THE YELLOW RIVER

THE YELLOW RIVER

A Natural and Unnatural History

RUTH MOSTERN

Maps and Infographics with the Assistance of
RYAN M. HORNE

Yale UNIVERSITY PRESS | NEW HAVEN & LONDON

Funding from the Asian Studies Center and the World History Center at the University of Pittsburgh is gratefully acknowledged.

Copyright © 2021 by Ruth Mostern.
All rights reserved.
This book may not be reproduced, in whole or in part, including illustrations, in any form (beyond that copying permitted by Sections 107 and 108 of the U.S. Copyright Law and except by reviewers for the public press), without written permission from the publishers.

Yale University Press books may be purchased in quantity for educational, business, or promotional use. For information, please e-mail sales.press@yale.edu (U.S. office) or sales@yaleup.co.uk (U.K. office).

Designed by Mary Valencia.
Set in Warnock Pro type by Newgen North America.
Printed in the United States of America.

Library of Congress Control Number: 2021932807
ISBN 978-0-300-23833-4 (hardcover : alk. paper)

A catalogue record for this book is available from the British Library.

This paper meets the requirements of ANSI/NISO Z39.48-1992 (Permanence of Paper).

10 9 8 7 6 5 4 3 2 1

CONTENTS

Graphics Style Reference Guide   vi
Acknowledgments   ix

Introduction
Seeing the Yellow River as a Whole   1

1 The Natural and Unnatural History of the Yellow River   22

2 Before It Was Yellow
The Great River from Neolithic through Medieval Times   62

3 Loess Is More
The Middle Period Tipping Point, 750–1350   121

4 Levies and Levees
The Engineered River, 1351–1855   179

Epilogue
The Yellow River in the Anthropocene   239

Appendix
Tracking Yu: Developing a Data System for Yellow River History, by Ruth Mostern and Ryan M. Horne   247

Glossary of Chinese Characters   267
Notes   273
Bibliography   293
Index   309

Color plates follow page 128.

GRAPHICS STYLE REFERENCE GUIDE

Administrative Geography Map Style. This map shows the style used throughout this book to depict prefectures, counties, settlements outside the administrative system, and places of interest.

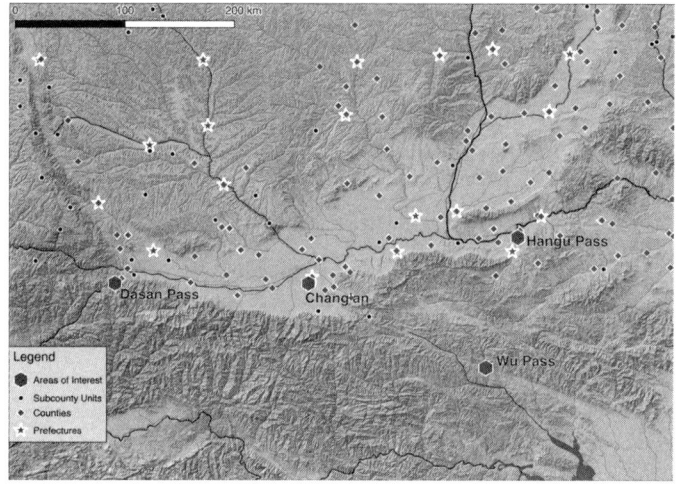

Event Map Style. This map shows the style used throughout this book to depict disaster events and infrastructure management events. Size gradients represent the number of events that occurred at each location during the time period depicted on a given map.

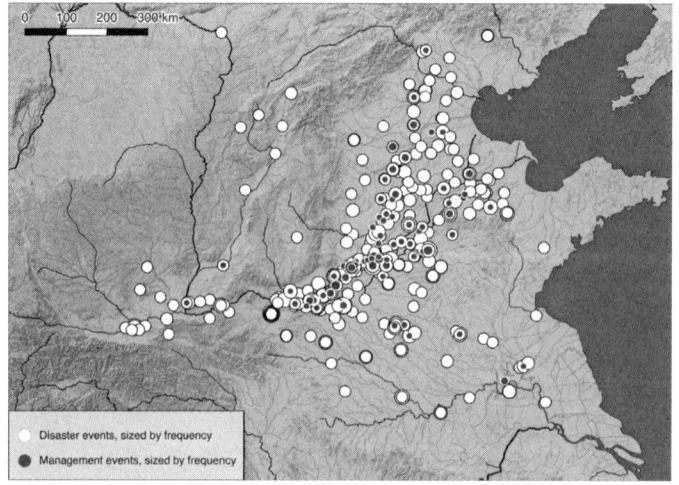

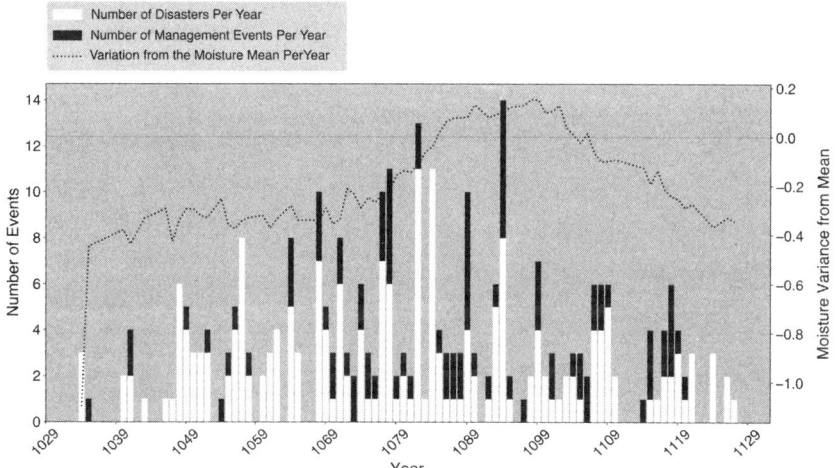

Timeline Style. This timeline depicts the style used throughout this book to depict disaster events and events of floodplain infrastructure management as they occurred in time.

ACKNOWLEDGMENTS

I have completed this book during the COVID-19 pandemic and amid the failures of governance and struggles for racial justice that have accompanied contagion itself here in the United States. I have written about the causes and impacts of environmental catastrophe while sequestered in a spacious and comfortable home office. This experience has caused me to reflect newly and deeply about the conditions that permit some people to remain sheltered from illness and death while others shoulder hazards on their behalf. My first and most profound acknowledgment is to the delivery and sanitation workers, the medical professionals, the education and child-care employees, the librarians and office staff, the cooks and shop clerks, and the protesters who have borne the perils that have permitted me to finish this book.

This has also been a troubled season in global geopolitics and in the relationship between the United States and China, as each country turns away from the other and toward a more repressive politics. I am moved to have written a book that has permitted me to spend so much time in dialogue with generous and erudite Chinese scholars and in intellectual engagement with Chinese scholarly traditions. The history of the Yellow River is a huge field in China. I will always be a student and a learner. Han Zhaoqing and Sun Jinghao have been my friends, my colleagues, and my hosts on numerous trips to Shanghai and Hangzhou. Chen Longwen and Li Denan have invited me on field trips to the river around Zhengzhou and Huai'an, respectively. Wang Yingjie hosted me in Beijing, and he and Su Yanjun provided me with an invaluable collection of data tables. Chen Yunzhen also provided essential data and advice. I am grateful to the Institute for Historical Geography and the International Center for Studies of Chinese Civilization at Fudan University, and to the History Department at Zhejiang University.

I have benefited from the periodic willingness of environmental scientists to help me understand my topic from the perspective of their disciplines. I am lucky to have embarked on this project while working in the exceptionally interdisciplinary culture of the University of California, Merced. There, interlocutors in the sciences helped to set my direction, steer me away from early mistakes, and let me know when I was on the right track. My colleagues in the sciences there included Asmeret Asefaw Berhe, Teamrat Ghezzehei, Josh Viers, Leroy Westerling, Valerie Leppert, and Shawn Newsam. More recently, I have benefited from conversations with Amy Hessl at West Virginia University, who sent me relevant content from the Monsoon Asia Drought Atlas and helped me understand how to use it (my thanks as well to Nicola Di Cosmo for introducing me to Amy), and with Mark Abbott at the University of Pittsburgh.

This book would simply not exist without immense amounts of labor, expertise, and creative contribution—far less well remunerated and more precarious than my own—on the part of numerous undergraduate and graduate students, technical staff, and postdoctoral fellows. Kaiqi Hua performed years of expert data collection, data entry, and data analysis. Ryan Horne is the lead author of the database, data management platform, data queries, and data visualizations that underpin this book. Shen Zhifeng was my coordinator and companion during two weeks of touring sites around the Yellow River basin. Shaobai Xiong, Erin Mutch, Rocco Bowman, Ed Lanfranco, and Elijah Meeks also played essential roles in the process. Elijah was the first person to launch me on this research and is the coauthor of my first article about it.

Two anonymous readers offered helpful readings of the draft manuscript of the book. I am humbled that four other extraordinary historians agreed to read the entire draft manuscript, to offer generous and bountiful advice, and to save me from numerous mistakes. My thanks go out to Brian Lander, Ling Zhang, David Pietz, and Ken Pomeranz. Ling Zhang was the colleague with whom I first discussed Yellow River history when I was initially scoping this project. She has offered innumerable connections to people and texts, she has helped me develop its intellectual framework, and she has been one of my main interlocutors throughout the project. Ian Miller shared prepublication excerpts of his important work on the history of deforestation in China.

I have presented this work in progress on literally dozens of occasions on multiple continents over the course of ten years. I have gained new insights each time. The conferences, workshops, and invited talks at which I have discussed Yellow River history are too numerous to name exhaustively. Some particularly memorable and generative events were a workshop on the history of the Grand Canal at Princeton University hosted by Paize Keulemans; a series of conferences on the global history of science hosted by Pat Manning at the University of Pittsburgh; a visit to Washington University hosted by T. R. Kidder, and a number of water history meetings and workshops. They were hosted by Ann Waltner and myself at University of Minnesota, by Rina Faletti and myself at UC Merced, by Philip Brown and Nicholas Breyfogle at Ohio State, by Ling Zhang at Harvard, and by Donald Worster and the Rachel Carson Center for Environment and Society at Renmin University. I also published work in progress on this topic in several journals and edited books, and I am grateful to the editors and reviewers of those publications: *Found in Translation*, *Nature and Culture*, and *Journal of the Economic and Social History of the Orient*.

James C. Scott expressed enthusiasm for this project at a key moment and discussed it with me on several occasions. I am endlessly grateful for his support and for introducing me to Jean Thomson Black, my editor at Yale University Press. Jean, and her colleague Elizabeth Sylvia, have been perfect project managers, cheerleaders, task keepers, and sources of wisdom about the publication process. Katherine Faydash has been an exceptional and meticulous copy editor. Charlie Clark at Newgen has overseen the production process.

In recent months various librarians and image owners have worked hard to ensure that I have access to the material that I need. At the University of Pittsburgh Library, Zhang Haihui and Chris Lemery responded to emergency requests that were made more numerous and more urgent during this season of library closure. The people who responded with generosity and alacrity to questions about images included Ian Teh, Sai Vanappali, Martin Heijdra, Ding Xiangli, Avigail Rotbain, Eva Myrdal, Bambi Xia, Shi Changxing, and Chien Fan-wen.

During my most intensive months of writing, I met weekly with a group of friends and fellow writers—Liz Arkush, Carla Nappi, Laura Lovett, and Mari Webel—who offered me just the right mix of encouragement and distraction.

As always, all remaining errors and infelicities are my own. To that traditional gesture, I ask for patience and forbearance from my readers, who will discover some footnotes that are less complete and precise than they would be if I had been able to visit a library during the past few months.

An ACLS Digital Innovation Fellowship funded a year of salary and project costs. I received institutional support and time off at both the University of California, Merced, and the University of Pittsburgh. I am grateful for the job security, staff support, and time off that my tenured position permits. This status is something else without which the book would have been impossible.

I was a fellow at the UC Merced Center for the Humanities during one year of its two-year water theme, and I served as interim director during the other year. The two water postdoctoral fellows, Rina Faletti and Kim De Wolff, were exceptionally influential on my thinking about the questions that this book explores. Kim was the first person to push me toward a science studies perspective that grants agency to water as well as people, and Rina helped me to think about the materiality of water infrastructure. Christina Lux was the able associate director of the center. The water theme launched under Susan Amussen's able leadership of the center.

At Pitt, Linda Howard, Kimberly Thomas, Cynthia Graf, and Ian McLaughlin have offered invaluable administrative support, which the COVID-19 crisis has made even more difficult than usual. The Pitt World History Center, which I direct, has offered both material and administrative support. Thank you to former center coordinator David Ruvolo, current center research associate Ali Straub, and associate director Molly Warsh for making that happen. The publication of this book has also been supported by the Evelyn and Thomas

Rawski Faculty China and Chinese Studies Research Fund at the University of Pittsburgh.

Throughout this long process, my family has often known better than I have myself when to distract me and when to leave me alone. They wish me well when I travel, and they welcome me home with love. No thanks are deep enough to acknowledge the contributions of my child, Pip; my mother, Wilma; and my husband, Kenny, who is the household chef, provisioner, and event planner as well as my brilliant and steadfast life partner.

THE YELLOW RIVER

INTRODUCTION

Seeing the Yellow River as a Whole

I was born a hundred years too late to be searching for the sublime in nature. To me, pursuing this would have just been an expression of nostalgia.... I decided that what was relevant for our times were pictures that showed how we have changed the landscape in significant ways in pursuit of progress.

—EDWARD BURTYNSKY, *Manufactured Landscapes*

According to a Chinese adage, "water flows downwards, but people stride upwards"—*shui wang dichu liu, ren wang gaochu zou*. The proverb signifies that the ambition of humans to improve their circumstances is as inexorable as the law of gravity itself.[1] There is a long tradition in China of using metaphors about flowing water as a symbol of human ambitions and ethics. As early as the fourth century before the Common Era, the thinker Mencius held that "the tendency of human nature to virtue is like the tendency of water to flow downwards. Among people, there are none who [by nature] are not virtuous, just as there is no water that does not flow downwards. Now by striking water and causing it to splash up, you may make it go over your forehead, and by damming and channeling it you can make it flow up a mountain. But is this due to the nature [*xing*] of water [itself]? It is, rather, the powers [*shi*] [of water] that enable this."² This book concerns environmental history, not moral philosophy, but the passage from Mencius introduces many of its themes nevertheless, for the book's central concern is what has happened over many centuries when people have sought to understand the capacities that inhere in water and the ways in which people have struck water and forced it uphill in the name of their own aspirations.³

The term that I have translated as *powers* in the quote from Mencius is a somewhat more complex and interesting concept than the English word implies. The expression *shi* describes a way for people to understand their relationship to phenomena in the natural world. The philosopher François Jullien translates the term as *efficacy*. Other terms for it might be *propensity* or *capacity*. The word reflects the idea that people who understand the inherent characteristics of some phenomenon can channel it in order to bring about a desired result.⁴ In the efficacy tradition, the official or engineer, faced with water and sediment in motion, seeks to understand hydrology as a dynamic system in which human activity is itself always implicated. In that tradition, statecraft and science are activities that require people to achieve a deep understanding of the potential power situated in any system, and only then to use that information to determine how to proceed.

In the context of Yellow River policy and practice during preindustrial times, *shi* represented the capacity of the river to unleash vast destructive potential. Good policy and engineering that accorded with the *shi* tradition often sought to diffuse the river's power and always acknowledged the risk and expense involved in any course of action. *Shi* was the keyword used in many debates about river engineering in historical times, but today, conflicts like these persist in other terms.

At the same time that Mencius was suggesting that the behavior of flowing water was a metaphor for human nature, other scholars were canonizing the story of Yu the Great (Da Yu). By the time of the Warring States era (475–221 BCE), when Mencius lived, the tale of Yu the Great was already ancient. The Yu legend explains that the ability to manage rivers and riparian environments in the interests of the state is what allows government to create territory, to tax it, and to exercise political authority.

According to myth, a great flood had begun during the reign of the sage-king Emperor Yao, inundating his entire domain. Yao appointed his relative Gun, the Count of Chong (Chongbo), to control the flood. Gun tried for nine years to tame the waters, but he did not accomplish that goal. As the legend goes, Yu the Great, Gun's son, succeeded where his father had failed. He traveled around the realm, channeling all the rivers and restoring them to stable courses. Unlike his father, Yu understood that the powers (*shi*) of water were such that he had to dig drainage canals to slow river currents, not simply trap waters behind high dams and levees, which would inevitably rupture. Once the rivers were channeled, Yu the Great used their courses to define the Nine Provinces (Jiuzhou). For each one, he set tax and tribute responsibilities. The Nine Provinces, the empire's first system of political geography, formed the basis for China's legendary founding regime, the Xia dynasty. The story of Yu the Great reveals the fact that in the Chinese tradition, water management begets social structure, prosperity, and political power.[5]

Figure I.1, the *Map of Yu's Traces* (Yujitu), completed in the year 1137, a millennium and a half after the Warring States era, enumerates place-names from Yu's Nine Provinces along with toponyms from the cartographer's own day. It situates both sets of names in a remarkably detailed rendering of East Asia's river network. The Yellow River, the subject of this book, is the prominent waterway that dominates the northern part of the map. This map, with its vivid description of rivers and its promotion of the enduring significance of the Yu the Great legend, is one of the inspirations for this book.

THE INVENTION OF RIVERS

By converting a wetland floodplain into a channeled network of managed rivers, the achievement of Yu the Great was an act of imagination as much as one of engineering. In *The Invention of Rivers*, the historian Dilip da Cunha avers that a river as we generally imagine it—"its source identified by a point; its course depicted as a stroke; and its propensity to flood imagined as the erasure of the boundary between water and land"—is a human invention. People create rivers, both as physical facts on the landscape and as objects of discourse and cartography.[6] In an active alluvial landscape, a river is not, in fact, a tidy channel of water traversing dry land. It is, as the landscape architecture scholar Anthony Acciaviati puts it, "a sediment sorting

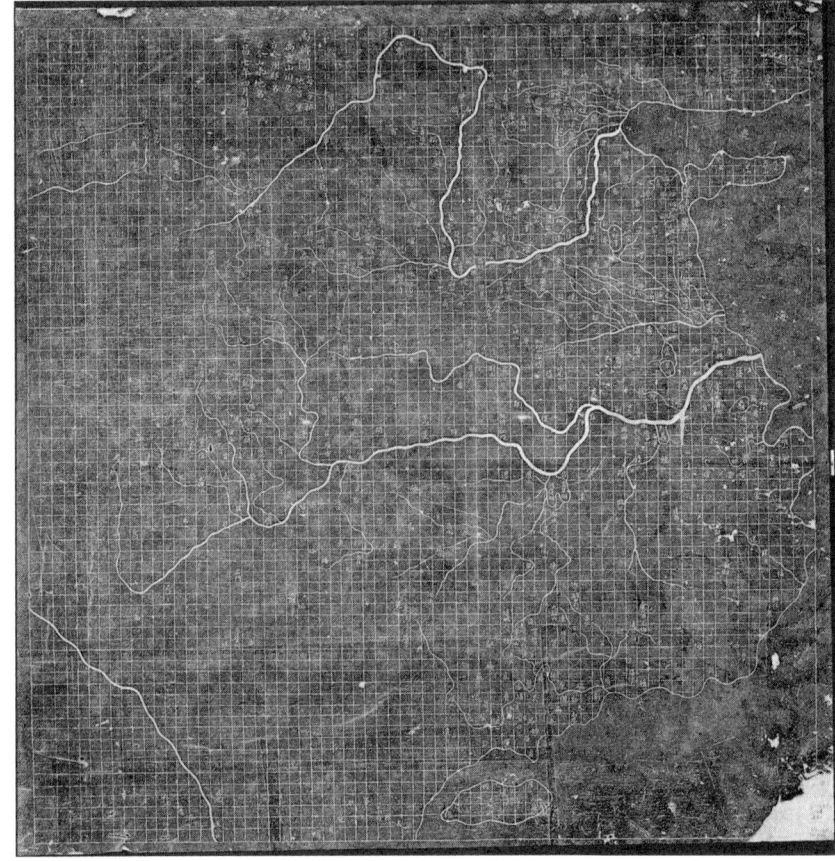

FIGURE I.1

Map of Yu's Traces. The 1136 CE *Map of Yu's Traces* is one of the works that inspires this book. The Yellow River is the major waterway in the north. Courtesy of the Library of Congress Geography and Map Division, Control Number gm71005080.

machine," an engine for transporting organic and mineral material downhill from higher-altitude locations to deposit on its floodplain or in its estuary.[7]

In the kind of river system that Jamie Linton and Jessica Budds refer to as hydrosocial, one in which people and water are fundamentally inseparable from each other, human activity that profoundly transforms the behavior of a river may transpire anywhere throughout its basin. Farming, mining, pasturage, or logging may occur on arid lands far from the sight of water, but if those actions alter the rate or character of sediment transport downstream, they may be an essential aspect of river history. Indeed, ocean currents, earthquake faults, and climate systems that affect precipitation and glaciation at watershed peripheries and river headwaters may be elements of river history as well.[8]

This book adopts this perspective—it is about how to *see* the Yellow River. Seeing is a kind of empowered illumination that has material consequences, here from the perspective of the river's entire basin and over the entire span of human habitation.[9] The book invites readers to expand their gaze upstream to take in the sources of sediment and, by doing so, to understand the alluvial plain as a territory that derives its character from events upstream. As the environmental scientist Josh Viers explains: "River systems are highly dynamic—they're always changing in time and space. As they rush from their headwaters, they break down mountains and transport a lot of sediment. By the time they reach the lowlands, they've started to deposit that sediment. It's that dynamic between water and land that creates floodplains." Throughout the world, states and powerful denizens have sought to separate land and water on the floodplains of major rivers in the name of controlling floods, as Yu the Great is purported to have done in China in the ancient past. This has been highly consequential, for societies and ecosystems alike. Viers continues: "Humans like to inhabit floodplains because of their rich soil and flat land. So we've engineered ways to protect people from floodwaters. But ecosystems evolved to have functions and processes that depend on floods. For example, flood flows are slower and warmer on floodplains. That creates good conditions for a lot of creatures—bugs, fishes, and birds, for example. When we cut off rivers from their floodplain, we're cutting off these essential habitat functions."[10]

Taking a similar perspective, in a classic 1942 work, the geographer Gilbert F. White asserted that "floods are 'acts of God,' but flood losses are largely acts of man." He advised against building dams, levees, and walls for flood management and against considering floods as natural adversaries against which society must perpetually battle. As he pointed out, victory in the war against floods has a high price: it is the cost of perpetually maintaining the engineering works necessary to confine the flood crest. Conversely, the price of defeat in that war is a continual sequence of flood disasters. People pay either way. Humans, White asserted, instead should encroach less aggressively onto floodplains, less aggressively exploit upstream lands, and less aggressively restrict river channels. He recommended adjusting human occupation to the floodplain environment so as to permit effective utilization of floodplain resources without the construction of large-scale physical structures.[11] Mencius might have recognized White's advice as an approach that recognizes the power of *shi*.

When people make floodplains habitable to dense human populations, and when they engineer levees and drainage canals that turn rivers into linear channels, they are actually increasing flood hazards because they are creating engineered versions of plains that have lost their capacity to store floodwater and sediment. Nutrient-rich sediments that are transported directly to estuaries through channeled rivers rather than being deposited on floodplains cannot be colonized by riparian plants and associated communities of insects, birds, and other animals. Floodwaters lost to the floodplain cease to recharge local groundwater supply.

Previous Yellow River histories, like most other histories of world rivers, have tended to focus exclusively on the floodplain and on particular short eras of human history. With these constraints, they presuppose the river has always been prone to flood and to change course, and they explain that this is precisely because it carries so much sediment. However, few river histories historicize the sediment itself, and few track flood frequencies over a very long term. Contemporary authors writing about the Yellow River in both English and Chinese thus replicate the blinders that many imperial Chinese hydrocrats wore themselves. These civil officials, experts in water science and engineering, took the sediment in the Yellow River as an inherent and permanent reality to manage rather than a manifestation of a particular history and geography.[12] This book, taking a whole-basin and sediment-centered approach to the Yellow River, works to historicize the processes of sediment transport and floodplain transformation and to make their long-term consequences visible.

Plate 1 depicts the entire hydrosocial landscape of the Yellow River. The Yellow River's history, the history of sediment as much as water, significantly concerns the Loess Plateau, which is the semiarid region that lies to the west of the floodplain. When people felled trees, grazed herds of animals, and cleared land for agriculture on the Loess Plateau, they generated high rates of erosion, which washed sediment into the river. Entrained in the water, the sediment traveled downstream until it reached the wetlands and slow currents of the floodplain and settled on the riverbed. Downstream, that alluvial deposit repeatedly caused the riverbed to rise higher than the banks that surrounded it and then to change course. Upstream, after a period of substantial erosion, the topsoil disappeared, and the land became sandy and fissured with gullies.

Absent levee construction, rivers, constantly settling sediment into their beds, are restless bodies of water that are always seeking new and straighter courses to the sea, and in doing so, sometimes making dramatic changes of course called avulsions. The Yellow River is currently the most sediment-laden river in the world. As such, it is prone to create particularly mutable landscapes. Plate 1 depicts the many occasions when the river has changed course. When levees prevent rivers from avulsing and spreading sediment across their floodplains, the result is that water tables, riverbeds, and riverbanks rise while land begins to subside. When artificially embanked rivers overtop or breach their levees, they disgorge onto land that lies lower than the river itself, causing catastrophic floods, and when their current is powerful during floods, they deposit entrained sand, gravel, and even boulders along with fertile loam.

It is possible to precisely track the changing relationship between human activity on the floodplain and the behavior of the managed river. Timelines reveal the gradual creation of a disaster-prone river basin, as well as the imperative to create the infrastructure that could control it (fig. I.2). Beginning in the late ninth century, the rising rate of erosion on the Loess Plateau initiated a millennium of frequent floods, followed soon thereafter by continuous mitigation efforts. Human activity created a disaster-prone river in another way as well. As Dilip da Cunha, Josh Viers, and Gilbert White have all pointed out, floodplain management itself is what creates both the material conditions and the language for disaster. Absent a densely populated floodplain and an engineered infrastructure of canals, levees, locks, sluices, and ponds, a river's meanders and avulsions are no calamity. Until people built a dense network of levees, historical sources did not attest levee breaches as a common disaster (fig. I.2a). Only after the alluvial plain was thick with earthworks did repair become a frequent element of infrastructure work (fig. I.2b).

These conclusions would not have surprised Yu the Great himself. The details of these timelines—perturbations up and down, and changes in the ratios of various activities—are the subject of the rest of this book. At the scale of decades and centuries, they help to reveal stories about climate and weather, war and colonization, taxation and demographics, and changing notions of state power and benevolent

FIGURE I.2

Breaches as a Percentage of Disasters. Until the tenth century, when levee building began in earnest, there were few disasters on the floodplain, and few of those that did occur were recorded as levee breaches. From the tenth century onward, historical sources report disasters on a near-annual basis, and half or more of all reported disasters were breaches. Figure (a) depicts attested levee breaches on the Yellow River as a percentage of all recorded disasters. Waterworks management was rare until the ninth century, and it increased to near-annual frequency only in the seventeenth century. Figure (b) depicts repairs as a percentage of all recorded events of waterworks management.

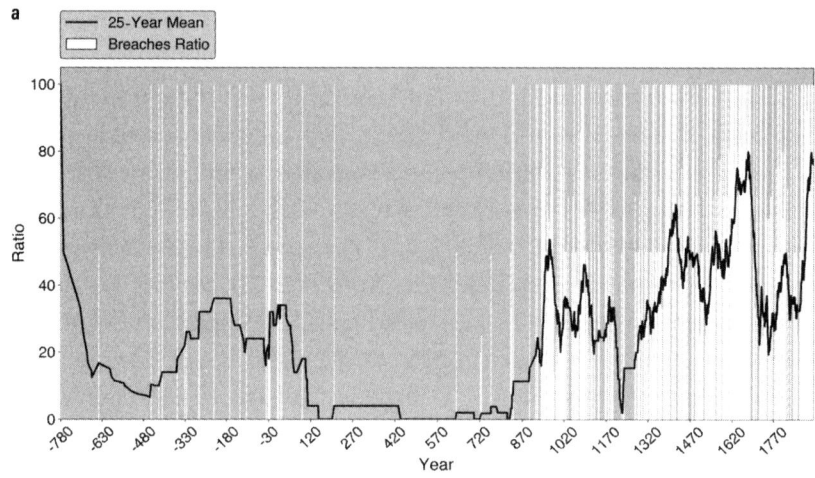

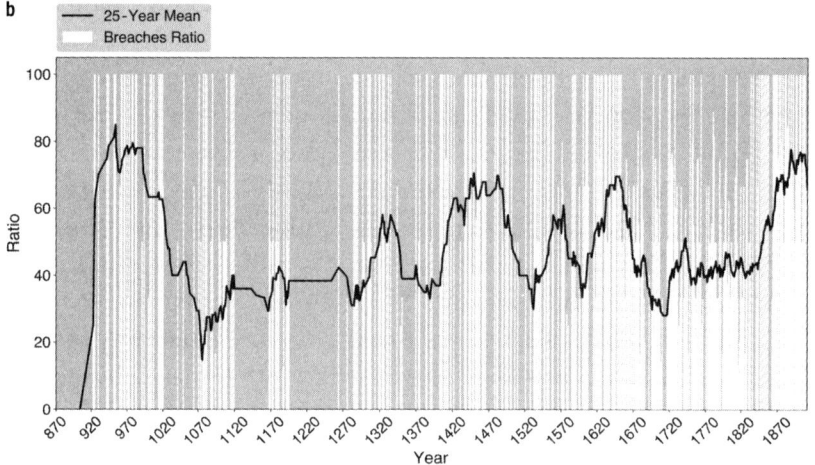

rule on both the Loess Plateau and the alluvial plain. In short, they reveal a series of biographies of the hydrosocial river.

The historian Richard White has played an important role in defining river history as a field. His important work *The Organic Machine: The Remaking of the Columbia River* introduces the idea of a river biography. In this genre, rivers themselves are the central protagonists. In a river biography, rivers and their stories have births and deaths; beginnings, middles, and ends. White explains how human activity creates new versions of rivers by making it so that "the human and the natural, the mechanical and the organic, [have] merged so that the two [can] never be ultimately distinguished."[13] He advises that the objective of a river history is to reassemble what has been broken into parts, because "our tendency to break it into parts does not work. For no matter how much we have created many of its spaces and altered its behavior, it is still tied to larger organic cycles beyond our control."[14]

In keeping with White's approach, I have structured this book around a sequence of life spans. By attending to the river's entire watercourse, its whole eventful history, and its slow transformations, I have identified exactly when it was that the lower course turned unruly and what happened, both upstream and downstream, that led it to be that way. Each chapter describes a turning point in world-making processes along the Yellow River and a span of time when the links between the middle course and the lower course were transformed.

Few of these turning points conform to the standard dynastic dividing lines familiar to historians of China. This book primarily concerns the history of the river during the two millennia of the imperial era, but it is not a work of political history. The first chapter, the first river life span, encompassed a period of many centuries during which a relatively small population lived upstream and during which writers attested only a small number of events on the floodplain.

That era ended in the ninth century of the Common Era, when Chinese armies fortified the Loess Plateau, installed colonist farmers, and attempted to expel the region's pastoralist residents. The next river life span resulted from the intensive period of frontier colonization on the fragile middle reaches of the river, which destroyed forests and grasslands and doubled the rate of erosion relative to earlier levels. The result was an unprecedented age of flooding, managed on the floodplain in only haphazard and sporadic ways. This is the era when the river began to be called Yellow. Historians of China have long recognized the end of the first millennium of the Common Era as a transition point in long-term history, resulting in new population distributions, new forms of economic and social organization, and new modes of governance. The beginning of the eventful era of Yellow River history is also an aspect of that transformation.

The final river life span of the imperial river is the age of the highly managed floodplain. The era began with proposals and experiments in the mid-fourteenth century and came into focus by the sixteenth century. Even as land degradation in the middle reaches of the river accelerated, with erosion there rising to unprecedented levels along with population, floodplain managers built levees ever higher and made earthworks and waterworks ever more elaborate. The system reached full buildout during the eighteenth century. Although efforts to curtail floods were successful in the medium term, the floodplain system, overwhelmed with silt, collapsed catastrophically in the mid-nineteenth century. When Chinese and global commentators refer to the Yellow River by the moniker of "China's Sorrow," they are unwittingly evoking only the river's final imperial life span, especially the nineteenth-century collapse and its aftermath.[15]

By claiming that the China's Sorrow life span of the Yellow River was short lived, this book challenges some deeply ingrained ideas about Chinese history and civilization. Scholars and commentators, domestically and globally, regularly link the fate of the empire with the condition of the floodplain. However, many such assumptions—that the North China Plain has long been ecologically degraded and that political legitimacy since time immemorial has necessitated extensive hydraulic management—may be of relatively recent origin.

In a now-classic article from 1993, the environmental historian Mark Elvin introduced the paradox of "three thousand years of unsustainable growth" in China.[16] On the one hand, China's ecological transformation can appear extraordinary, even impossible, in both its intensity and its historical span. On the other hand, despite all the evidence for landscape destruction, species extinction, environmental disaster, and human misery, imperial China also possessed the world's longest continuous state and civilization, its largest population, its most ingenious engineers, and its most dynamic engine of global prosperity. These phenomena are inseparable from one another. The history of the Yellow River is not a simple story of environmental decline.

It is not instructive to periodize Yellow River history according to the rise and fall of regimes or the fluctuations of climate. Nevertheless, political history and climate history braid together with river history in significant ways. Drought on the Loess Plateau is one example. With the final imperial Qing ruling house (1644–1911) as an exception, regime changes made relatively little difference to the direction of policy on the floodplain. However, large-scale political changes were often quite significant on the Loess Plateau, where

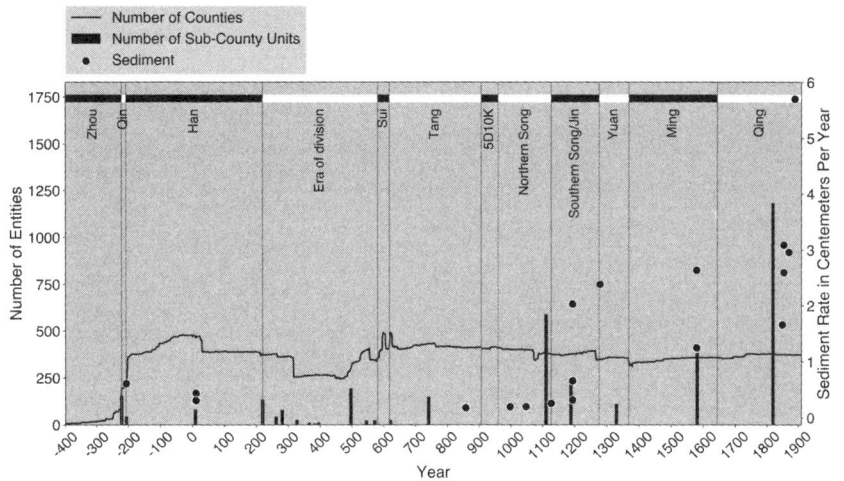

FIGURE I.3
Erosion and Settlement History. There is a correlation between the density of settlement on the Loess Plateau and the rate of sediment accumulation. Both settlements and sediment expand on three occasions: during the last two centuries before the Common Era, around the turn of the first millennium of the Common Era, and in the nineteenth century.

particular regimes sponsored settler colonization or fomented military contestation. Many factors entwined, for instance, on occasions when harsh weather on the Loess Plateau increased conflict over arable land. Particular governments reacted to those conflicts with their own characteristic policy tool kits, and erosion was the result. This river biography is populated by peoples and their ambitions and by dynamic climate conditions, not only by water and sediment.

WATER FLOWS DOWNWARD

It is possible to pinpoint with some certainty when two kinds of human activities—those that accelerated erosion upstream and those that reduced resilience to floods on the alluvial plain—fundamentally changed the propensities of the hydrosocial Yellow River. Figure I.3 depicts the inflection points in Yellow River history from the perspective of the Loess Plateau. On each of the three occasions when the number of settlements on the plateau significantly exceeded previous levels—the third century before the Common Era, the eleventh century of the Common Era, and the eighteenth century—the rate at which sediment accumulated on the floodplain doubled anew. Each of these phenomena, in turn, correlates with the event timelines in figure I.2. This series of charts reflects the entire argument of this book in a simplified form. Generally speaking, whenever the Loess Plateau population became denser and settlements became more numerous, erosion increased; and whenever erosion on the Loess Plateau rose, so too did catastrophe on the floodplain.

The number of settlements on the erosion-prone Loess Plateau, at the frontier between agrarian and pastoralist modes of subsistence, increased rapidly just over a thousand years ago, an event that occurred in tandem with erosion rates. The number of events related to the river that people attested in historical sources (disastrous events like floods and management events like levee construction) also traced a similar trajectory. As the environmental scientist Chen Yunzhen and her coauthors put it, "The co-evolution of river dynamics and Chinese society is remarkable, especially farming and soil erosion in the middle river, and central authority and river management in the lower river."[17]

That inflection point around a thousand years ago is also the time that imperial China's regional concentrations of wealth and revenue, the locations of its northern borderlands, and the zones of its waterworks spending diverged from one another and from the geography of the watershed itself. At around the same time, the imperial capital moved east, away from the Loess Plateau and onto the floodplain, and the prosperous core of imperial culture and society shifted even farther south and east, out of the Yellow River basin altogether. From this time onward, it ceased to be conceptually possible for the Chinese imperial state to "see" the politically and spatially distinct regions of the river basin as a single ecological system. The results of that failure were disastrous for people and nature alike. The historian Randall Dodgen calls this "the conflict between the geologic and the fiscal," and Pierre-Étienne Will refers to the "hydraulic contradictions" between the upper and lower reaches of the river.[18]

In the Yellow River watershed, at the northwestern periphery of the East Asian monsoon system, the record of human activity extends from Neolithic times to the present. Beginning in the mid-Holocene, farmers, herders, soldiers, and woodcutters began to disturb the ecologically fragile grasslands of the Loess Plateau in the middle course of the watershed. This accelerated soil erosion on the grasslands and flooding on the plains downstream. Over centuries, the human challenge of coexisting with increasingly desiccated grasslands and turbulent waters stimulated innovations in water engineering, farming, and state financing. It transformed demographic structures and modes of subsistence. It also provoked debate and struggle about which people and which regions to inundate when the water was high and the levees failed.

The river's middle course, traversing the Loess Plateau, passed through a multiethnic, multispecies, and multinational world at the edge of the East Asian monsoon system, one that brought together agriculturalists, pastoralists, sheep, warhorses, traders, and salt miners who were beholden to various state and clan systems. When Chinese and other states sought to extend their reach there, they did so by voluntarily or forcibly resettling herders and farmers and by building, ultimately, hundreds of towns and forts in the region, constructed out of pounded earth and often-scarce timber. In this ecologically fragile zone of friable soil and unpredictable precipitation, the long-term result was devastating erosion and a river that carried an increasingly large amount of sediment.

On the floodplain, the river traversed a populous farming region. By the middle imperial era, the river also transected numerous transportation canals, ultimately including the famous Grand Canal. The canals bifurcated the river, creating a human hydrology distinct from that of the river and its tributaries. They were the state's engineered lifelines between the semiarid north and the humid south. They transported people, grain, and commodities from the wealthy and urban rice-growing regions of the Yangtze delta to the capital cities and military frontiers of the north. To keep the canals in working order, governments sponsored reservoirs, levees, and sluiceways that pushed sediment past the canals and out to sea. Farmland near the river became saline and waterlogged as tributaries and drainage systems collapsed. The frequency and violence of catastrophic floods increased.

By the eighteenth century, government spending to maintain the system consumed between 10 percent and 20 percent of all government spending, an amount higher than the entirety of all government spending by most European regimes at that time.[19] Historical documents and environmental science research confirm that during imperial times the river experienced thirty major course changes, approximately 1,500 floods, and, over the last millennium, almost perpetual projects of construction and repair to maintain the engineered river system. Domesticating the floodplain required a "technology complex," a set of tools that included forms of administrative organization and a cultural imaginary as well as earthworks and dams.[20]

Dams and earthworks "define social reality" by remediating human relationships with the natural world. These "ubiquitous and uncontested elements of material life ... construct the order of things."²¹

The middle course and the lower course signified different things in the imperial imagination, and they had different roles in the realm. They were always managed separately. The unruly river and its fractious denizens thwarted the state's dreams of classification and simplification. A series of regimes destroyed human and fluvial systems that linked the middle and lower course of the river and replaced them instead with a spatialized Chinese empire consisting of an often militarized frontier and an ecologically precarious center—a process with unintended consequences for people and river alike. Human imagination and engineering split the river into a middle course and a lower course that were conceptualized separately, even as the water continued to flow from mountains to ocean.

The point is that the state "saw" the middle course and the lower course separately, and the court saw the river differently than local officials did, let alone the human and animal denizens of the watershed. The state simplified and rationalized river management by splitting its watershed into multiple parts under different jurisdictions, but this had disastrous environmental consequences. Management of the Yellow River played a significant role in China's "developmentalist project" to exploit land and resources and to make nomadic populations sedentary, a long-term trend with both an environmental and imperial geography.²² However, although developmentalist ambitions drove activity on all parts of the watershed, they did not do so in the same direction on the lower reaches of the river as they did on the middle reaches. Even on the lower course, vigorous management ended at the confluence of the Yellow River and the Grand Canal. Downstream from there, flooding and waterlogging was a matter of concern only insofar as it degraded river and canal transportation elsewhere.²³

By attending to the river's entire watercourse, its whole eventful history, and its slow transformations, I identify when and how the alluvial plain became flood-prone. The singular length and consistency of northern China's historical record means that the Yellow River is the world's best-documented example of the creation of a waterscape, a long-term entanglement of diverse, fractious, and impactful human agents arrayed unevenly across an entire continent, with water embedded in a complex and open-ended human-oriented environment, and in the dynamic and unpredictable natural worlds upon which people depended. The three-thousand-year history of this book offers a unique perspective on matters of urgent significance today: enduring patterns and gradual transformations in historical ecology, sudden ruptures that upend them, and unintended environmental consequences of human action that reverberate back to the societies that engendered them.

THINKING BIG

Despite the excellent reasons to attempt it, holding together an entire continent and three millennia of time in an environmental history narrative is not easy, and few authors have tried. This book does so by combining the historian's craft with analysis, maps, and infographics generated from data and from environmental science literature. Indeed, the project is necessarily unique to the Yellow River, for there is no other watershed that can claim more than two millennia of consistent documentation, and it is impossible to manage this material within the confines of the traditional scholarly monograph.

The book is based on an innovative historical geographic information system that depicts geomorphology, waterworks, settlements, and flood locations. The geographic information system is also underpinned by a database of 3,754 spatially and temporally referenced events related to the Yellow River as recorded in historical sources: floods and droughts, construction and repair activities, policy arguments, and funding requisitions. Using the concept of the spatial and temporally referenced event as a basis for reasoning and visualization brings long-term trends into focus in a way that spatial analysis alone cannot accomplish.[24] From the sources in the database, and in collaboration with my data manager Ryan M. Horne, I have generated visual material that I pair with imperial-era maps and artworks that depict the river system, diagrams from environmental science articles, and contemporary photographs. The appendix describes the data system in more detail.

Although the book is replete with maps and timelines that suggest a veneer of objectivity, it is important to remember that they are simply based on collations of attestations about the river and about settlement geography that come from historical sources. They are biased toward the elite state actors who produced them, they reflect different emphases from one era or regime to the next, and there are more surviving sources from later periods than from earlier ones. They offer a heuristic, a guide to broad trends and a way to reengage with the sources themselves, rather than the final word on what "really happened." The ratio of reported river management events and river disaster events during a given time period offers a snapshot of ideas about success and failure at that time. Comparing the number of such events over time is suggestive, but the results are not perfect, as the number of extant sources and the amount of reporting varied between regimes. They can serve as a road map back to close reading, and they can be correlated with records from other corpora of historical sources and with data from the climate and sediment record of environmental science. Creative queries also help to reveal different pathways through the sources and different gaps and elisions in them. For instance, it is possible to isolate major events from the noisy records of later time periods: events recorded in the official dynastic histories, events recorded in multiple modern handbooks, events that affected a large number of places. It is also noteworthy that the rapid increase in attested river events that began in the ninth century actually preceded the print revolution that began in the twelfth century. This reinforces my conviction that the occurrence is not merely an artifact of the historical record.

Such approaches make it easy to see a river at many temporal and spatial scales, from individual events to trends over centuries. The concept of the event is key to this book's methodology, one that I adopt from the work of the historical social scientist William Sewell. Sewell defines events as the sorts of occurrences that disrupt social structures and transform them into something new. For the purposes of my Yellow River database, I define an event as any instance of river management or disaster that is attested in the historical record.[25] In Sewell's sense, disasters, which displaced people from their homes and destroyed their crops, are such events. In the sense that the author and activist Naomi Klein uses the term, they are frequently also "shocks," which permit power holders to consolidate authority and enact preferred policies in the name of offering solutions to the problems of vulnerable people who might otherwise be able to resist these changes.[26]

Sewell distinguishes historical events from historical processes. In his taxonomy, processes are slow changes that occur within stable structures. The event database focuses on episodes that destroyed lives and livelihoods, and those that mobilized massive spending and marshaled laborers by the tens of thousands. It is not equipped to record how processes unfolded at the rate of slow time. Sediment slowly erodes upstream and accretes downstream. Population gradually rises and its distribution shifts. The narrative of the text describes these processes, many of which are the kinds of phenomena that the literary critic Rob Nixon refers to as slow violence, which affects people who are poor, vulnerable, and subjugated more profoundly than it affects the population as a whole.[27] Generations living precariously on sand fields in the shadow of the great levees did not inevitably experience catastrophic flooding events, but they were affected by slowly unfolding processes of declining soil fertility and food insecurity. Environmental history necessitates work at multiple temporal and spatial scales simultaneously. A well-written river history integrates a whole watershed and even a whole climate system, but it also invokes individual wells, sites of erosion, and zones of flooding. This narrative resists declensionism, the notion that human intervention has led inexorably to ecological degradation. Loess Plateau vegetation repeatedly rebounded between episodes of erosion, and the floodplain was a site of remarkable innovation as hydrosocial conditions changed. As the story of Yu the Great attests, the standards of successful river management varied over time as different rulers and hydrocrats articulated a range of ideas about what kinds of places to protect and what kind of activities to prioritize.

As for geographical scale and spatial production, ambitious preindustrial states, faced with population pressures and seeking economic growth and military victory, have only two options. They can encourage or compel people to intensify extraction and production in densely settled regions of significant infrastructure, or they can cause them to settle imperial frontiers and to farm less ecologically desirable territory.[28] These, too, are among the important processes in Yellow River history. They resulted in an "environmental mode of consumption" in which the state commandeered staggering amounts of political capital, finances, labor, and timber to maintain erosion zones and flood zones as governed territory.[29] The Chinese imperial state routinely circulated natural and agricultural resources throughout the Yellow River basin and between it and the rest of the empire, and in doing so, it created self-consciously constructed zones of ecological precarity as well as wealth, both of which served the interests of the empire as a whole.[30] Maps throughout the book illustrate these structures and processes.

PEOPLE STRIDE UPWARDS

Shifting baseline theory suggests that one challenge of protecting the environment is that, given the scale of human life spans, people have a poor conception of the environmental impact of social activity because expectations shift with every generation. We see as pristine what our ancestors might have taken to be degraded.[31] Geoarchaeology and environmental science provide a clear history of soil erosion on the Loess Plateau. Commentators over the centuries refer periodically to unprecedented quantities of sediment inundating the floodplain. However, the change in the sediment rate was not necessarily visible at the scale of a human lifetime, and writers typically discussed many aspects of the complex hydrosocial system together, rather than focusing on sediment as such. One advantage of the *longue durée* and event-modeled approach of this book is that it invites readers to step back from historical documents written by individuals who may have taken their present circumstances as normative and everlasting.

However, the historical record also offers glimpses of river managers and hydrocrats who realized what was happening around them. As long as two thousand years ago, some commentators recognized that erosion upstream caused flooding downstream. Zhang Rong (fl. 4 CE), who served as commander in chief (*da sima*) at the imperial court around the turn of the Common Era, was the first observer whose writing is still extant to identify each dynamic element in the interlocking system of farmers, rulers, soil, water, and weather.[32] He explained how state-sponsored colonization drove farmers onto hillslopes, initiating land clearance that caused erosion, and washing sediment into riverbeds. He also wrote about how water diversion for floodplain irrigation reduced water flow and velocity of currents, causing sediment to accrete in the riverbed instead of washing out to sea. He noticed that, in addition to irrigation canals, tributaries were being blocked by levees, thereby damaging floodplain dynamics and accelerating riverbed elevation. Finally, he recognized that once human activity caused erosion of hillslopes upstream and extinguished floodplain tributaries downstream, it was specific heavy rainstorms that were the proximate trigger for occasions of flooding and embankment collapse. Zhang's insights influenced floodplain engineering in the decades that followed his commentary, but they did not alter subsequent approaches to the colonization of the Loess Plateau. Nevertheless, his work remained well known.[33]

A millennium later, during the eleventh century, a time of rapidly increasing erosion and intensive human activity throughout the middle and lower course of the river, two insights came into focus once again. One was that deforestation on loess hillslopes was causing erosion. The other was that the amount of sediment on the floodplain channel was rising, and that it was for this reason that the river traversed the floodplain with increasing instability. The renowned essayist and reformist statesman Ouyang Xiu (1007–1072) spent time on both the Loess Plateau and the floodplain. In a series of three petitions drafted between the late 1040s and 1050s, he explained that the Yellow River had begun to carry such a large quantity of sandy sediment that it would inevitably block the river course.[34]

Until the late sixteenth century, floodplain managers, often explicitly invoking the principle of efficacy, tended more toward establishing drainage canals and overflow basins than high levees. The kind of floodplain lock-in that Gilbert White decried in the twentieth century and that Yu the Great was said to have rejected in primordial times began to predominate only after the late sixteenth century. Once the maintenance of high earthworks barriers became standard policy, however, it was difficult to undo. In 1743, an inspection official (*jiancha yushi*) named Hu Ding (1709–1789) traveled to the Loess Plateau to investigate the situation there. Upon his return, he sent the emperor a proposal to manage erosion by creating small check dams far upstream rather than vast structures on the floodplain. The court rejected this solution because the routines for floodplain management—personnel, expertise, and landscape modifications—were already so entrenched.[35] Budget structures and bureaucratic incentives, which enforced the distinction between upper reaches of the river chosen for agricultural colonization and a floodplain designated for water management, were remarkably persistent.

In turn this suggests an unsettling conclusion: that two thousand years ago, as now, governments and other empowered decision makers have charted ecologically consequential courses of action not out of inertia or ignorance but because environmental protection has been a lower priority than economic growth, social stability, political power, or imperial conquest. One lesson of this book, therefore, is to offer a historical case about the limits of what is politically possible for environmental politics.

As late as 1934, Li Yizhi (1882–1938), a hydrologist, author of *Opinions about Yellow River Management*, and a native of the Loess Plateau, could still erupt in frustration about the problem of thinking about the Yellow River in terms of its floodplain alone: "The great illness of the Yellow River is an agonizing tumor: it is sand. Without excising the agony caused by sand, the river will never have a day of governance. The fundamental layer of the Yellow River is its soil, which is loess, and which washes away more easily than any other. The soil follows the flow of the river, and that is why the river cannot be secured."[36]

Li's commentary would not have surprised his imperial-era counterparts except insofar as they may have wondered why the problem had persisted for so long. The purpose of this book is to chart, map, and piece together the causes and consequences that Zhang Rong, Li Yizhi, and their peers identified, and to offer some lessons about the consequences of shortsighted political regimes ignoring scientific facts about long-term environmental change.[37]

Li Yizhi was a member of a loosely connected generation of Chinese intellectuals in the early and middle twentieth century who paid close attention to erosion on the Loess Plateau, the precarious livelihoods of the people who lived there, and urgent challenges that erosion posed for life downstream and for modernist ambitions of monumental dam construction.[38] Their flagship journal, *The Tribute of Yu* (*Yugong*), invoked the classical text that tells the story of Yu the Great. Ji Chaoding (Chi Ch'ao-ting) (1903–1963), who described the large-scale spatial economy of water management in China in the 1930s, influenced Anglophone commentators about Chinese history including Karl Wittfogel and Joseph Needham.

After the founding of the People's Republic of China in the 1950s, generations of historians and historical geographers, many of them associated with the Fudan University Historical Geography Institute (Fudan daxue lishi dili yanjiu zhongxin)—Cen Zhongmian (1886–1961), Tan Qixiang (1911–1992), Shi Nianhai (1912–2001), and Zou Yilin (1935–2020)—conducted extraordinary and urgent research about the history of the Loess Plateau of their day and its relationship to the Yellow River floodplain. Today, a server named for the *Tribute of Yu* hosts the website of the Historical Geography Institute, which remains a center for lively and innovative scholarship on Yellow River history. Although an in-depth discussion of the intellectual lineage of Yellow River scholarship is beyond the scope of this book, I stand humbly behind generations of scholars for whom the legacies of the Yellow River define the prospects of a nation.

CHAPTER 1

The Natural and Unnatural History of the Yellow River

The spatial form of the Yellow River has never tidily mapped onto the political and social structures of human geography, and neither does it lie neatly within a single climatological region. Its channels have always traversed ecological and political territory that is very diverse indeed. Upstream, the river's middle course, which traverses the Loess Plateau, loops around the semiarid and arid Ordos region, which lies at the periphery of summer monsoon precipitation. Historically a multiethnic frontier, the Ordos is a borderland between multiple modes of subsistence, a focus for agrarian colonialism, and a site of frequent military contention. Downstream, the lower course traverses its floodplain, a moist continental climate, a historical wetland expanse where farmers tilled land made fertile by sediment that had been deposited throughout the Holocene era. The entire North China plain was the Yellow River's floodplain until farmers and the regimes that directed them locked the river into a narrow and stable channel by means of earthworks and drainage systems. Periodically, the structures failed catastrophically, but otherwise, they held the river in place.

Regional climate and topography are such that the story of the Yellow River extends far beyond China, involving distant tectonic plates as well as global wind patterns and pelagic currents that shift as the climate changes. Much of this chapter follows the Yellow River downstream, from its headwaters on the Tibetan Plateau to its estuaries on the Yellow Sea; to where it passes through the Loess Plateau, the source of the Yellow River's famous sediment, as one of the objectives of this chapter is to explain that soil is just as important as water to understanding the history of this river; and through the province of Henan, which is the ecological and cultural pivot point between the Yellow River's many worlds. This project offers a perspective on the whole river basin all at once: its headwaters and its floodplain, all its tributaries, and the larger systems of terrain and climate within which it is constituted. Much of this chapter is about monsoons, mountains, marshes, and deserts. Rather little of it concerns water flowing through a stable and clearly banked channel, because the Yellow River, like many other rivers, has not spent most of its life span behaving that way. River history benefits from a millennial-scale perspective. The chapter ends with a section about the first span of the river's life, which bridged millions of years of changing geology and climate and supported numerous forms of flora and fauna, including, in the end, humans.

Plate 2 introduces the entire Yellow River watershed. This vast expanse encompasses most of North China, the eastern part of the Tibetan plateau, and the northeastern deserts of Central Asia. It includes the river and all its tributaries, and the watersheds of the Hai and Huai Rivers as well as the Yellow River itself. The vegetation cover today does not reflect the extent of primeval forest cover, but it does reveal the ecological possibilities that persist under intense human stewardship: a vast expanse of cropland on the floodplain and on the southern part of the Loess Plateau, upland forests, and grassland and desert throughout much of the Ordos region.

CLIMATE, TOPOGRAPHY, AND THE RIVER

Plate 3 shows that eastern Asia follows a generally downward slope from the Tibetan Plateau, where the Yellow River originates, to the East China Sea, where it disgorges into the ocean. Like almost all the rivers that pass through the territory of China as we know it today,

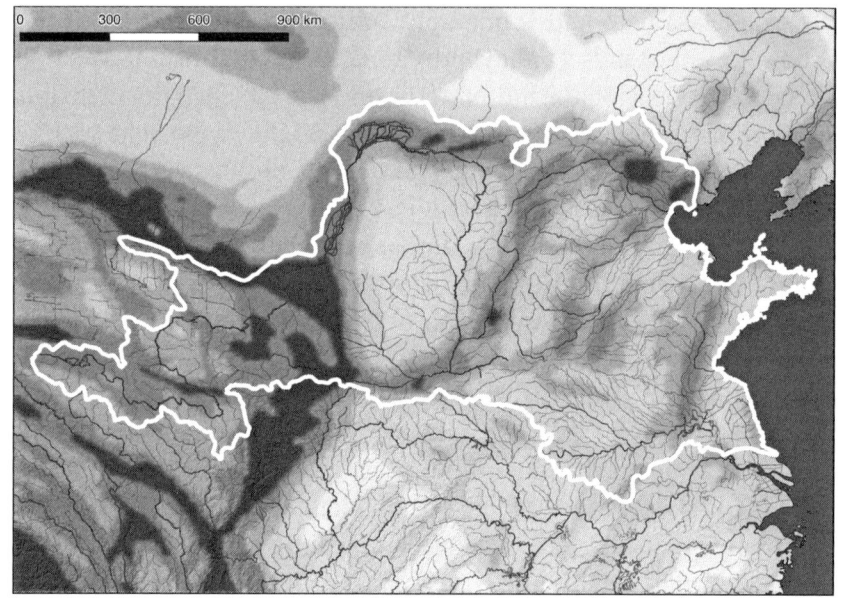

FIGURE 1.1
Seismic Regions and Historical Earthquakes in North China. The entire Yellow River is a seismically active zone. The Ordos Block is offset by faults, the Wei and Fen Rivers follow fault lines, and the Tibetan plateau is extremely seismically active. Based on Parara-Caryannis, "Historical Earthquakes in China."

it flows from west to east. The river transects the Loess Plateau—the world's largest formation of loess, a porous and friable soil. The Yellow River watershed lies at the midpoint of a steep longitudinal moisture gradient. The vast Gobi Desert lies to the north of the Yellow River watershed, and the Qinling Mountains and other ranges separate the Yellow River watershed from the subtropical Yangtze River watershed to its south.

The Tibetan Plateau is extremely seismically active (fig. 1.1). So too is the periphery of the Ordos Block that underlies much of the Loess Plateau. Its gradual uplift is the reason the river course makes a characteristic "Great Bend"—flowing north, then east, then south again through its middle reaches. Two major Yellow River tributaries, the Wei River in the southwestern part of the Ordos and the Fen River to the east of the Ordos, follow earthquake faults. Earthquakes have formed and re-formed the courses of the Yellow River and its tributaries, and they have upended human activity as well.[1]

The gradient of the downward slope of the Asian landmass toward the Pacific Ocean declines as the Yellow River and its tributaries flow across the Ordos Plateau. As the slope decreases, so does the rate at which the river transports sediment from west to east. As the geographer Charles Greer explains: "The geologic role performed by the

Yellow River is the removal of material from the western uplands and the deposition of that material on the slowly subsiding [North China] plain. To accomplish this, the river must range freely over the length and breadth of the area of deposition. The difficulties presented by this process for permanent human settlement are obvious, and they are exacerbated by the presence of loess in the middle portion of the basin."[2]

Topography is one important factor in sediment transportation, and rainfall is another. The most important characteristic of the Yellow River precipitation regime is that it is complex and difficult to predict. The river's western reaches traverse landscape features that are so large that they influence regional climate. In addition to the Tibetan Plateau (sometimes referred to as Earth's third pole because of its concentration of ice fields and the major control that it exerts on regional atmospheric circulation), the Loess Plateau is ringed by deserts and mountains, places of very diverse climate and topography that exert an influence on the weather of the Yellow River basin.

The other major climatic influence on the Yellow River basin comes from the warm and moist Asian and Indian summer monsoons, the westerlies (the prevailing winds that circle the globe from west to east in the middle latitudes), and the dry and cold Central Asian winter monsoon. Air flows toward the east throughout the year, and it also flows toward the south in the winter and toward the north in the summer (figs. 1.2a and 1.2b).[3]

The history of the Yellow River, which entrains water, sediment, and wind from thousands of kilometers away, is the history of an entire hemisphere. During the winter, the cold and dry winter monsoon that blows from Central Asia carries large quantities of aeolian (wind-blown) sand and dust, which settles in gullies and dry creek beds. During the summer, generally in July and August, humid maritime air arrives from the southeast, bearing 80 percent of the year's precipitation. Thirty-five percent of the rain falls in heavy downpours.[4] When ground cover is intact, the summer storms foster plant growth, accelerate soil formation, and stabilize sand dunes, but when grasses and trees are absent, the heavy rains herald a season of erosion and flooding. Although the total annual runoff is not large by global standards, it can cause severe flooding because of its seasonal concentration. Even a single major downpour concentrated in one region of the Loess Plateau is sufficient to cause flooding downstream, and this can occur even during a year of generally low rainfall, if a single

FIGURE 1.2

Monsoons in East Asia. The winter monsoon (a) carries sand and dust into the Yellow River basin from the northwest, and the summer monsoon (b) carries moisture to the Yellow River basin from the South China Sea. These alternating processes are essential to the natural history of the Yellow River. Based on Yi, "Holocene Vegetation."

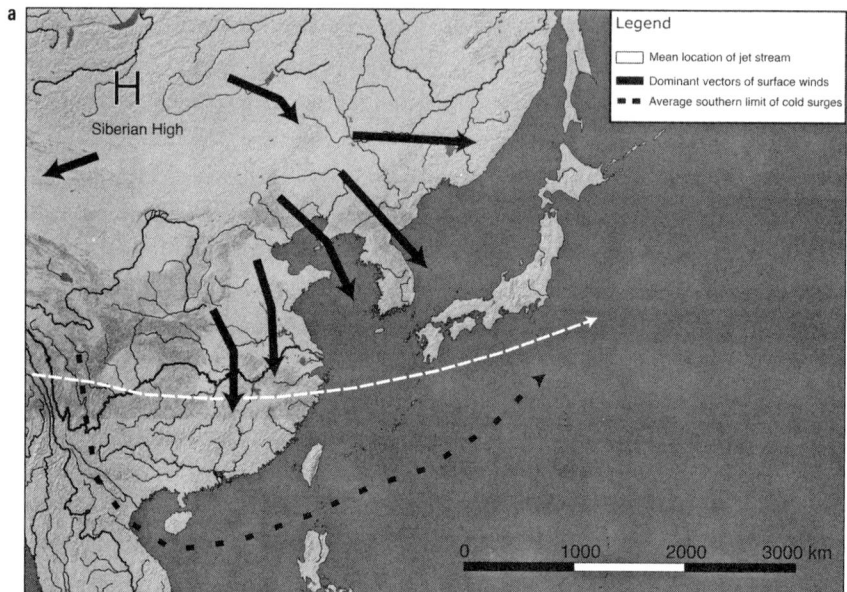

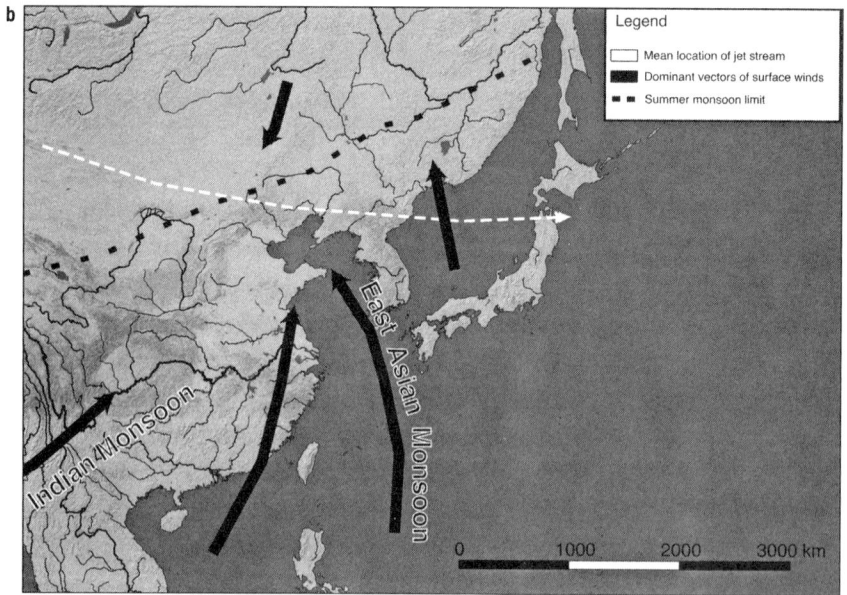

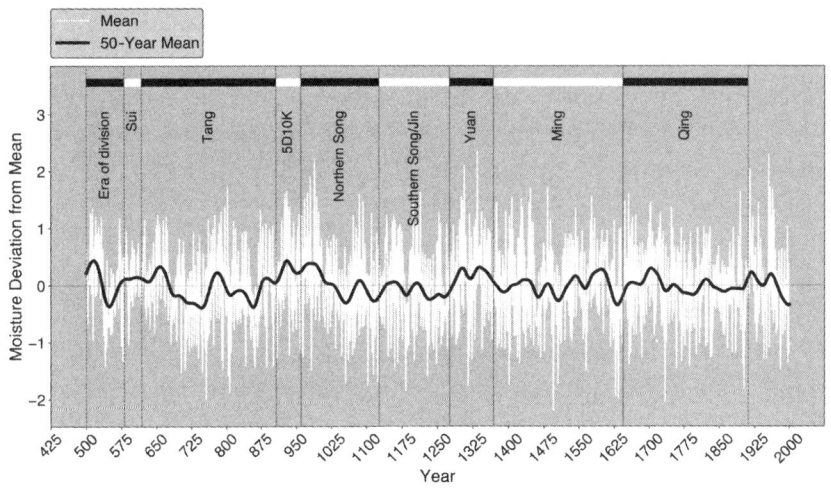

FIGURE 1.3

The History of Moisture on the Loess Plateau. This data, from Cook's Monsoon Asia Drought Atlas, records annual variation in moisture on the Loess Plateau since 500 CE based on tree-ring records. The unsmoothed narrow white line shows the dramatic variation in annual trends. The thick black line is a fifty-year rolling mean.

violent storm falls onto dry earth that has a low moisture-retention capacity.[5] The well-being of flora and fauna on the Loess Plateau, as well as of the humans who rely on its plants and animals for survival, hinges upon the amount of rain that falls during a single season each year.

The timely arrival of a handful of summer storms, during a few weeks of the year at the peak of growing season, determines how much of the previous winter's windblown sand and dust is incorporated into loam soil, the extent of vegetation growth, and the well-being of those humans and other animals that depend on plants and water. At the scale of the Pacific Ocean, unpredictable El Niño events influence monsoon behavior. During strong El Niño events, when warm ocean water wells up in the central equatorial Pacific, central China becomes extremely arid.[6] But regardless of specific weather variations, northwestern China always lies at the far periphery of the summer monsoons, in a complex precipitation regime of its own, and the amount of precipitation that falls in any particular location on the plateau varies dramatically and unpredictably from year to year.[7]

Figure 1.3 depicts the history of moisture (a composite of temperature and rainfall, a proxy for the rate of vegetation growth) on the Loess Plateau. The index is derived from a collection of tree rings, the oldest of which date to the year 500 CE. Although throughout history people could often store enough surplus grain to sustain themselves through one or two lean years, this timeline reflects the fact that periods of drought sometimes lasted for decades. During moist phases, agriculture could move toward the north. When it was dry, farmers retreated south with the monsoon perimeter. There is some correlation between the sunspot cycle and flood events, but this remains an "erratic precipitation regime which produces conditions marginal to unirrigated agriculture throughout much of the basin."[8] The whole Yellow River basin is situated at a climatic frontier, and the transition zone between the Mu Us Desert and the Loess Plateau, within the Ordos, is particularly sensitive to climate variability.

Although weather in the Yellow River basin is unpredictable and varies annually, it does so within the framework of large-scale climatic patterns. For instance, the summer monsoon tends to be more forceful when Pacific temperatures are warm. High-precipitation rainy seasons are most likely to occur during periods of rapid climate transitions between cold and warm spells, when violent storms follow arid periods during which wind-borne sediments and dead vegetation accumulate in upstream gullies.

Temperature is one factor that correlates with precipitation: cooling is associated with drier conditions and warming is associated with wetter conditions.[9] Annual and decadal precipitation totals affect phenomena like population density and modes of subsistence on the Loess Plateau. However, concentrated rainfall and high-intensity rainstorms occur more often during years that are dry overall, and periods of drought and sediment accumulation punctuated by heavy storms are the conditions most conducive to catastrophic floods.[10]

However, summer precipitation does not reliably reach the whole Ordos region under any climate regime. The northern extent of the summer monsoon determines the human and animal carrying capacity of the Loess Plateau, the potential mix of agriculture and pastoralism, the locations occupied by nomadic bands and their herds, and the likelihood of conflict over fertile land and scarce water.

The large-scale climate picture presented thus far is critical for understanding Yellow River history. During the past millennium, human activities have had more impact than climate and weather on desertification and erosion. Soil cores at the desert-loess margin, where sandy loess layers alternate with vegetation-rich soil, reveal that at the beginning of the Holocene era as temperatures warmed, the latitude of the desert margin moved northward by about three degrees, the summer monsoon margin moved north, and precipitation increased. However, today the desert margin has shifted close to its most southerly position, even while the monsoon is near its most northerly margin. It is unprecedented for the desert margin to diverge from the summer monsoon margin. There is no way to explain this from climate models. Human activity is the driving mechanism—though the pressures of weather variation and climate trends influence social behavior as well.[11]

At a large scale, Yellow River history is a story about how global weather and climate patterns affect a transnational eastern Asian region, a river basin that sits in a seismically active region of unstable climate and even more unstable annual weather, and astride a region of transitions between ecological zones. These facts have an impact on Yellow River history, even though people forge the events that populate historical sources.

THE RIVER'S COURSE: THE UPPER REACHES

The Yellow River is a single river basin, although historically it has seldom been treated that way in Chinese political culture. The river emerges from headwaters on the Tibetan Plateau, traverses the Ordos Plateau and the Loess Plateau, and disgorges to its alluvial plain on the North China Plain before entering the Pacific Ocean at the Yellow Sea. The regional designations of upper, middle, and lower course are convenient terms for indicating ecological and topographical distinctions between parts of a single fluvial system, but of course they are not sharp dividing lines. For instance, Chinese provincial borders, which have been quite stable across history, do not map onto any of these ecological divisions. In human imagination and in imperial policy and engineering, these regions of the river have often been treated as if they were disconnected in spite of the fact that upstream events affect downstream phenomena.

The headwaters and upper reaches of the Yellow River basin gather at the northern foot of the Bayan Har Mountains. This range, a branch of the Kunlun Mountains in Qinghai province, is a watershed between the Yellow River and the Yangtze River, separating the drainage area of China's two major rivers. The region of the watershed that is conventionally designated as the upper reaches extends to the northeasterly point in the Ordos Loop in Inner Mongolia, where the river turns southward in its Great Bend. Over the course of this journey, the river's elevation drops by a total of 3,496 meters (11,470 feet) and the river adds forty-three tributaries. This region, much of it featuring fast-flowing current through rocky gorges, accounts for only 8 percent of the river's contemporary silt load.[12] Plate 4, an overview of the upper reaches, oriented toward the northwest, depicts the five provinces that incorporate this region of the river basin and shows that, although many tributaries join the river on the Tibetan Plateau, little of the region is arable farmland even today.

The sparsely populated, high-elevation headwaters region is a land of lakes, wetlands, and grasslands. The river descends from there into a series of more than twenty gorges with rapidly falling elevation where many tributaries join the turbulent and rapidly flowing stream. After emerging from the gorges, the river reaches the city of Lanzhou, the provincial capital of Gansu, before commencing its great turn north to circumscribe the Ordos region.

In 2014, the photographer Ian Teh traveled along the Yellow River to chronicle environmental changes and social transformations along its upper course. Plates 5, 6, and 7 are among the gorgeous pictures he produced. At Ngoring Lake at the Yellow River headwaters (plate 5), glaciers are melting rapidly and water levels are rising. Around Gyaring Lake in Qinghai province (plate 6), grasslands have degraded into gullied and treeless expanses as a result of overgrazing after communal pastures were fenced into household plots in the 1990s. At Bayin, in Gansu province (plate 7), sand dunes are evidence of desertification that is presently consuming a million acres of grassland each year.[13]

East of Lanzhou, in Gansu province, the river turns north into its Great Bend and curves around the periphery of the Ordos Plateau. Here, the Liupan Mountains and the western extension of the Qinling Mountains separate the Yellow River from the headwaters of the eastward-flowing Wei and Jing Rivers. These three rivers converge hundreds of kilometers downstream at the southeastern edge of the Ordos.

The Yellow River's 1,500-kilometer-long loop flows north, east, and south before turning east once again. This Great Bend, also known as Hetao, or "riverbend," takes the waters around the periphery of the Ordos Block, a solid section of Earth's crust that is bounded by faults. When the Yellow River first formed, between fifty-six million and thirty-four million years ago during the Eocene epoch, it drained

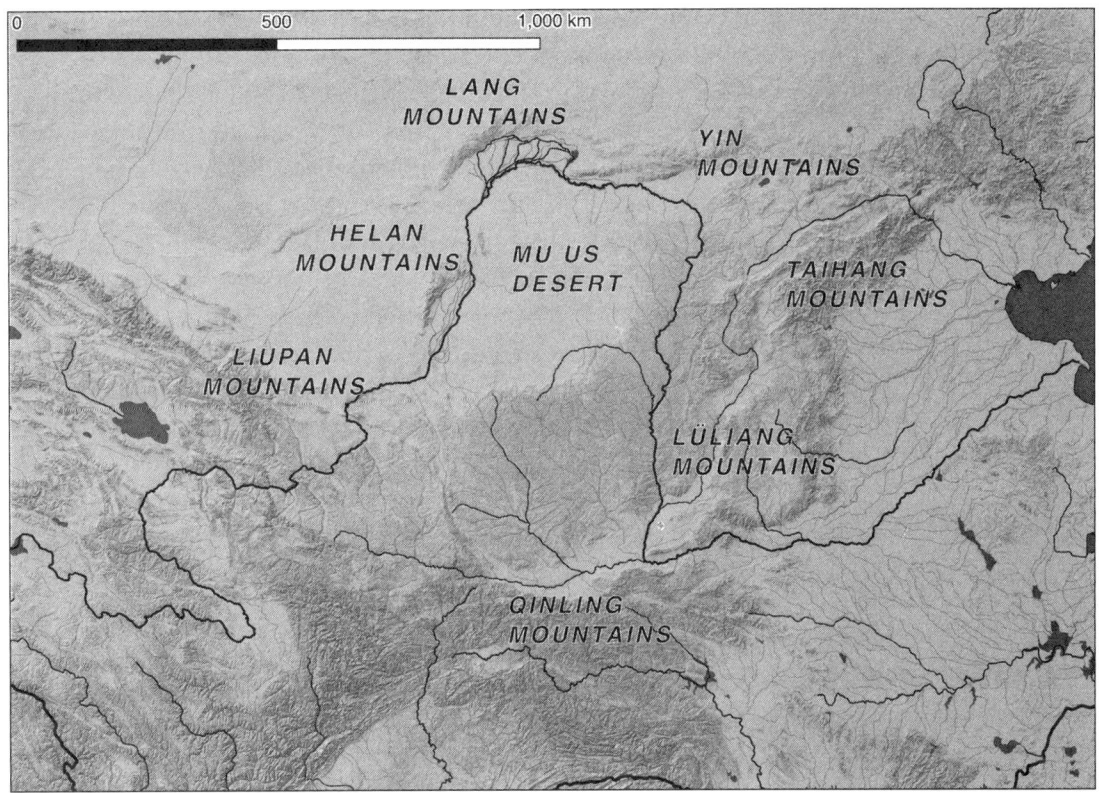

FIGURE 1.4

The Hetao Region. The Hetao region, circumscribed by the Yellow River's Great Bend, includes the Mu Us Desert in its northern portion and is ringed by mountain ranges that affect both rainfall and human history.

due east. It began to flow around the Hetao bend around five million years ago as the Ordos Block folded and uplifted into place, a result of the tectonic collision of India into Eurasia as well as local rifting.[14] The river flattens out as it circles the Ordos. Ringed by mountains and deserts, the Ordos is relatively isolated from the rest of China, which has had a significant impact on human history.

Hetao, a semiarid and arid region of hills, grasslands, and deserts that is ringed by mountain ranges, contains the Mu Us Desert (Maowusu shamo) as well as a large portion of the Loess Plateau (fig. 1.4). The Mu Us occupies the northern half of the Hetao region. In the grassland regions to the south of the desert, along with the Yinchuan wetlands and Liupan Mountains to the west of the desert, more than in the Mu Us itself, people have contended for territorial supremacy.

The Mu Us lies in the rain shadow of the Lüliang Mountains, which divide the Yellow River from the Fen River, one of its major tributaries. The Lüliang Mountains force the Yellow River southward on the eastern side of Hetao before tapering off and permitting the Fen to turn west and join the Yellow River as it flows south. To the east of both the Yellow River and the Fen River, the Taihang Mountains form the eastern perimeter of the Loess Plateau. These mountains and their passes are historically strategic places that isolate the Ordos in all directions; they were frequently fortified during historical times, with significant consequences for erosion.

Although Hetao is predominantly semiarid and arid, the floodplains of the southeast-flowing Yellow River tributaries that transect it—the Wei, the Jing, the Luo, and the Wuding (a name that translates as "the Never Stable")—all have substantial arable land. When Chang'an, modern Xi'an, located on the Wei River plain (a region known as Guanzhong), was the capital of the medieval Tang regime, it supported a population of a million people who ate Ordos grain and burned Ordos timber. The Ordos region also contains two irrigated alluvial plains, regions that catch moisture at the windward foot of mountain ranges: the Yinchuan Plain on the western side, and the Hetao Plain on the northern side. Saline and waterlogged regions of poor drainage, they were nevertheless large and strategically situated oases. Both of them, the Yinchuan Plain in particular, were historical sites of grain production and settlement, repeatedly contended among multiple cultivators and regimes.

Plates 8, 9, and 10 are photographs of the Yinchuan landscape, and plate 11 is a map of the landscape's eighteenth-century settlement geography and vegetation zones. A narrow strip of farms and forests, traversed by the Yellow River and numerous tributaries and canals, the Yinchuan Plain is situated between the Mu Us sand dunes to the east and the rocky slopes of the Helan Mountains to the west. Even this far upstream, the river's motion shaped human settlement. The historical city of Lingzhou ("the miraculous prefecture"), situated on the southern part of the Yinchuan Plain, began as a sandbar fortress in the middle of the river, which, according to legend, remained "miraculously" free of flooding. However, by the eighth century, the sandbar on which the city was situated joined the eastern embankment as the increasingly muddy river shifted west. By the eighteenth century—after numerous changes to both the city site and the river's course—the river lay 10 kilometers away.[15]

THE RIVER'S COURSE: THE MIDDLE REACHES

A traveler following the river from its headwaters would turn north with it along the western side of the Ordos, then east along the northern side of the Ordos—a region that includes the western and northern sides of Hetao—and then would next track it along the southward-flowing part of the Hetao bend. At the southeastern periphery of the Ordos, the river turns east, passes through a final set of gorges, and debouches onto its floodplain. This relatively short portion of the river's course—only about 1,200 kilometers, or 750 miles, less than a third the length of the upper course—is the segment at which the river traverses the highly erosive Loess Plateau, where its sediment-engorged tributaries join the Yellow River.[16] This site is where the Yellow River, at least for the past millennium, has taken on the hue that provides its name.

The entire Ordos hydrological system flows southeast, so this segment of the river drains the entire Ordos Plateau and its sediment-choked tributaries. With a total elevation drop of only 890 meters, just under 3,000 feet, the Yellow River and its tributaries travel slowly through this area, forming a distinct hydrological region that extends downstream to the city of Zhengzhou, in Henan, at the western edge of the North China Plain. Thirty large tributaries enter the Yellow River in its middle reaches, increasing water volume by almost 50 percent and contributing almost all of the river's sand and silt.[17] Figure 1.5 depicts the middle course of the Yellow River, which runs from Hekou, a village of Toghtoh County in Inner Mongolia, to Zhengzhou. The map also depicts the major tributaries, all of which drain the loose soil of the Loess Plateau.

Downstream from Hekou, just past the southern turn of the Great Bend, lies the Jinshaan Grand Canyon (Jinshaan daxiagu). The canyon, and the Hukou Waterfall (Hukou pubu) beyond it, flow yellow from sediment. They are also barriers to river transport, and thus the Yellow River is not a transportation artery in its middle reaches. Boats could transit parts of the Loess Plateau by way of major tributaries, but they could not travel to the upper reaches of the Yellow River itself. The Hukou Waterfall marks the edge of the Ordos Block, and it is also the edge of the Mu Us Desert, with its sandy and alkaline soils. Downstream from there, the river widens and slows as it cuts through the heart of the Loess Plateau. This region, with numerous tributar-

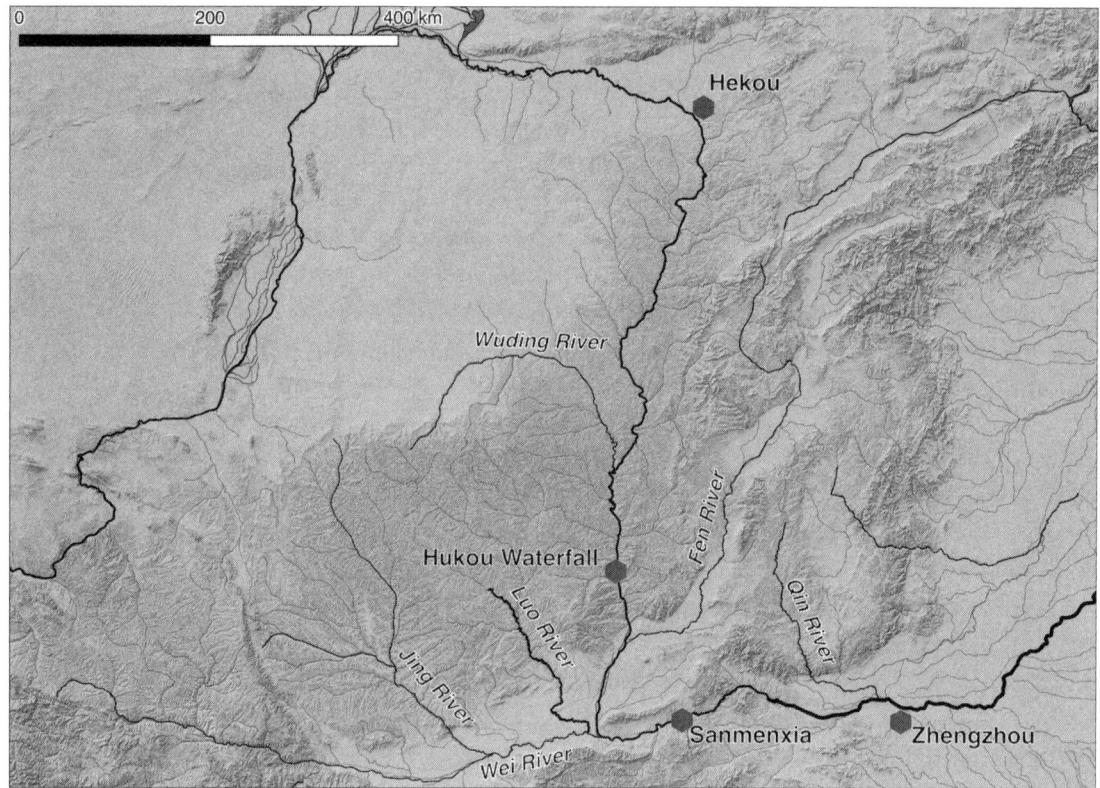

FIGURE 1.5
The Middle Course Region.
The middle course of the Yellow River, from Hekou to Zhengzhou, drains most of the Loess Plateau and includes all the river's major tributaries.

ies and annual rainfall of less than fifty centimeters (twenty inches), is where the river begins to accumulate massive quantities of sediment—up to 90 percent of its total sediment—as almost all the Yellow River's middle-course tributaries funnel into this small region.

One-third of Yellow River sediment comes from the Wei basin.[18] Bounded by mountain passes and loess hills and known as the Guanzhong Plain, the lower reaches of the Wei River is one of the most important regions of the Yellow River watershed. As long ago as Neolithic times, Guanzhong was known for fertile soil, forest cover, and excellent water and drainage.[19] Its relative peripheralization from the center of political and economic activity in China after the Tang era may be both cause and effect of the ecological deterioration of the region, which became more distant from state scrutiny and security than it had been before Chang'an (Xi'an) lost its viability as an imperial capital in the ninth century. Figure 1.6 depicts the dense settlement geography of Guanzhong and its hinterland during its heyday

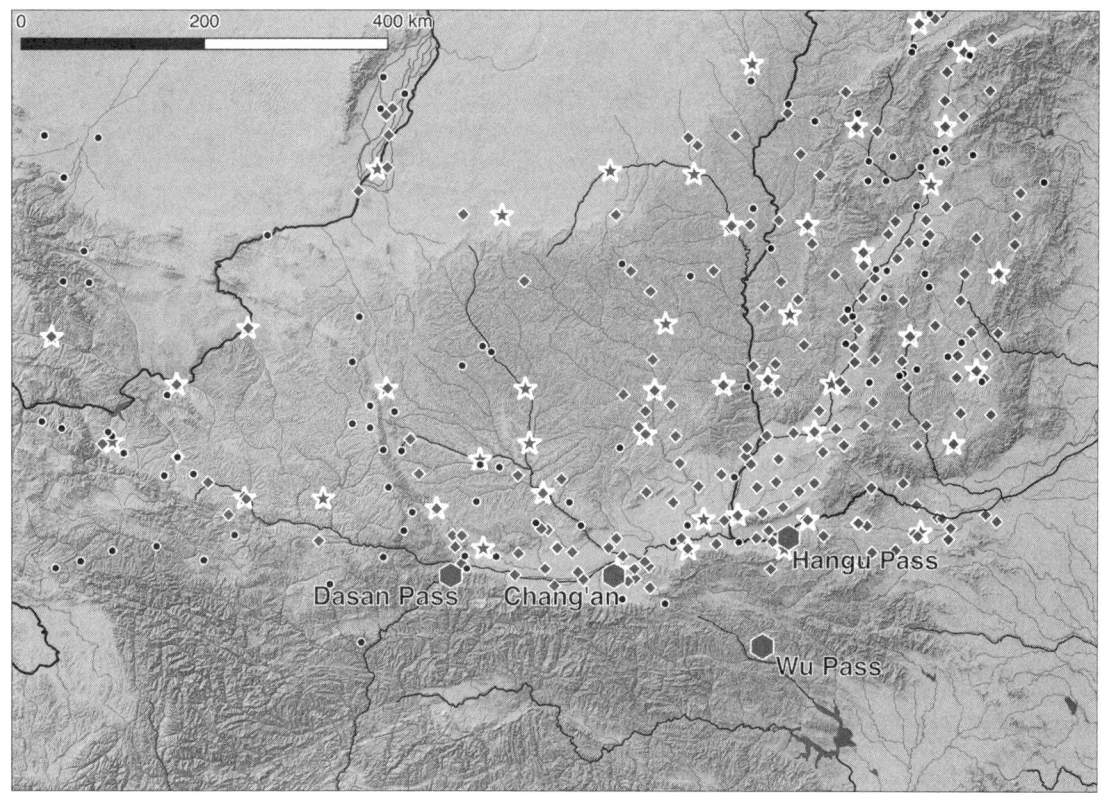

FIGURE 1.6
The Guanzhong Basin. The Guanzhong region, the Wei River valley between Dasan Pass and Hangu Pass, was a frequent site of regime capitals until middle imperial times. During the eighth century, as shown here, the capital was at Chang'an, and Guanzhong was densely settled.

as the Tang capital around the turn of the eighth century. In addition to the metropolis of Chang'an, the Guanzhong Plain, approximately the size of the state of New Jersey, was the site of seven prefectural capitals, more than fifty county seats, and numerous small settlements.[20] Until the ninth century, Guanzhong was often the site of an imperial or regime capital. With deteriorating ecological conditions in Guanzhong and throughout the Loess Plateau, the region came to depend on imported supplies, but these were difficult to transport into the relatively isolated area.[21] With increasing erosion, irrigation works on the Guanzhong Plain became more difficult and costly to maintain while also yielding ever-smaller results. Plate 12 depicts the ruins of irrigation works at the western edge of the plain. Dating to around the turn of the Common Era, they are located today far from any water source.

The rugged Qinling Mountains extend from east to west to form the southern boundary of the Yellow River basin. They serve as a

FIGURE 1.7
Sanmenxia before the Dam. Before dam construction in the 1950s, the narrow channels, rapids, and hairpin turns at Sanmenxia were a major impediment to river transportation. Courtesy of the Sanmenxia Local History Office (Sanmenxia difang shizhi bangongshi), item number 2783. Thank you to Xiangli Ding for identifying this image.

watershed between the arid north and the humid south. They turn the Yellow River, engorged with sediment and tributary waters, sharply eastward at Tong Pass (Tong guan), which divides the Wei Valley from the North China Plain. East of the pass, the river is constrained by rocky gorges and mountains that create another transportation bottleneck, forming the Sanmenxia rapids as well as famous cliffside sites such as Longmen Grottoes (Longmen shiku). This unnavigable stretch of the river made it difficult and expensive for grain and other resources produced on the floodplain to be easily transported to Chang'an and the rest of the Ordos. Hetao's resulting self-sufficiency, then, has been a major driver of deforestation on the Loess Plateau.[22] Sanmenxia today is the site of a massive hydroelectric dam. Figure 1.7 depicts the river's former hairpin curve through the denuded hills of the Sanmenxia Gorge and the islands that formed the narrow "three gates" that gave the site its name and that forced the current to rush rapidly downstream through the site.

At the eastern edge of the middle course, the river flows through a narrow valley punctuated by a final group of several gorges. The tributaries that enter here drain the eastern edge of the Loess Plateau. This region receives more precipitation than the northwestern portion of the plateau. It is a site of dense population, intensive agriculture, and many ancient historical sites. The region contributes significant water and sediment runoff to the main course of the river.

LOESS SOIL AND THE LOESS PLATEAU

Having followed the river to its floodplain, we now take a lengthy pause from our downstream journey to focus on soil rather than water. The Yellow River today is the most sediment-laden river in the world. It carries about 1.5 billion tons of silt per year, 90 percent of which comes from the Loess Plateau. That sediment—and its complex of shifting interactions with weather, water, terrain, and society—has a history, just as the river itself does. The most important fact about loess landscapes is that they retain moisture very well when they are intact, but they deteriorate rapidly when ground cover is removed. Throughout history, when ground cover has been removed, all the soil carried off the plateau by wind and water has made its way into the Yellow River and onto the floodplain.

The loess region is not one single ecological zone. The region is dominated by a single type of soil, loess, but it spans forest and steppe vegetation, covering parts of Gansu, Inner Mongolia, Shaanxi, Shanxi, and Henan provinces. Average annual precipitation transitions sharply from over 600 millimeters, or 24 inches, in the southeast to under 300 millimeters, or less than 12 inches, in the northwest (fig. 1.8).[23] The northern part of the Loess Plateau is dominated by dry grasslands and shrubs, with sand dunes in the more arid areas but forested areas along river corridors. Forests flourished along the littoral zones of streams, and vegetation grew more densely during warmer and wetter periods, but according to pollen studies, shrubs and grasslands were the dominant vegetation types outside of river zones.[24] On the southern part of the plateau and in rain shadows like the Yinchuan Plain, vegetation is relatively lush. Plates 13–16 depict the terrain in four different regions of the Loess Plateau. The arid terraces and gullies at the peripheral agricultural zones of Lüliang (plate 13) and Yulin (plate 14), the tree-lined riverbanks of Yinchuan (plate 15) and the lush subtropical vegetation of Pingliang, punctuated by muddy loess cliffs and spires (plate 16)—all of these places (indicated by stars in figure 1.8) are Loess Plateau landscapes.

Climate and weather variation also play an important role in understanding the diversity of the Loess Plateau. Rainfall numbers are highly inconsistent annually, sensitive to climate variability and to the reach of the monsoon in any given season. During wet summers, soil and vegetation form on grasslands and dunes, and during typical winters, a layer of aeolian sand blows in to cover them. During

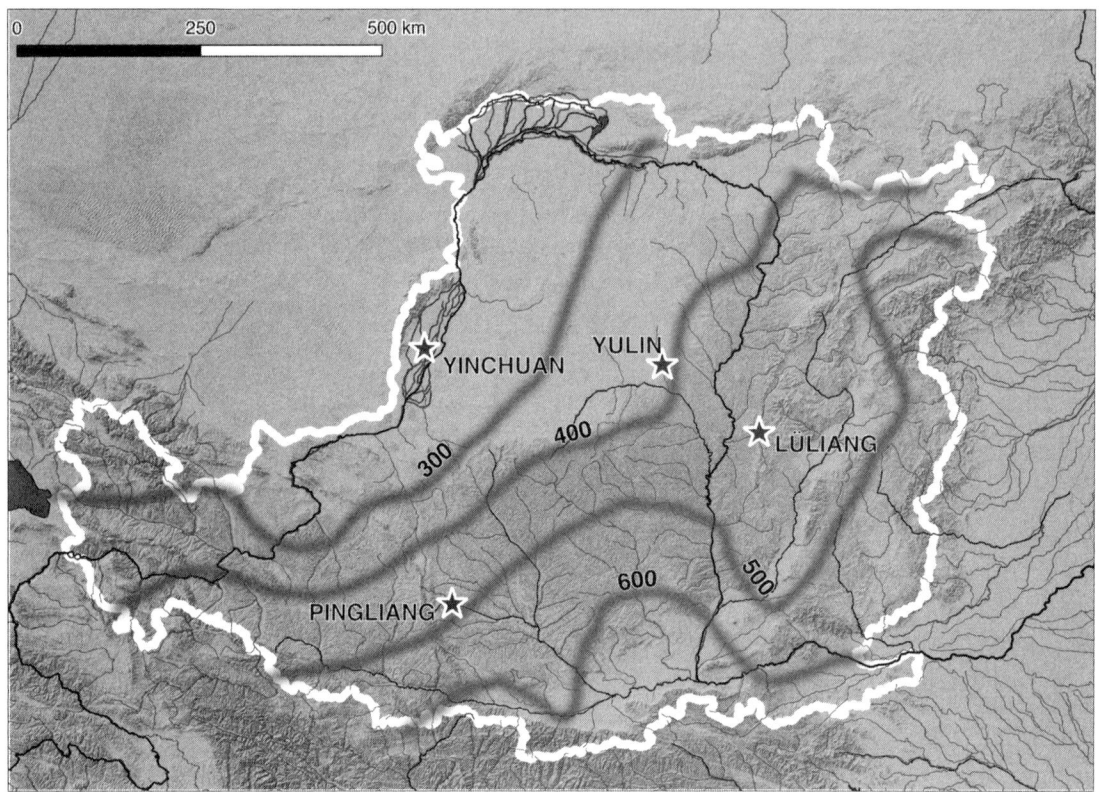

FIGURE 1.8
The Loess Plateau Rainfall Gradient. There is a steep moisture gradient on the Loess Plateau, ranging from an average of more than 600 centimeters per year in Guanzhong to less than 300 centimeters per year in the Mu Us Desert. Stars mark the locations of the four photographs in plates 13–16. Based on Wang et al., "Precipitation Gradient Determines the Tradeoff."

dry periods, forest degrades to steppe and becomes less resistant to erosion.[25]

The Loess Plateau, which is approximately the size of the state of Minnesota, is ecologically diverse, and it also includes regions of rocky mountains, alluvial soils, sand dunes, and other formations that are not characterized predominantly by the presence of loess soil. Its elevation gradients extend to summits over 3,300 meters, or 11,000 feet, and its varied climates and terrains historically supported a wide range of subsistence modes, including farming, forestry, salt mining, pastoralism, hunting, trade, and soldiering.[26] Herd animals enhance the health of soil and ground cover when their numbers are moderate, but overgrazing can contribute to serious erosion. Xi'an, the largest city on the Loess Plateau, the historical capital city of Chang'an, is situated at an ecotone, a transition zone between two historical biological communities: grasslands to the northwest and woodlands to the east.

In recent years, improved access and analysis of soil cores and pollen samples is reducing the amount of controversy about the extent of forest cover on the Loess Plateau.[27] Historically, forests were much more widespread than they are today. Their exact extent and location depended on the proximity of water sources and the aspect of slopes (whether they face north, south, east, or west) as well as variations in rainfall and evaporation rate at any time. What is certain is that there has always been a "mosaic of vegetation" on the plateau, including thick forests around rivers and other water sources, plains with deciduous woodlands, coniferous trees at higher elevations, savannas with sparse trees and brush, and dense woodlands on south-facing foothills.[28] For most of the Holocene era, the western Loess Plateau's ground cover was steppe; the north was sparsely wooded steppe, and the south was more densely wooded steppe, but forests persisted even when the climate fluctuated. The Loess Plateau became predominantly dry steppe during cold and arid glacial periods.[29] Grass-covered and forested lands still constituted more than half the plateau's total land area as recently as 2,500 years ago, a time for which historical documents, pollen records, and archaeological sources all reflect considerable forest as well as shrubs and grassland. Today, however, grasses and trees cover only 3 percent of the plateau.[30]

Forest cover on the Loess Plateau was particularly extensive in riparian regions, around lakes and springs, at higher elevations, especially on slopes with south- and west-facing aspects, and on more southerly parts of the loess region. Groups of people, as well as trees, plants, and other living things, assembled themselves near water sources, in regions of greater rainfall, and in less rugged terrain. People and their crops and livestock competed with plants and wild animals for water and soil. Grasslands and forests interlocked with each other, with one or the other predominating on the basis of slope and aspect, elevation, hydrology, and other characteristics. Few vast stretches of terrain were devoted to only grasses or trees. Many animals, including early humans, occupied both forests and grasslands and moved freely between them.[31]

Although there is anecdotal evidence in the Chinese historical record about people managing forests to provide sustainable access to fuel and building material, the preponderance of the evidence reflects periodic large-scale destruction of forests for conversion to agricultural land. Before the seventeenth century, this often resulted from government-sponsored colonization of remote places by civilian and military settlers. Later on, local officials were evaluated in part on their success in expanding cultivated acreage. As populations grew, the demand for fuel could also outstrip the carrying capacity of the land. Forests and grasslands were often destroyed in warfare as well, cleared to deprive adversaries of cover and of fodder for their animals, and when they were burned, it affected soil as well as vegetation.[32] Construction, especially for large Loess Plateau cities like Chang'an, also required a lot of timber. By the turn of the first millennium, there was a commercial timber market along with government requisition.[33] Soldiers stationed on the Loess Plateau did not receive adequate pay or provisions to support themselves, and they often turned to timber sales as a kind of sideline employment.[34] Under ideal conditions, even intensively exploited forests can grow back, but forestry on the fragile slopes of the Loess Plateau often degraded the soil and created steep ravines that prevented this from occurring.

Vegetation insulates the soil by reflecting sunlight and trapping moisture. Plant roots retain water in the ground, and forest canopy prevents water vapor from escaping to the air. Therefore, deforestation and ground-cover disturbance themselves cause droughts and exaggerate seasonal temperature variations. When vegetation is destroyed, water vapor that rises into the atmosphere may dissipate or be transported hundreds or thousands of kilometers away rather than locally providing a cooling and moistening effect. Less water in the soil also means less evaporation to the air. Over an area that is hundreds of square kilometers or more, this leads to less regional rainfall and less soil moisture. When rain does fall, without soil and vegetation, it does not penetrate the ground and results in runoff. New soil ceases to form. Temperatures rise and drought stress increases. Today, even in the tropical Amazon, forest destruction is creating a new regional prairie ecosystem.

Human activity caused deforestation and permanent ecological changes. However, changing social activity also permitted periods of regrowth. The Loess Plateau population periodically contracted as a result of rebellion, epidemic, famine, and regime change and collapse. Governments sporadically enforced laws against forest poaching, particularly those regimes that favored herding over farming, and forests and grasslands had a chance to rebound in particular locations. Nevertheless, the overall trend was toward decreasing areas of forest cover. When erosion occurred and when streams desiccated, forests could not grow back easily.[35]

The Loess Plateau was never glaciated, so its soils have been evolving throughout the 2.5 million years of the Quaternary era, alternating between moist periods of alluviation, when soil was formed and distributed, and dry periods of incision, when local soil and wind-blown loess eroded and rivers cut channels through layers of sediment. The deposition of today's loess began around forty-five thousand years ago.[36] The soil on the plateau today, one hundred meters thick, or over three hundred feet, consists of bands of silty and sandy loess transported by wind from the northwestern Gobi and modern Xinjiang and Mongolia. These bands alternate with clay-rich layers that formed locally. Dust storms during the dry and windy fall and winter have always been a feature of the Yellow River watershed—even the earliest historical sources refer to dust storms. Sand and loess blew into the river and across the North China Plain eons before human activity had an impact.[37] Indeed, Asian loess, borne by wind across the Pacific, is also a component of soil in California's Sierra Nevada mountains.[38]

Loess sediment is fine dust, typically in the 20–50 micrometer size range, less than half the width of a human hair, and its formations are loose, friable, and highly porous. As figure 1.9 depicts, loess comprises three differently shaped particles that leave interstitial voids when packed together and is therefore quite porous. The fine powder is easily kicked up by the wind, but once it is blown in and deposited, its microscopic crystals link together to create sturdy landforms. Plate 17 is a segment of the loess-constructed Great Wall at Yulin, which still stands where it was constructed in the fifteenth century using the pounded-earth technique.

Loess was easy to work in preindustrial agriculture and also prone to erosion. Plate 18 shows how easily a clod of loess soil, which holds together when undisturbed, breaks down when lightly manipulated. Owing to these characteristics, loess soil is very permeable to moisture when it is covered, but it is also prone to erode when it is not. Roots enhance its porosity. Its fine grains, loose texture, and high mineral content make it "perfectly arable."[39] When it is covered with vegetation, it is highly permeable and has a high capacity for water storage, so there is no runoff even during the heaviest rainfall. A healthy loess landscape is "a water reservoir in the soil."[40] As the soil

FIGURE 1.9

A Microscopic View of Loess Soil. Loess soil comprises fine grains of various shapes that are separated by interstitial voids. This characteristic contributes to excellent drainage when ground cover is intact and a high propensity for erosion when it is not. Courtesy of Li Ping, Sai Vanapalli, Tonglu Li, and *Journal of Rock Mechanics and Geotechnical Engineering*.

scientists Ren Mei-e and Zhu Xianmo conclude, "Loess is very resistant to erosion under vegetation cover but readily erodible without it. . . . [U]nder forest or grass cover, slope and rainfall intensity have relatively little effect on soil erosion."[41] One text on erosion worldwide typifies China's Loess Plateau as "the most highly erodible soil on earth."[42]

Erosion is influenced by the nature of the loess itself, the slope of the ground, rainfall intensity and type, vegetation cover, and soil conservation. Many of these factors are influenced by human activity. Beginning more than five thousand years ago, land conversion for agriculture began to strip away soil faster than natural forces of erosion did, making humans into geomorphological agents. By the fourth century, loess layers that had accumulated during geological time were exposed to air as a result of anthropogenic erosion.[43] In the past three centuries, the erosion rate has been ten times faster than natural erosion rates.[44]

Although there are stony mountains of various sizes throughout the plateau, the loess-dominant parts of the plateau have three characteristic landforms that have largely been formed through erosive processes. Highland plains (*yuan*) are large areas with slopes of less than 5 percent and abrupt edges descending to valleys. Long ridges with arched tops and smooth slopes (*liang*) lie between two valleys. There are also round mounds with steep slopes and many gullies (*mao*). Erosion on the Loess Plateau proceeds in a sequence through these landforms. A *yuan* dissected by valleys becomes several *liang*. Gullies cut *liang* into multiple *mao*. Today, *mao* dominate the Loess Plateau, especially in the north and northwest. During the Bronze Age, the early historical Zhou era (1046–256 BCE), *yuan* occupied at least three times as large an area as they do today.[45] Valley plains, like the Fen River plain and the Wei River plain, occupy less than 15 percent of the plateau.[46] Plates 13, 14, and 16 depict a range of *mao* landforms.

Because of the arid and semiarid climate, vegetation on the Loess Plateau grows slowly and takes a long time to recover from disturbance, and erosion there creates a feedback cycle from which recovery is difficult. As erosion increases, there are more regions with steep slopes, which in turn exacerbate erosion further. In fact, the rate of erosion is five to ten times as rapid on a twenty-eight-degree slope as on a five-degree slope, so hillside settlements were far more destructive than settlements on plains and valley bottoms, and erosion accelerated as *yuan* turned into *liang* and *mao*. People moved uphill under pressures of military defense needs and population growth, and this process fed on itself. The prehistoric topography of the plateau was quite different from what it is at present. Only after vegetation destruction did the gentle and flat ground of the plateau become dissected into gullies and hills, and that landscape, which is characteristic of much of the plateau, is only between 1,000 and 1,500 years old.[47]

The Loess Plateau is sensitive to climate variation, but especially to anthropogenic change, which has been the main controlling factor of vegetation dynamics over the past one thousand years or more.[48] Tianchi Lake, located in the Liupan Mountains in the southwestern part of the Loess Plateau, is today a disturbed grassy steppe landscape with some pine trees and plants near the lake. However, pollen core samples from the lake reveal that it was a coniferous and deciduous mixed forest until about 2,200 years ago, when tree cover began to give way to herbaceous plants in the artemisia family, such as tarragon, wormwood, mugwort, and sagebrush, which thrive in disturbed landscapes in dry or semiarid habitats.[49]

Sediment transportation involves multiple regions of the middle course, and multiple entrained materials. Coarse sediment is sand and residue from sandstone and weathered bedrock that comes from semiarid non-loess areas in the western and northern part of the Ordos region. In winter, when the weather is dry, the wind from the northwest blows sediment into gullies and channels and sometimes forms dunes. There it sits until the summer wet season or for the duration of a drought.[50] Thereafter, it is carried downstream slowly, season by season, depending on rainfall intensity. It is this seasonal alternation that makes this process and its effects so powerful.

Summer rainstorms cause runoff, which contains fine loess material that is no longer protected by the roots of trees and grasses, as well as the previously deposited coarse material. Summer rainfall is heavy, and there is extremely little moisture the rest of the year. During rainfall events, the runoff becomes a syrup of water and fine loess sediment that allows particles of coarser windblown sand from non-loess sources to remain suspended rather than settle out to the bottom. These thick, hyperconcentrated flows, characteristic of the Yellow River basin, are the source of high-intensity erosion and high sediment yield on the plateau as well as flooding downstream.[51] The fact that there is this variation in sediment size is a significant reason Yellow River flood events can be so intense.[52]

Because loess is so prone to erosion, and because much of the rain on the Loess Plateau falls in downpours, often after very dry conditions, the river carries proportionately more silt relative to the volume of water during flood periods. High water on the flood plain in effect takes the form of mudslides. The heavy sediment load also speeds up alluvial action, as its deposition on the riverbed forces the river to change its course much more frequently—and sometimes suddenly—than it would otherwise. This is what has caused the river's rampages throughout history. When the sediment discharges as the flood crest reaches the plain, the river can change course very rapidly.[53]

Plate 19 shows the approximate size and location of the deltas that the Yellow River has formed at the ten locations that span approximately 500 kilometers, or 300 miles, from north to south from which it has disgorged into the East China Sea over the past six thousand years. Each of these is a site of paleosediment deposition and they are one of the signals of changing sediment accumulation rates.[54] The largest delta by far is the one that the river occupied between 1128 CE and 1855 CE. Figure 1.10, based on analysis of soil cores in historical river courses by the environmental scientist Xu Jiongxin, confirms that sediment rates doubled above past maxima at around the time the river commenced its southern course, then doubled again around the time that it turned to the north once again in the nineteenth century. Box 1.1 offers a more detailed explanation of the techniques that he and other environmental scientists use.

Xu Jiongxin has used soil cores from Yellow River paleocourses and other locations on the North China Plain to estimate the human contribution to erosion from the Loess Plateau. Between eleven thousand and six thousand years ago, according to his research, the Yellow River deposited approximately 0.54 centimeters of sediment on its alluvial fan each year. Between six thousand and three thousand years ago, the rate was approximately 0.75 centimeters per year. At present, the rate is an average of 8.2 centimeters per year, more than fifty-eight times the rate of the early period. The difference between the baseline and contemporary rates cannot be explained without human activity.[55]

Another research group estimates that human activities have augmented natural soil erosion by about 41 percent in the Yellow River basin.[56] This is a conservative estimate that reflects the assumption that erosion rates would have risen more rapidly even without human intervention as the plateau became more dissected as the climate became drier in recent millennia. The current rate of erosion on the Loess Plateau is mainly due to humans' land use and destruction of the natural vegetation, and it correlates directly with population size.[57] By contrast, yet another study compared a snapshot of sedimentation three thousand years ago with that of the mid-twentieth century to conclude that human activity accounts for 52 percent of the total present erosion rate, and natural erosion

Box 1.1
How Do Scientists Study Changes in the Erosion Rate?

Scientists use several methods to study the changing rate of erosion and to determine the likely human contribution to erosion. They may take cores from loess-paleosoil sequences, which allow them to date eras of soil formation and soil loss. They may also analyze dated assemblages of pollen in soil cores from rivers, lakes, ocean deltas, and undisturbed land in order to determine how land cover changed over time. Soil is a "geoarchive" of past conditions. There are well-developed and preserved soil layers that reflect times of little erosion, truncated layers that denote eras of significant erosion, and discontinuous layers that mix soil and loess and reveal the most intense erosion of all. Geoscientists can study the history of the change of coastal and terrestrial landforms and analyze coastal sediments, and also conduct laboratory and field experiments to study which factors influence the rate of soil erosion in real time, because there is enough evidence from past soil records to expect that well-established patterns from the two-million-year Quaternary period would predict recent dynamics if erosion were not caused by human activity.[1] Since the 1950s, the Yellow River Conservancy Commission has made detailed cross-sectional measurements in the seven hundred kilometers (over four hundred miles) of the lower Yellow River, usually twice a year, in fifty locations. From this, it is possible to determine the volume of sedimentation at different periods. All seven major Yellow River paleochannels have also been identified and excavated, and soil cores from them reveal the amount of sediment that accumulated on the bed of each one during their duration, the beginning and end dates of which are clearly dated in the historical record. Moreover, thirty-seven viable data points (of varying accuracy and resolution) show sedimentation rates over the past thirteen thousand years on the North China Plain. Hydrologists have studied the alluvial fan, marine sediments, and paleoshorelines as well.

---

1. Xu, "Naturally and Anthropogenically Accelerated Erosion."

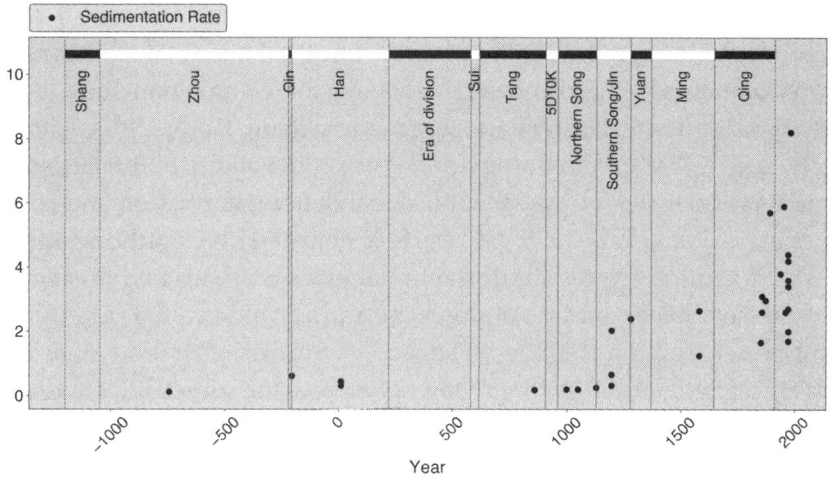

FIGURE 1.10
The History of Sedimentation on the Yellow River. Sedimentation rates remained low until the twelfth century. They maintained a plateau until the nineteenth century and rose rapidly thereafter.

for only 48 percent.[58] A fourth attributes only 30 percent of erosion to human activity.[59] One environmental scientist, He Xiubin, along with his coauthors, concludes unequivocally that "serious accelerated soil erosion has occurred during the last 2,500 years because of man-induced devastation of vegetation and other anthropogenic disturbance of the environment."[60] According to their study, the modern erosion rate is four times as high as one would expect from the long-term geological norm. Another group of scientists, focusing on data from marine sediments and paleoshorelines, discovered that, when averaged over the past six thousand years, sediment discharged into the Yellow River at a rate that was only about one-tenth of the present level of 1.4 trillion kilograms, or 1.6 billion tons, per year. Their data shows an "abrupt" and "sudden" increase in the erosion rate about a thousand years ago. This was a consequence of human activity on the Loess Plateau: land cultivation and particularly deforestation.[61]

As these summaries suggest, estimates vary, as do the methodologies for disentangling distinct causes in complex and integrated systems.[62] The point is that all authors attribute a significant amount of erosion to human activity; all authors recognize that today's hydrosocial system of the Loess Plateau has been profoundly influenced by farming, pasturage, and settlement there; and all identify a similar collection of inflection points. Furthermore, all authors agree that there is essentially a one-to-one relationship between rates of upstream erosion and downstream sedimentation. Almost all the sediment in the Yellow River comes from the Loess Plateau. As the environmental scientists Ren Mei-e and Zhu Xianmo conclude: "Although . . . flood disasters are governed by many factors, it is generally agreed that the predominant factor is the volume of fluvial sediment load which, in turn, is highly dependent on vegetation and land use in the Loess Plateau."[63] And as He Xiubin and his coauthors put it, "The documentary record of the frequencies of floods and breaches of the riverbank can be employed as a proxy indicator of intensive surface erosion on the Loess Plateau. . . . Studies of the rate of sediment accumulation in the Yellow River provide direct information about soil erosion on the Loess Plateau, since the sediment delivery rate remains a constant value close to one."[64]

According to most worldwide studies on soil erosion, sediment fluxes are highly sensitive to changes in local land use. Climate change plays only a secondary role, although it is still important because it is one of the most significant drivers of land use change.[65] Droughts, locusts, and frost, for instance—natural effects that can alter erosion rates on an annual or decadal basis—trigger long-term systemic change only when they occur in regions that have lost resiliency as a result of human activity. The long-term history of the Loess Plateau exemplifies this process.

HENAN: BETWEEN HEADWATERS AND FLOODPLAIN

We now return to our trip downriver. Before we turn to the floodplain and the delta, we must focus on Henan, a historical province and a distinct region in the middle of the Yellow River watershed that straddles the middle course, the lower course, and the Loess Plateau (plate 20). Here, hydrology, climate, soil, and political economy map poorly onto one another. In the Henan region, environmental history and political history produce quite different accounts of the same territory. The fact that Henan persisted as a very stable political entity despite its ecological diversity also helps to explain the challenges that ecological and watershed-scale policy posed for the imperial state. Eastern and western Henan experienced their proximity to the Yellow River in quite different ways.

Henan is a modern province that is also a historically ancient designation. Bisected by the Yellow River, hilly in the west, and flat in the east, Henan, a name that literally means "south of the river" (even though a quarter of it lies to the river's north), straddles the Yellow River's middle course and floodplain. The relative stability of its provincial boundaries in an ecologically diverse region reflects the fact that cultural and political geography does not fit neatly into hydrological boundaries. Henan is the site of some of China's most ancient archaeological sites, its oldest Buddhist monuments, and four of its ancient capitals. From deep antiquity until about a thousand years ago, Henan was a center of Chinese civilization and urban life. For the past millennium, however, deforested and frequently inundated by floods, its prosperity and cultural prominence have declined.

Western Henan is where the most easterly of the Yellow River's sediment-laden tributaries join the river's main course. The largest among them, the Luo River, drains Mount Hua (Huashan), a peak in the eastern part of Qinling range; Mount Hua is known as the westernmost of the Five Great Mountains of China (*wuyue*). After the Luo is joined by the Yi River, it is known as the Yiluo. It flows northeast to empty into the Yellow River at Luoyang. East of Luoyang and south of the Yellow River lies Mount Song (Songshan), a massif of seven peaks designated in Chinese cosmology as the Central Great Mountain (*zhongyue*) and the easternmost peak that feeds the Yellow River. Its northern-flowing streams debouche into the Yellow River, and its southern flowing ones flow into the Huai River and the East China Sea. East of Mount Song, the Yellow River disgorges to the flat North China Plain.

Today, the Xiaolangdi Dam bisects the Yellow River about 20 kilometers, or 12 miles, northwest of Luoyang. Plate 21 depicts the silt release that occurs there every summer, "a ritual cleansing so violent that it can look as if the earth just exploded" as "clouds the color of doom ascend beyond the dam's walls." For up to two weeks before the start of the summer monsoon flood season, officials open two portals in the dam. One releases clear water from the top part of the water column of the reservoir behind the dam, and the other releases muddy silt. The muddy cascade evacuates sediment that would otherwise accumulate in the reservoir and render it unusable. Meanwhile, the clear water scours out sand from the channel below to reduce flood risk. Every year, it lowers the riverbed by almost two

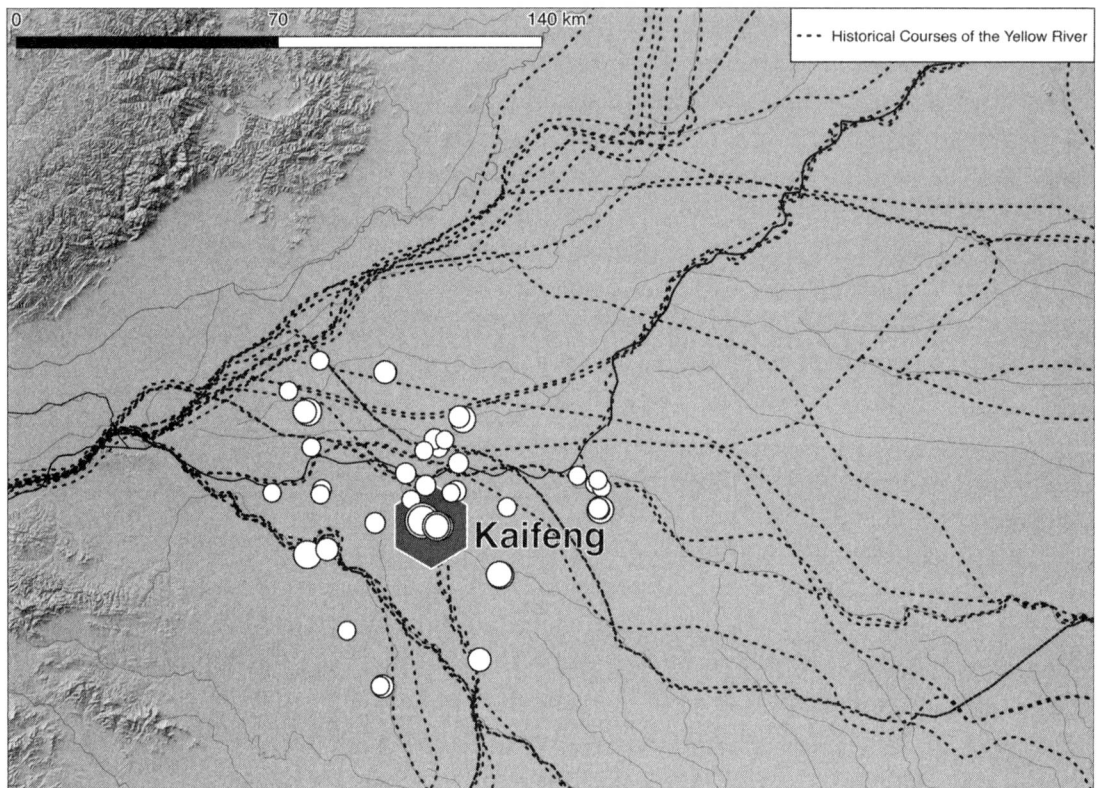

FIGURE 1.11
Historical Floods around Kaifeng. Kaifeng, the major city of the floodplain and sometimes an imperial capital, lay in the middle of the flood region.

meters, or about six feet. This water and sediment discharge project is the latest version of Yu the Great's imperative to use the river's power to regulate its own course.[66]

The Yellow River reaches its floodplain past the confluence with the Yiluo, the Wei, and the Su. A distinctive region of convergence between multiple ecologies and topographies, the low-lying basin east of Mount Song has been the site of many major floods and course changes as well as smaller course adjustments that occurred even during periods of overall stability.

Situated in a low-lying part of the Yellow River basin and crisscrossed by streams and historical wetlands, as well as the Yellow River itself, the city of Kaifeng was a grand metropolis and imperial capital of the Northern Song dynasty (960–1127). However, as depicted in figures 1.11 and 1.12, there were 279 floods within 50 kilometers of the city during the imperial era.

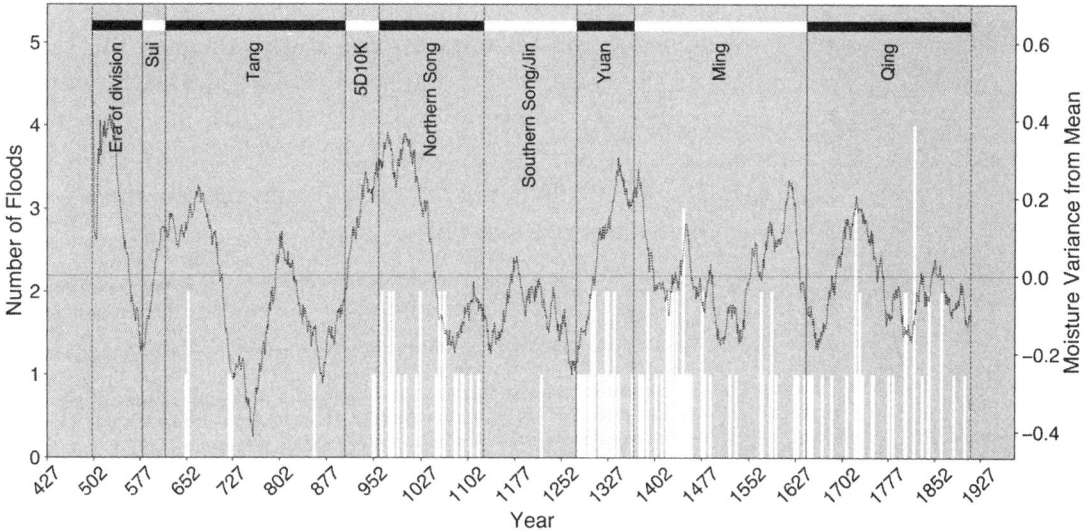

FIGURE 1.12
A Timeline of Historical Floods around Kaifeng. Around Kaifeng, as everywhere else on the floodplain, the turn of the first millennium of the Common Era marked the beginning of the disaster era.

When the Yellow River overtops its banks as it enters its floodplain and disgorges into the basin, it captures southward-draining Huai River tributaries that originate from Mount Hua, Mount Song, and the eastern part of the Qinling range, as well as the historical canals that transect the basin. Many of these rivers and canals have become historical courses of the Yellow River. Plate 22 shows that almost all of the historical courses of the Yellow River have avulsed from a small triangle of less than 200 kilometers (124 miles) per side, a low-lying basin just east of Mount Song that occupies only about 3 percent of the total floodplain.

Repeatedly inundated by sand and silt, the great cities of Henan—Luoyang, Zhengzhou, and Kaifeng—lost population, stature, and wealth in the second millennium of the Common Era. Agriculture declined precipitously during the final centuries of the imperial era in eastern Henan locales such as the historical county of Lankao, which is depicted along with Kaifeng and Zhengzhou in Plate 22. Poorly drained, Lankao became a site of saline wetlands interspersed with sand and gravel deposited during floods. Its population came to depend for subsistence on farming lotus and other pond commodities and on foraging for reeds and herbs rather than farming. In the twentieth century, decades of large-scale reclamation to compost organic matter have created new soil there.[67]

Hydrologically and tectonically, one can distinguish the middle course from the lower course of the river with some clarity, but culturally, climatically, and politically, Henan province extends from Sanmenxia in the mountainous west to Lankao in the sand fields of the east, about 400 kilometers, or 250 miles, as the crow flies, a day's journey on modern roads or rails.

THE RIVER'S COURSE: THE FLOODPLAIN

The Yellow River floodplain occupies approximately 568,000 square kilometers, which is a bit bigger than France. The average slope of the river course gradient declines suddenly after the river exits the mountains in central Henan for the last time. It slows down and disgorges onto the plain, depositing sediment and overtopping its banks during times of high water flow. Given the sharp and sudden decrease in rate of flow, most of the silt carried by the river is deposited in this flat, low-lying, poorly drained region, and all the major course changes have originated from there.[68] Plate 23 is a map of the floodplain. It shows Zhengzhou at its western edge and the line of cities linked by the Grand Canal across its center. The Mount Tai massif directs course changes to either the north or the south of the Shandong Peninsula. Streams and canals crisscross the floodplain flatlands.

On its modern course, the river's passage across the floodplain runs 786 kilometers (almost 500 miles) to the sea. Its elevation drops only 93.6 meters, or just over 300 feet. Because the plain is so flat, the velocity of the river is very slow. Sediment settles on the riverbed, floodwater drains slowly, and there are no barriers to course changes. Entirely channeled between engineered levees, the modern drainage area covers only 23,000 square kilometers, less than 15,000 square miles. Only one remaining tributary, the Dawen River, flows into the Yellow River, which it does via Dongping Lake, the only extant remnant of the "800-*li* Liangshan Marsh" (*babaili Liangshan shuibo*) famous to readers of the classic sixteenth-century novel *Water Margin* (*Shuihu zhuan*).[69] Figure 1.13 is an illustration from a late imperial edition of *Water Margin*, a seventeenth-century or eighteenth-century image from a historical novel about twelfth-century life in

FIGURE 1.13
Liangshan Marsh in the Late Imperial Imagination. The artists who created this woodblock print sometime between 1621 and 1800 depicted the twelfth-century floodplain as a locale of water and wetlands. Image courtesy of the Harvard-Yenching Library of Harvard University.

the floodplain wetlands. They provide an impression of the expanse that these marshes occupied at the time, when they were sufficiently extensive to provide shelter beyond the perimeter of state scrutiny for bands of outlaws.[70] In the north the floodplain takes in the Hai River basin. In the south the floodplain connects to the Huai River and in turn to the Yangtze (Changjiang) delta. Course changes have had a fairly regular pattern over time, gradually sweeping from north to south and back again across this vast plain.[71]

The Yellow River, depositing the sediment that it accumulated upstream, is the most active geomorphological agent of the alluvial plain other than humans. However, the landforms, soils, and groundwater of the plain—as well as its aquatic, agricultural, riparian, and other ecosystems, and its engineered human landscapes—all have feedback and response effects of their own, human and climatic alike, and short and long term in scope. The environmental effects of human river regulation are simply one factor in a system that has always been complex and active.

Even before human intervention, the river carried a load of sediment and routinely overtopped its banks, depositing the sediment along channel banks as natural levees. Over time, this created landforms known as perched channels, in which the riverbed, protected by the natural levees, rose to an elevation higher than the surrounding topography. In these, surface water elevations within the channel can be considerably higher than the land surface elevations immediately outside the levees, and the levees can be rather distant from the channel itself.

Human activity came to emulate and finally overtake the natural processes. During the high-sediment era, the riverbed rose about one meter, or three feet, per century.[72] By the seventeenth century, the river often ran many precarious meters above the surrounding floodplain, exacerbating effects of flooding. The only way that a perched river can be resolved is for the river to breach or overtop its levees and inundate the adjacent lower-lying land.[73] After a perched river ruptures its levees, it is almost impossible for it to return to its previous course, which fills entirely with silt, sand, and gravel as the water drains out.

Within levees, whether natural or human engineered, the river slowed and formed great loops known as meanders (*zuowan*). The current on the outside of the meander moved rapidly, crashing into the levee and carving the soft soil of the riverbank with great erosive power. During flood season, in a matter of hours the river could swerve ninety degrees and strike a perpendicular dike.[74] This was a frequent cause of levee breaches. Many river engineering projects entailed shoring up the weak spots on levee banks and using bricks and stones to reinforce the most vulnerable sites and the ones most closely proximate to population centers. Plate 24 depicts meanders within far-set levees.[75] Plate 25 depicts the multiple tiers of highly reinforced embankments that were typical of the floodplain by the nineteenth century, these ones surrounding the city of Ji'nan.[76] Plates 24 and 25 reveal the fact that even an engineered river remains an active and restless geomorphological force. Levees, likewise, are dynamic structures, not just inert earthworks.

Human impact on the Yellow River floodplain was limited before about 350 BCE. For centuries after that, there were still numerous lakes and wetlands, rapidly emerging and disintegrating micro-landforms, and many short tributaries that flowed out of Mount Tai, the eastern Great Mountain in the *wuyue* cosmology. High water levels, sediment deposition, course shifts, and landform changes did not register as disasters because they did not inundate, dislodge, or destroy large-scale human settlements or disrupt human property arrangements. The river generally had numerous dynamic interlocking channels rather than a single embanked course, and it meandered freely across a marshy plain that people claimed for settlement and agriculture only gradually. The Huai River did not originally take a clear channel to the sea. Only with intensive engineering did this flat, swampy, and poorly drained terrain of wetlands and dynamic and unpredictable river courses come to favor dense human settlement. By the turn of the Common Era, as we will see in the next chapter, the floodplain was the most densely populated part of the empire, although it became depopulated after floods soon thereafter.

As the plain developed over the course of the hundreds of thousands of years of the Pleistocene and Holocene eras, the natural migration of the river channel produced U-shaped oxbow lakes at the sites of former meanders, as well as sloughs, wetlands, and sediment deposits disconnected from subsequent channels. There were numerous historical large lakes and wetlands, all of which have since disappeared, and there were countless smaller ponds and marshes as well. Dalu Marsh, the largest body of water in ancient China, persisted until at least the twelfth century. Daye Marsh, also known as Liangshan Marsh, situated outside effective state control at the boundary of several administrative frontiers, was home to the twelfth-century Song Jiang outlaw band (fl. 1121) immortalized in the previously mentioned novel *Water Margin*. The marsh shrank considerably after the Yellow River course change of 1279. Repeated inundation from sediment-filled floodwaters further desiccated the wetlands region, and the marsh was entirely lost to silt deposition by the eighteenth century.[77] At least one lake disappeared as early as the earthworks construction initiatives of the first century of the Common Era.[78] Plate 26 depicts the lake district between the Yellow River and the Yangtze River today, with the Huai River flowing into Hongze Lake at its center. Even after millennia of efforts to create drainage canals, reservoirs, and embankments, this remains a marshy region. Its contemporary hydrology is a hydrosocial accomplishment. Despite the precariousness of life there, the region has long supported a dense population and numerous cities, with lifeways that included fishing and birding as well as farming.

As lakes and streams disappeared, inundated by sediment, floodplain drainage declined even further, land became waterlogged, and even heavy local rains could be catastrophic. Floodplain tributaries, repeatedly scoured and inundated by the silty Yellow River during centuries of rising sediment rates, became shallow, and some dried up completely. Lakes became marshlands that were gradually reclaimed into farmlands, which, as low-lying tracts, always remained highly vulnerable to flooding. The extinction of lakes, tributaries, and wetlands caused the climate to become drier and more prone to water shortages as silt and sand deposits raised the level of the plain. Wells had to be dug deeper and deeper. The silt and sand transported from the Loess Plateau also caused dust storms during the dry season, and the same extinction of bodies of water led to salinity and waterlogging.[79]

As the Yellow River course shifted, as the channel moved across the floodplain and deposited sediment along its many beds and banks, it created micro-landforms, low mounds of deposited material on the flat plain. The traces of historical floods and course changes remain visible on the land today. After a course change, former perched riverbeds became elongated micro-highlands that still rise as much as three to ten meters above the land surface. These raised bands of loess and sand mark the paleocourses of the Yellow River. On today's floodplain, there are also traces of splay fans—triangular deposits of sand, gravel, and boulders that reveal where coarse sediment and debris shot out from dike breaks at high velocity in floods that occurred hundreds of years ago. Micro-landforms in turn control subsequent distributions of new deposits of sand, loess, and moisture. To this day, the bands of soil on ancient micro-highlands and the splay fans that are the landscape records of historical floods still suffer from low groundwater and sometimes form sand dunes. They still require mitigation to encourage soil formation.[80] The depressions that lie on either side of the sites of old earthworks are full of fine silty clay, and the groundwater is high and saline. In some places, there are standing salt marshes.

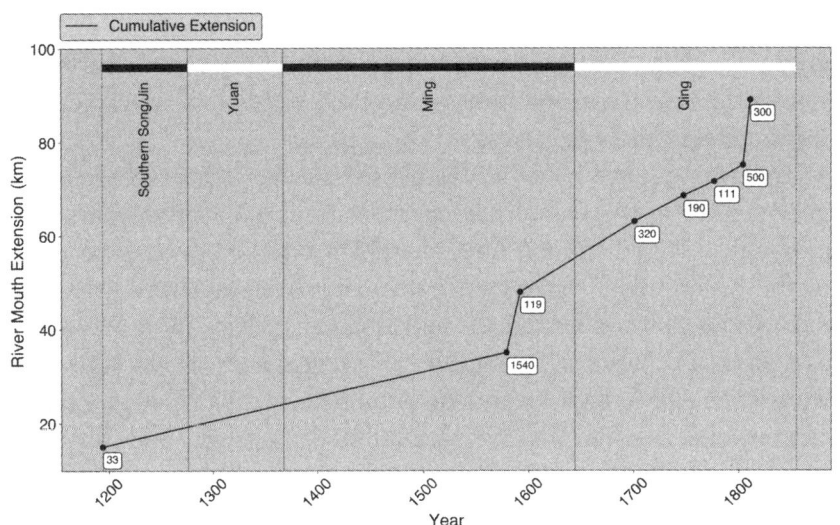

FIGURE 1.14
The Rate of Extension of the Yellow River Delta. The mouth of the Yellow River extended rapidly between 1194 and 1855 CE. The boxed figures are the average rate of river mouth extension as a result of sediment deposition, in meters per year, during each time period. Based on Xu Jiongxin, "Naturally and Anthropogenically Accelerated Sedimentation."

The Yellow River delta is downstream from the lake district. As plate 19 depicts, the Yellow River has had numerous deltas in human and geological time, each one reflecting the duration and sediment accumulation rate of one of its paths to the sea. These muddy, saline estuary districts have generally been sparsely populated regions of little state scrutiny. As figure 1.14 reveals, and as we will see in Chapter 4, water engineering and water policy could have a profound effect on the rate at which land formed on the delta. The hydrosocial system of the Yellow River produced an immense, growing quantity of silt in historical times. People managed it in ways that determined where on the floodplain it settled, but they did not mitigate it at its source on the Loess Plateau.[81]

A RIVER IN MOTION: FROM PLEISTOCENE TO HOLOCENE AND THE END OF THE HOLOCENE CLIMATIC OPTIMUM

The river system has experienced dynamic changes at a geological time scale. It never occupied a position of primordial peace and quiet. Even before intensive human activity began to shape the landscape in the Neolithic era and thereafter, the river was always an agent of its own landscape, never a passive recipient of human activity. Throughout the Cenozoic era, which is the geological period that has occupied the past sixty-six million years, the Indian tectonic plate has been colliding with the Eurasian plate, causing the Tibetan Plateau to gradually uplift. That process has created the characteristic eastward slope of the East Asian landmass and the morphological framework for its eastern-draining rivers.

The course of the Yellow River formed during a relatively brief cool period early in the Quaternary period about 2.5 million years ago, although the Ordos Loop took shape later.[82] The drainage patterns of the Yellow River basin achieved their historical shape gradually during the Pleistocene, which ended about eleven thousand years ago. Prior to that, during the Pleistocene ice ages, when sea levels were lower, the Yellow River delta extended into what is now seabed and cut multiple, various, and simultaneous northerly and southerly channels to the sea, channels that are still visible under water. As temperatures warmed at the end of the Pleistocene and sea levels rose, the planetary system transitioned into the Holocene epoch, the focus of this book. The current coastline took a recognizable modern form around seven thousand years ago.[83]

Loess soil took its current form on the Loess Plateau during a series of cold and dry intervals of the Quaternary period. There were eight episodes of extreme natural erosion on the Loess Plateau prior to the anthropogenic era. These generally occurred during times of transition from lengthy dry and cool periods to warmer, wetter times. The erosion eras punctuated epochs during which soil, rich with organic matter, formed locally. The soil formed during four moist phases: between 10,000 and 9,000 years ago, between 7,500 and 5,000 years ago, between 4,000 and 3,000 years ago, and between 2,700 and 2,000 years ago. The paleosoil profile of the Loess Plateau today consists of soil layers interleaved with the wind-borne sandy loess layers deposited during cold, dry periods.[84]

The Holocene Climatic Optimum, the warmest, moistest period of the Holocene, was the most pronounced era of soil formation. It was the period during which the Loess Plateau reached its maximum forested extent. The summer monsoon gained strength with rising temperatures. Soil formation was high, and erosion and sand transport were low. Creeks, wetlands, lakes, and springs emerged throughout the Loess Plateau, surrounded by dense vegetation; their traces in historical times are salt ponds and salt flats. Human communities, their populations growing, began to practice agriculture along with herding, hunting, and foraging. The Mu Us Desert retreated northward by about three degrees latitude. Numerous Paleolithic and Neolithic sites in places that are now desert show evidence of fishing, hunting, animal husbandry, and dry farming along extinct rivers and lakes.[85]

The global shift to a cooler and drier climate at the end of the Holocene Climatic Optimum is well documented on the Loess Plateau and in northern China generally by multiple proxies: oxygen isotopes, changes in the prevalence of various plant and tree pollens, and the deposited sediment throughout the Yellow River basin and in the ocean. All of these sources document a decline in deciduous broadleaf plants and trees and a rise in grasses. The "abrupt" change, which occurred in north China around 5,800 years ago, was unrelated to human activities.[86]

At that point, the monsoon began trending toward a less northerly extent. East Asian summer monsoon precipitation now extends only to the southern and eastern parts of the Loess Plateau.[87] As the northerly extent of the monsoon receded over time, Neolithic agriculture gave way to pastoralism on the northern part of the Loess Plateau, and nomadic pastoralists migrated in search of better pasture.[88] Nevertheless, the Loess Plateau has always supported multiple modes of subsistence from year to year and from place to place, and in ways that have fluctuated with smaller shifts in weather and climate.

The end of the Holocene Climatic Optimum about five thousand years ago marks the birth of the modern Yellow River ecosystem. Soil began to deteriorate, some marginal forests transformed to grasslands, and some grasslands to desert. Springs and ponds began to dry out. Erosion increased. As we have seen, between 30 percent and 50 percent of the sediment in the Yellow River can be attributed to the effect of cool and dry conditions in the past two millennia, not to anthropogenic factors. As the climate cooled and dried and forests contracted after the Holocene Climatic Optimum, the quantity of water in the Yellow River diminished, but its annual variability increased, as did the increasing unpredictability of the monsoon. This was the setting in which recorded history and complex society emerged in the Yellow River basin.

Still, as recently as three thousand years ago, during the Neolithic era and the early Bronze Age, considerable forest remained on the

Loess Plateau along with amble shrub and steppe cover. Downstream, the periodically inundated wetlands of the plain remained more sparsely populated than the hospitable tributary valleys throughout the river's middle course.[89] At the same time, as temperatures became cooler and drier on the Loess Plateau, as forest retreated and desert expanded, Loess Plateau lifeways began to diverge and population distribution changed. Some people moved south and east and closer to water sources. Farming remained a dominant mode of subsistence, although farmers had to practice more intensive and invasive methods than those employed by their ancestors. Other people moved north, and pastoralism became the most important part of their subsistence strategy.

Charcoal particles, which reflect people's burning of biomass to clear forest and brush for farming and to heat homes and to fire pottery kilns, join the geological record after the end of the Holocene Climatic Optimum. The evidence from the Fen and Wei valleys of the southern Loess Plateau reveals a rapid rise in the levels of biomass burning during the late Holocene as the climate became drier and human land use became more intensive. About 3,100 years ago, human fires became more common than naturally occurring ones, with profound effects upon previously forested landscapes and even regional climate.[90]

The charcoal history dovetails with four inflection points of population increase and land-use intensification on the Loess Plateau, each of which created intervals of intensive soil erosion. The first occurred around about 7,000–7,500 years ago, and it coincided with the initial emergence of Neolithic agriculture in the region. The second happened as a result of state-directed Iron Age agricultural colonization around 300 BCE to 0 CE. The third, driven by population growth, military defense, and timber commodification, occurred between about 800 and 1100 CE and featured the most abrupt, sudden increase of sediment discharge of all. A final turn toward high erosion rates coincided with intensive dryland farming of maize and tubers in the seventeenth and eighteenth centuries and the population boom that accompanied it. The following three chapters of this book focus in turn on each of these three inflection points.

That is to say, it is possible to pinpoint with some accuracy the moment at which human activity became a dominant feature of Yellow River ecology, even though it is still, to this day, a complex ecosystem with numerous and dynamic intersecting components. As we will see in chapter 2, just as the climate began to generate conditions on the Loess Plateau that were particularly conducive to erosion, early farmers began to engage in precisely the activities that most accelerated those pressures.

CHAPTER 2

Before It Was Yellow: The Great River from Neolithic through Medieval Times

The first life span of the hydrosocial Yellow River lasted a very long time indeed. It began in the middle of the Holocene era, roughly 7,500 years ago, and ended around 1,300 years ago. Only for the final thousand years of this lengthy epoch did people produce extensive written records documenting their ideas and activities. Moreover, only in that same period did effective and widespread iron tools permit people to intensively modify the environments of the Loess Plateau and the alluvial plain. Although humans did alter landscapes during the early Holocene, about 11,500 to 7,500 years before the present, their modifications were local and site specific. This chapter begins during the middle Holocene, 7,500 to 5,000 years ago, when the effects of human landscape modification become visible at the scale of tributary watersheds. By the late Neolithic and early Bronze Age, 5,000 to 3,500 years ago, population growth and intensive agriculture spread the human footprint across the entire Yellow River watershed, and by the middle to late Bronze Age and into the Iron Age, beginning about 3,500 years ago, even larger populations, equipped with better technology and propelled by more centralized governments, were altering land throughout the region and fundamentally shaping the landscape through their activity. Although by the standards of later imperial times that impact was still limited, human land management became as consequential as natural forces during the dominion of China's first unified states, the Qin (221–206 BCE) and the Western Han (206 BCE–9 CE). Thereafter, the impact of human activity on the Yellow River basin receded for more than half a millennium.[1]

During the first eight millennia of the Holocene, the Yellow River maintained a northerly course across its floodplain. It meandered through wetlands and routinely overtopped its low banks during times of high rainfall. It did not experience any avulsions—sudden changes in course—and at a large scale, the floodplain landscape was stable.[2] Erosion and sediment transport began in earnest around five thousand years ago. The erosion rate increased dramatically between three thousand and two thousand years ago as growing populations on the Loess Plateau used new and more efficient technologies, supported by ambitious state-level infrastructure, to expand the amount of land under cultivation and the intensity with which they worked their fields. Flooding became more common. The flood frequency timeline correlates well with human activity but does not align with climate change. The strong monsoons of the Holocene Climatic Optimum do not correlate with turning points of major increases in erosion. By contrast, signs of hillslope farming and deforestation do correlate with the erosion and flooding rate.[3] Emerging anthropogenic transformations began to alter the topography of the North China Plain itself as the alluvial fans of Yellow River tributaries encroached onto the edge of the floodplain, tilting the basin to the south and the east and magnifying the gradient of its slope.[4]

As people started to build levees on the floodplain to control the increasingly unstable river and to allow larger populations to settle on the alluvial plain, they exacerbated the risk of floods even more. Once the river began to be fixed in place between levees, sediment accumulated only on the riverbed and not on the surrounding plain. When floods did occur in the managed floodplain landscape, they were catastrophic events that transpired amid a dense population.

Three timelines offer a summary introduction of the first fifteen hundred years of documented human activity in the Yellow River basin. During the centuries covered in this chapter, the colonization of

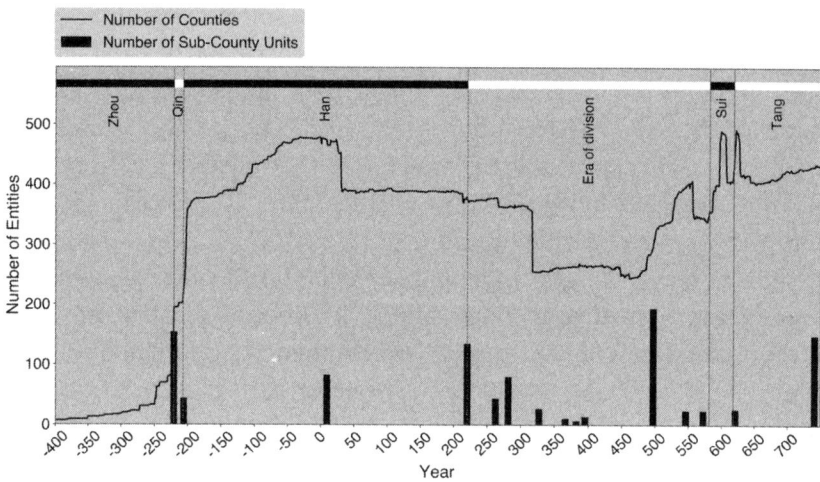

FIGURE 2.1
Loess Plateau Political Geography, 400 BCE to 750 CE. Settlement density on the Loess Plateau varied from earliest recorded times in the fifth century BCE until the middle imperial period in the eighth century CE. Records of the number of counties are continuous. Settlements outside the administrative system are recorded only sporadically and should not be considered complete. The density of settlements rose until the turn of the Common Era before declining, beginning to rise again only after the 500s. The solid line depicts the number of counties that existed anywhere on the Loess Plateau during each year between 400 BCE and 750 CE. The vertical bars are the number of settlements that were documented by historical sources but lacked formal administrative status.

the Loess Plateau, the stability of the floodplain, and the succession of imperial regimes all align with one another. Figure 2.1 depicts the changing number of settlements on the Loess Plateau.[5] Data about them exists only for sporadic dates. This timeline reveals a sudden rise in the number of settlements on the Loess Plateau around the turn of the third century BCE. This was the time when China's initial unified states established suzerainty over large realms, which for the first time encompassed the entire Yellow River basin and also territory that extended to the west and far to the south of the Yellow River. The density of settlement on the Loess Plateau peaked around the turn of the Common Era and then declined continuously for five hundred years before rising once again.

On the floodplain (figs. 2.2–2.4), historical sources depict only sporadic disasters during the first fifteen hundred years of documentation. In particular, it is noteworthy that that the most peaceful periods on the floodplain coincide with the centuries when the Loess Plateau was less intensively settled. The two disaster-prone intervals on the floodplain—the decades around the turn of the Common Era and the period that began in the early seventh century—both align with recorded political history. The former case, which accompanied the transition from the Western Han to the Eastern Han regime, is documented in figure 2.4. The latter one corresponds with the founding of the Tang regime. Over half of the events attested on the floodplain during the early centuries of record keeping were disasters—

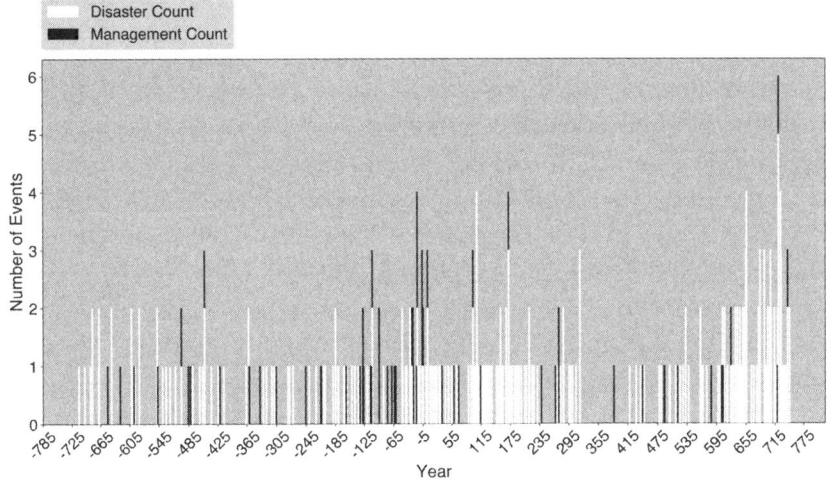

FIGURE 2.2
Event Timeline, 750 BCE to 750 CE. Historical sources attest relatively few disasters and infrastructure management events during the first 1,500 years of recorded history on the Yellow River. The most active disaster periods cluster around the turn of the Common Era and the period from the seventh century onward. The active management period is around the turn of the Common Era.

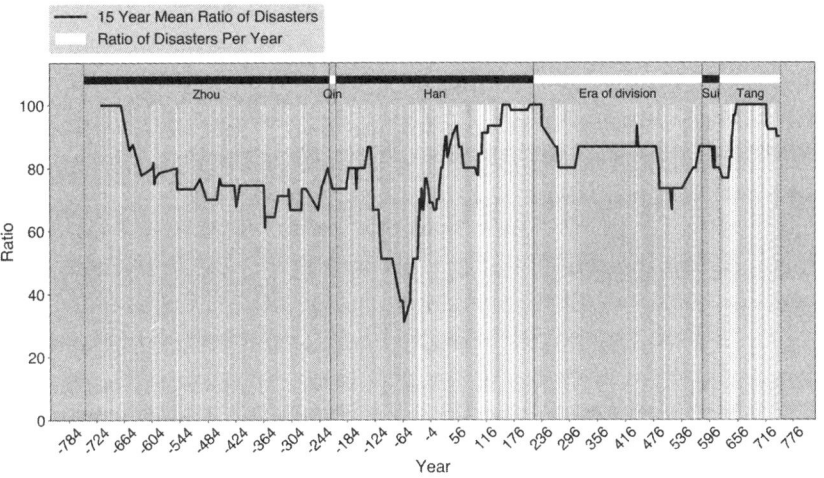

FIGURE 2.3
Disaster to Management Ratio, 750 BCE to 750 CE. The most active management period in the early recorded centuries of Yellow River history transpired around the turn of the Common Era.

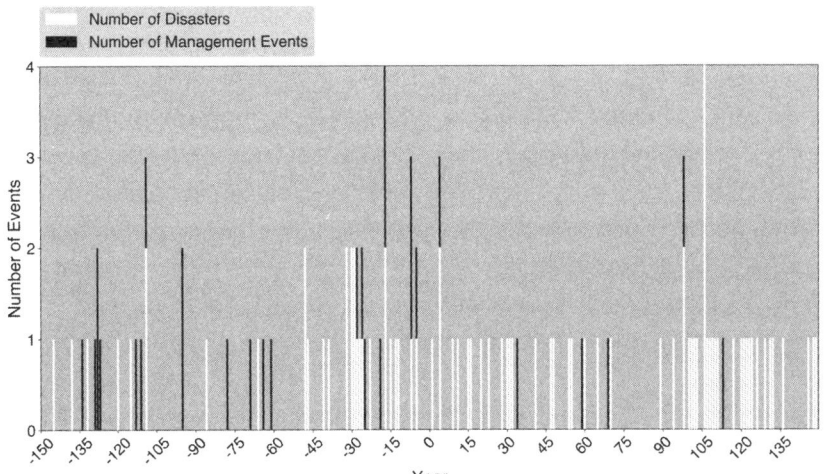

FIGURE 2.4
Event Timeline, 150 BCE to 150 CE. There was an unprecedented disaster rate immediately preceding the Common Era, followed by significant attention to infrastructure management.

floods and levee breaches and so on. However, the events recorded during the earlier part of this time span were more likely to be related to floodplain management events—the development and repair of embankments and canals—than were those than occurred during later times.

This chapter spans such a long time period because, even as human footprints on the river basin became more evident, this lengthy time span—relative to later times—represents a single coherent era of low erosion and limited flooding. During most of the period discussed here, the increase in the rate of erosion was gradual. Excepting two centuries of high erosion and alluviation (the process of sediment deposition on the floodplain) during the last two hundred years before the Common Era, the river was relatively stable compared to later times. Even the events of those centuries were of limited impact compared with those of middle and later imperial times.

By asserting this I am also engaging with a lively intergenerational debate about Yellow River history between eminent Chinese scholars. In an influential 1962 article, published soon after the completion of the controversial Sanmenxia Dam, the historical geographer Tan Qixiang established himself as an early advocate for the idea that human-caused erosion on the Loess Plateau directly correlated to the rate of flooding on the alluvial plain. He proposed that historians who investigate the history of the river identify environmental turning points in middle-course settlement and agriculture rather than prioritizing the importance of feats of engineering genius on the floodplain. He focused on the Qin and Western Han regimes to make his case.[6] Conversely, in a 2012 article, the historical geographer Xin Deyong argued that Tan, excessively influenced by his alarm over twentieth-century deforestation, overstated the case for ancient erosion and discounted the successes of floodplain engineering.[7]

In fact, although Xin Deyong has written a monumental account of late classical floodplain engineering, archaeology and environmental science tend to bear out Tan Qixiang's conclusions. In addition to the work of both Tan and Xin, this chapter is indebted to recent scholarship in geoarchaeology and landscape archaeology by T. R. Kidder, Arlene Rosen, and their collaborators. This work, surveying sediment deposits themselves, confirms a close correlation between deforestation, erosion, and flooding. It demonstrates than engineering followed floods rather than preceding them. As Kidder and his coauthor Yijie Zhuang put it, "massive" landscape transformation occurred between late Neolithic and early dynastic times, especially on the Loess Plateau, which increased sedimentation in the Yellow River and required massive investment in flood control measures, that "human activity in the mid- to late Holocene contributed to large-scale changes in the behavior of the Yellow River and [that] . . . these changes were of sufficient magnitude to bend the arc of China's history."[8]

Still, although I broadly follow these conclusions, I am struck most by the fact that the rate of Neolithic and Iron Age erosion pales in comparison with that of the middle and later imperial periods, which are the topics of the following two chapters. I want to invite three thoughts simultaneously. First, changes of epochal scope in the relationship between people and the Yellow River transpired during the Western Han. Second, these two centuries were an anomaly that was followed by a long stable period. Third, the most significant changes in the life of the river were still to come. Kidder and his coauthors cite a change in average annual sedimentation rates that expanded from 0.02 to 0.20 millimeters per year (still only twice as tall as a sheet of paper) during the period this chapter covers. However, as we will see, during later imperial times, the rate continued soaring, far above that ceiling, depositing quantities of sand and silt that overwhelmed people who made plans based on precedents in the Han era.[9] Figure 2.5 documents that sediment history.[10]

In the final millennium before the Common Era, farmers and city dwellers peopled and transformed the floodplain, drained wetlands, and built dikes, chronicling as they went the catastrophes that occurred when the levees broke. Their ambitious and literate Iron Age states moved people, cut trees, produced documents, built walls and garrisons, contended territory, and built canals and levees. They created a language about the relationship of people, water, and state power. Following the collapse of a centralized agrarian empire in the third century, military and civilian colonization of the Loess Plateau retreated. At that point, the grasslands revived, erosion diminished, and floods abated, not resuming until almost a millennium later in the tenth century.

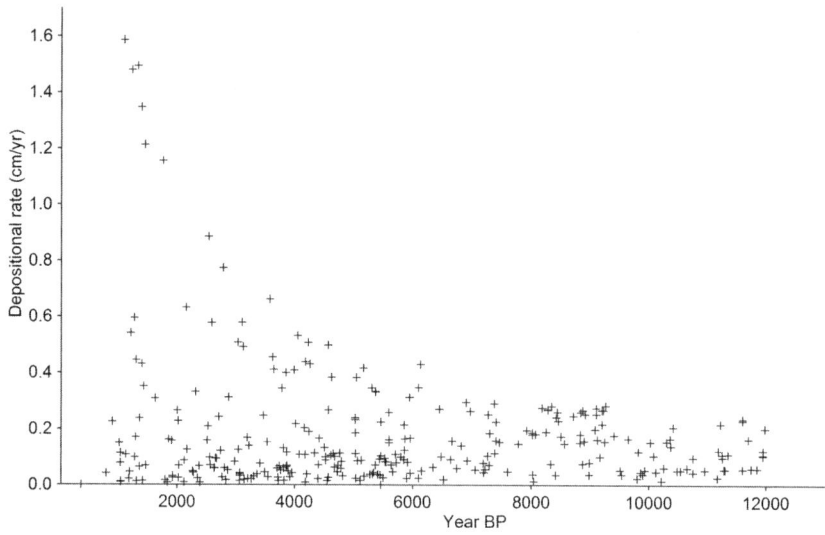

FIGURE 2.5
Yellow River Sediment Deposition at the Scale of the Holocene. Baseline early and mid-Holocene erosion rates are between 0.2 and 0.4 centimeters per year; 0.5 and 0.7 in the Neolithic and Bronze Age; 0.8 and 0.9 in Warring States; and 1.0 and 1.2 in Han, which was just the beginning of the doubling that occurred (according to this data set) during the course of the imperial era. Reprinted from *Geomorphology*, 46.3–4, Changxing Shi, Zhang Dian, Lianyuan You, "Changes in Sediment Yield of the Yellow River Basin of China during the Holocene," 267–283, 2002, with permission from Elsevier.

THE NEOLITHIC AND THE BRONZE AGES

The next few pages cover 3,500 years of history. The narrative begins with the emergence of agriculture, the era when human activities began to go "against the grain" of both ecological stability and relative social equality.[11] Long before they documented their activities in written records, people channeled small streams to irrigate crops and grazed animals and cultivated crops in ways that increased sediment deposition above the baseline rate of under 0.02 millimeters per year. Compared with later times, these impacts were relatively small and local. People were limited by the technologies of stone and bronze tools. Nevertheless, the impacts of their activities appear in geoarchaeological strata. Communities tended to occupy the valleys of Yellow River tributaries. Settlements were sparse on the marshlands of the alluvial plain and the grasslands of the central Loess Plateau.

Growing populations of humans, working the land with stone tools, did have a demonstrable effect on the environment. The table of sediment deposition in figure 2.5 depicts a rise in sediment accumulation (according to a different form of measurement than the sedimentation rate referenced in the previous paragraph) from about 0.3 centimeters per year to about 0.8 centimeters per year during these millennia. However, the change was small and gradual relative to the standards of later eras, and human impacts coexisted with

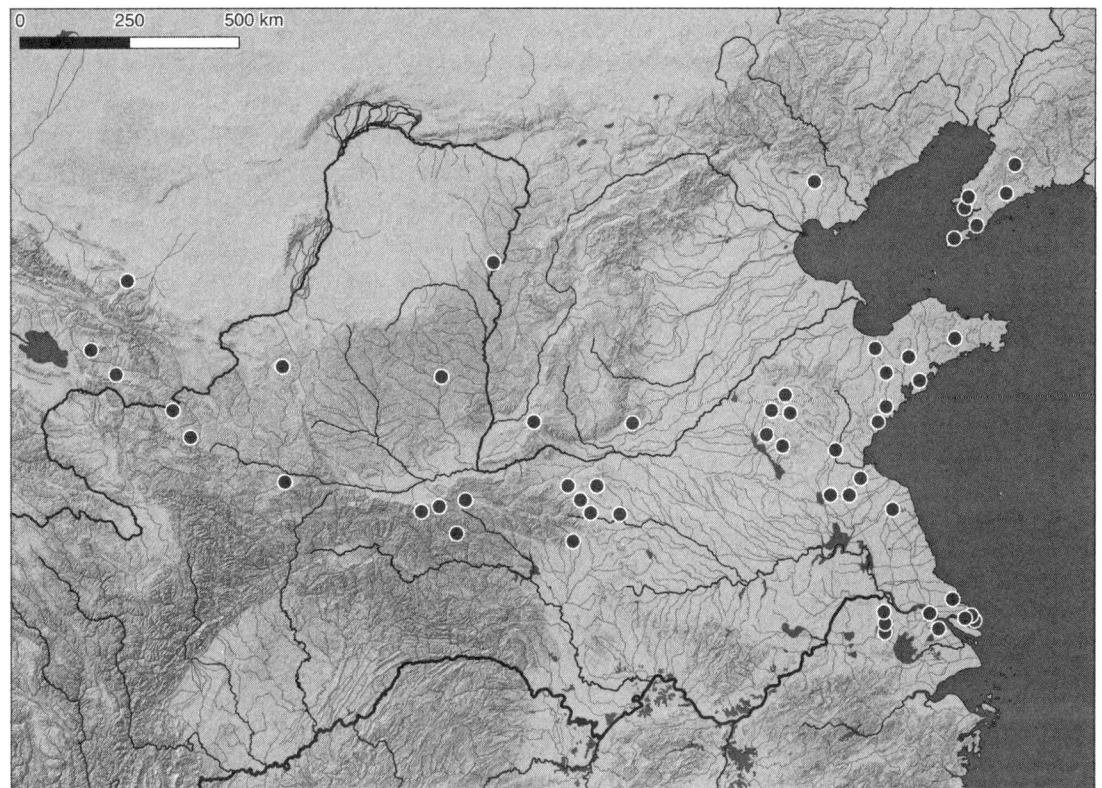

FIGURE 2.6
Overview of Major Neolithic Sites in North China. Few Neolithic sites in north China were located on the Yellow River floodplain, excepting those clustered at higher elevations on its perimeter and on the western slopes of Mount Tai. Based on Li, "The Products of Minds as Well as Hands."

planetary climate changes. Neolithic-era hillslope erosion primarily affected tributary valleys, not the vast Yellow River alluvial plain wetlands. Figure 2.6 maps major Neolithic sites in northern China. The sites are clustered around the highlands near the Wei and Fen Rivers, the Yiluo River valley, the Mount Tai and Mount Song foothills, and the coastline of the East China Sea. Few appear on the floodplain or the Loess Plateau.[12]

Until about eight thousand years ago, the climate was colder than later times, human populations were small, people practiced hunting and foraging, and their impacts on the land were limited. As the climate warmed, the impacts of farming and animal husbandry become visible in the archaeological record in the form of traces of charcoal and deposits of fertilized soil that contain more organic matter than surrounding plots do. However, forests and grasslands remained largely undisturbed. Farming, foraging, and herding coexisted for

BEFORE IT WAS YELLOW   69

thousands of years. Dominant activities at any given time and place depended on the amount of rainfall in a given year, and many people lived in mobile bands amid plains, forests, and hills.[13]

Precipitation increased during the Holocene Climatic Optimum, about seven thousand to four thousand years ago. The number, size, and geographic area of settlements in the Yellow River basin rose dramatically as temperatures and precipitation peaked. As the climate became warm and moist, new sites in the Ordos emerged on the desert-loess transition zone. People fished, hunted, and raised animals, sometimes in locations that today have become semiarid or desert climates. They practiced dry farming along rivers and lakes that have since disappeared.[14] People farmed rice on floodplains and millet on dry lands and raised herds of goats and sheep on grasslands. Signs of landscape transformation appear in the bioarchaeological record at only a few sites.[15] During these millennia prior to the construction of large-scale earthworks, agriculture remained sparse on the shifting wetlands of the North China Plain, although fishing and hunting peoples established some permanent settlements there. Agrarian society intensified instead at the southeastern edge of the Loess Plateau, especially along large Yellow River tributaries like the Wei, Fen, and Luo.

The landscape archaeologist Arlene Rosen has identified the earliest signs of large-scale and long-term Neolithic modification of drainage systems in the valley of the Yiluo River, a Yellow River tributary region at the southeastern edge of the floodplain. During a process that transpired between 7,200 years ago and 3,600 years ago, sediment and gravel accumulated at the valley's bottom, a sign of substantial deforestation, soil erosion, and gully formation in the river's upper catchment area.[16] As the V-shaped valley filled in with eroded sediment from the surrounding hills, it became a moist floodplain, allowing for irrigated rice paddies to come into production.[17] The relationship between farming on slopes, upstream erosion, and downstream flooding is clear in individual tributary valleys during the Neolithic era. These are the same hydrosocial dynamics that would transform entire watershed landscapes in later eras.

About 4,200 years ago, the Holocene Climatic Optimum began to wane. The Yellow River watershed climate became cooler and more arid.[18] The monsoon retreated and became more unpredictable. At around the same time, paleoflood deposits reveal a series of massive flood events on the Jing River, which drains the central Loess Plateau. They occurred during an era of periodic severe droughts, and they herald the abrupt arrival of an unstable and catastrophic climate regime typified by both droughts and floods. Extraordinary floods recurred throughout the following two hundred years. At the sites of the earliest of the floods, pottery shards, charcoal, and burnt earth commingle with flood deposits, but eventually the signs of human habitation disappear. The earliest versions of the story of Yu the Great, the legendary ancient ruler said to have channeled rivers and drained wetlands, may echo these real-world events.[19] Generally attributed to climate change, the floods are also collocated with evidence of human-induced erosion, and they should probably be attributed at least in part to anthropogenic landscape change.[20]

These floods, and the end of the Holocene Climatic Optimum, also coincide with the transition from the Neolithic to the Bronze Age. The new reality of harsher, less predictable climate conditions required more specialized modes of subsistence. Foraging, agriculture, and pastoralism came to dominate different parts of the watershed. Settlements declined in number but increased in population density. Societies evolved distinct practices and modes of survival. Populations that had grown relatively high during a warmer period with more predicable precipitation had to engage in more intensive, diverse, and ingenious practices in order to produce more food on less land and under conditions of flood and drought. More specialized occupations and more social inequality were among the results of these transitions. This era, the dawn of the Bronze Age in North China, saw the emergence of wheat cultivation and other new types of crops, crop rotation techniques, and the construction of waterworks such as irrigation canals and reservoirs. Populations grew rapidly as formerly dispersed groups gathered into fewer but larger centers that contained sites of ceramics and metallurgy production and complex political formations that incorporated written documentation of their activities.[21] Although the population grew in the long run, communities collapsed, fought, or migrated when resources were scarce. They learned to store harvest surpluses from good years to hedge against future crop failures. Droughts became more frequent on the arid and unpredictable Ordos Plateau. To the north, settled regions contracted in semiarid lands, whereas to the south, farming society became more intensive around the lowland

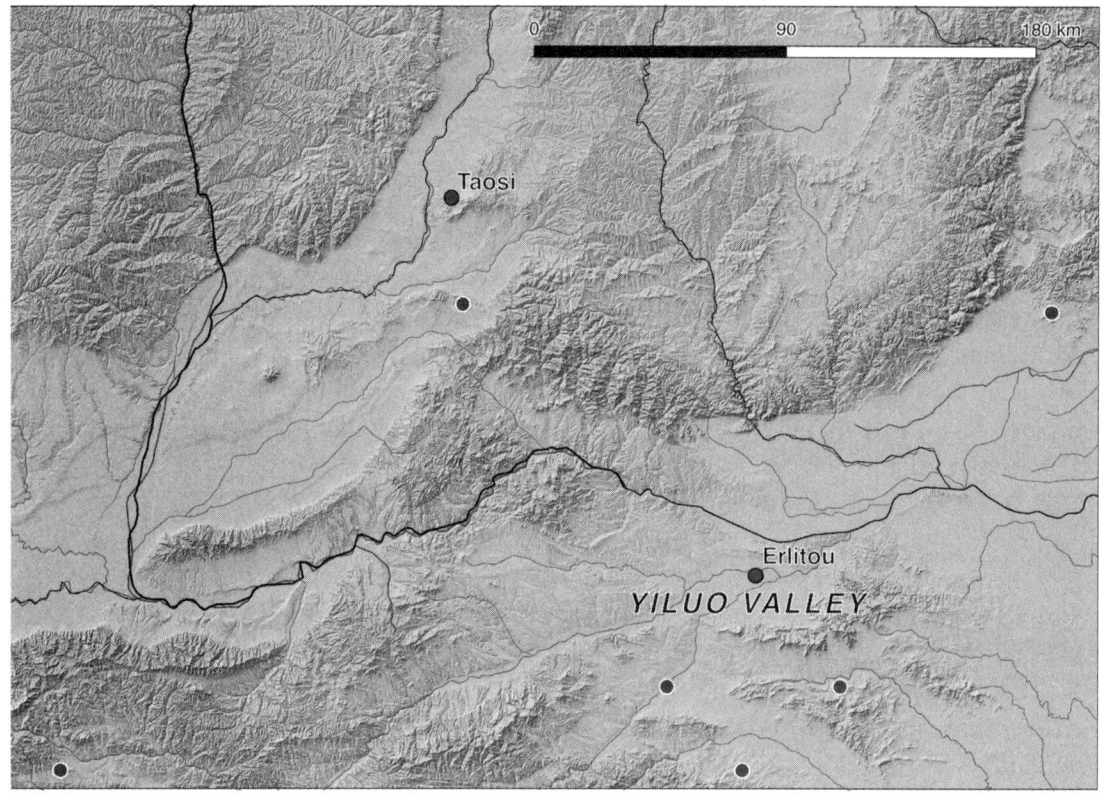

FIGURE 2.7
Taosi, Erlitou, and the Yiluo Valley. The largest early settlements were situated on Yellow River tributaries, not on the river itself. They were located on well-watered plains at the foot of mountains, where residents had access to resources from multiple ecosystems and farmed on slopes. Unlabeled points are other Neolithic sites.

riverbanks of the Guanzhong basin and the southeastern reaches of the Loess Plateau.[22] Pastoralists occupied the grasslands. Multiple proxies—isotopes, pollen, and sediment found in soil cores throughout the Yellow River basin and in the Bohai Sea into which it drained at the time—reveal an "abrupt" decline in deciduous broadleaf plants and trees and a rise in grasses around 3,800 years ago, an indication of a drying and cooling climate.[23]

Taosi, a late Neolithic city surrounded with pounded-earth walls in the Fen River valley, was occupied between about 4,400 and 3,900 years ago (fig. 2.7). It had industrial and ritual neighborhoods, at least fifty-four associated suburban sites outside its walls, charcoal remains from at least twenty-five tree species, and cemeteries of about ten thousand graves. Around Taosi, there is evidence of significant erosion and gully formation, which coincide in the archaeological record with rapid population growth in the vicinity.

Major floods deposited gravel and boulders around Taosi. Meanwhile, on the floodplain, people began building levees to control the wetlands hydrology. Geoarchaeological research reveals that episodes of landscape stability became shorter as floods became more frequent.[24]

The intensification of land use in the millennium after the end of the Holocene Climatic Optimum marks the onset of what archaeologist T. R. Kidder and his coauthors refer to as an early Chinese "great acceleration." Pollen data reflect a considerable shift in vegetation from natural ground cover to food crops. The charcoal record reveals widespread burning for land clearance and other activities. Soil cores disclose the fact that sediment modification was occurring at a greater rate and higher intensity than ever before.[25] Even as the climate became more arid, technology and social organization continued to become more complex, and populations kept expanding. Some people practiced metallurgy and lived in cities, surrounded by rammed-earth walls. Others were farmers and herders. Nevertheless, people occupied a world still populated by elephants, rhinoceroses, monkeys, and numerous species of deer, in addition to their favored domesticated animals.[26] By around four thousand years before the present, the pine forests of the southeastern Loess Plateau disappeared. Human activity, not climate change, seems to be primarily responsible. There is no precedent in the paleoclimate record for ground-cover change at this geographic scale, pace, and magnitude. Moreover, the locations of deforestation correlate with dense human populations and with the charcoal concentrations associated with those settlements. In the Yellow River delta, around the same time, there is a significant reduction in arboreal pollen and a dramatic increase in nonarboreal pollen, and on the Loess Plateau there are the first signs of erosion, gullying, and increased rates of sediment transportation.[27] As figure 2.5 shows, the sediment deposit rate doubled from about 0.3 centimeters per year in the early Neolithic to about 0.6 centimeters per year by the end of the Bronze Age. Written evidence joins the record around this time as well. The divination records known as oracle bones and the narratives inscribed on cast bronze vessels both document large-scale, state-managed irrigation works and levee construction as well as the use of fire to clear land for farming. The archaeological record reveals several measures of landscape change. Large charcoal deposits and decreasing arboreal pollen suggest deforestation. Floods that began to carry large pebbles rather than topsoil are a clear sign of erosion.[28]

Erlitou, a major Neolithic and Bronze Age cultural area centered in the Yiluo valley (see fig. 2.7), was the most densely populated region of the Yellow River basin at the time. Occupied between about four thousand and thirty-five hundred years ago, Erlitou straddled the emergence of literate culture in northern China. The lowest geoarchaeological stratum there is a silty, sandy deposit that is characteristic of waterways meandering across a broad floodplain. Increased sedimentation and frequent flooding began there during the late Bronze Age and the Iron Age, almost surely triggered by deforestation, and accompanied in the archaeological record by signs of dense population, intensive millet production, rice agriculture, and levee and wall building.[29] As the Neolithic gave way to the Bronze Age, the Erlitou floodplains shifted dramatically from soil formation to erosion. As the valley bottom filled in with sediment, agriculture shifted away from rice and other wetland crops and toward wheat and millet.[30]

During the harsh and unpredictable millennia of the Neolithic and Bronze Ages that followed the end of the Holocene Climatic Optimum, farmers and, to a lesser extent, herders, engaged in activities that intentionally and unintentionally remodeled the landscapes they inhabited. They caused floods and erosion—of that there is no doubt—but they also learned to live with the environments they occupied.[31]

PLOWING WITH IRON

Documentary history joins the archaeological and paleoenvironmental history of the Yellow River about three thousand years ago. As cities and their hinterlands evolved into states, the earliest regimes in the Yellow River basin organized efforts to feed the populations of farmers and herders who congregated in regions of fertile soil, adequate rainfall, and low flood risk. They waged war on one another. They exploited coerced laborers and extracted agricultural surplus in return for some measure of stability. Population and state power expanded together, and along with these phenomena, so did intensive and ingenious civil engineering on the middle and lower courses of the Yellow River. Throughout centuries of migration, warfare, urbanization, and state formation, the extent and scope of cultivated land, waterworks, and earthworks expanded as well.

During the last millennium before the Common Era, iron technology advanced toward the production of economical, interchangeable, and widespread iron tools and weapons. People hewed forests to feed iron forges, drove iron plows to cut through heavy soil and tough roots, and wielded iron shovels to move earth to build canals and levees. Large infantries, armed with iron weapons, overwhelmed the characteristic Bronze Age charioteer combatants. Warfare had direct and indirect environmental impacts. Soldiers chopped forests and trampled grasslands around campsites and battlegrounds. Mining and smelting required timber and compromised ecosystems. Early states enforced and taxed intensive agriculture and created reliable supply chains, permitting them to equip, feed, and manage their troops and warhorses. Large armies with full bellies won battles. Intensive and extensive agriculture drove population growth and generated the surpluses that fed soldiers and horses, industrial and construction workers, scribes and bureaucrats who managed complex social structures, and artisans and other new kinds of specialists.

The Yellow River's annual sediment accumulation rate nearly doubled again during the last fifteen hundred years before the Common Era, although it remained below one centimeter per year. A sudden spike in the volume of Yellow River sediment occurred around three thousand years ago.[32] The frequency and amplitude of both droughts and floods began to increase around the same time.[33] In part this resulted directly from the continuing shift to a cooler and drier climate. Climate change prompted new agricultural and demographic patterns that were the proximate causes for further transformation of hydrosocial landscapes.[34] On the southern Loess Plateau around three thousand years ago, fires set by farmers for land clearance, a sign of first cultivation, became more common than naturally occurring fires.[35] At some sites, the intensity of human disturbance by fire and cultivation increased continuously from that time onward. At others, local fires occurred most frequently between 3,100 years ago and 1,500 years ago, a period of intensive land reclamation for cereal cultivation. Human burning of the landscape then declined once all the land was continuously under cultivation. At sites in the middle Yellow River, soil moisture diminished and aeolian (wind-driven) silt content increased, signifying more frequent and intense dust storms. On the northern Loess Plateau, some vegetation cover transformed to steppe.[36] People cleared forests for agriculture. They built and occupied large cities and other structures, using timber for construction and fuel, and the trees floated downstream from the Luo, Wei, and Fen Rivers.[37] The changes were not limited only to the floodplains. At the desert-loess margin, natural vegetation was "severely affected by an increase in human population, dry farming activity, forest clearance, and frequent warfare. Wind and water erosion became a problem, and desertification became intense. The grassland ecology in some places was changed into semi-desert or desert, and mobile sand dunes developed."[38] Between three thousand and two thousand years ago, on the hillsides overlooking Tianchi Lake at the southwestern margin of the Loess Plateau, a dense forest of mixed coniferous and deciduous trees transformed into a landscape in which those trees intermingled with plants and shrubs.[39]

Meanwhile, herds of domesticated animals were becoming larger. These herds were more likely to overgraze grasslands, and they needed to move from place to place more frequently. Around the eighth century BCE, pastoralist communities throughout inner Asia took up horse riding, introducing a large new population of herbivores into a fragile landscape. Written documents begin to designate horse-riding peoples as adversaries to sedentary farmers. Agrarian regimes defined territorial borders that separated their lands from one another and from their pastoralist neighbors. The large armies and large settler populations characteristic of Iron Age regimes began to subjugate indigenous peoples who practiced hunting, foraging, and swidden and pastoralist modes of subsistence. The iron-wielding regimes modified landscapes at their peripheries by constructing roads and fortifications and by establishing farming settlements in their midst.[40]

In 770 BCE, the Zhou kingdom, the dominant regime of the Yellow River basin at the time, moved its capital downstream. The court embarked from Haojing on the Wei River (the city later known as Chang'an and finally as Xi'an), a location that could be provisioned only from Guanzhong. The new capital was at Luoyi (later Luoyang), east of the Sanmenxia Gorge, situated on the Yiluo River near its

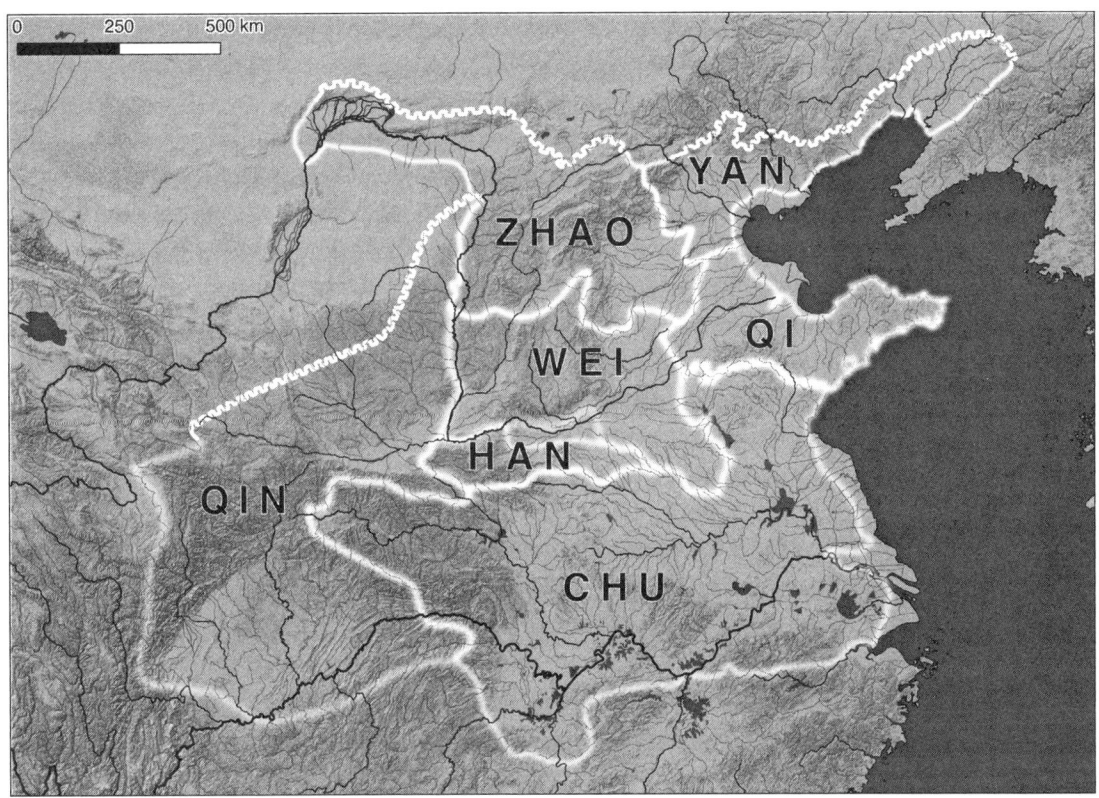

FIGURE 2.8
The Warring States Period.
By the third century before the Common Era, the number of independent states had contracted to seven from a total of 148 named in earlier sources. Control over the Yellow River basin was divided between six of the seven regimes, each of which built its own flood-control and irrigation works, and several of which built walls, including the Qin fortifications across the central Ordos.

confluence with the Yellow River. This was the first of many occasions when the capital would cycle between Guanzhong, which was well-protected from floodplain adversaries and easily accessible to the pastoralist frontier but had limited resources, and disaster-prone Henan, which was easy to provision but hard to defend from floods and invaders alike. This cycle is a theme in Yellow River history.

Around the same time, various aristocrats began struggling with one another over access to the power of the throne. Throughout these centuries, the king of Zhou ruled as a figurehead while aristocratic kingdoms contended with one another for suzerainty and waged war with one another and with their pastoralist neighbors. Not only did they fight battles; they also moved earth for walls, canals, and fields, and cut trees for forges and to clear land. These centuries are known as the Spring and Autumn Period (Chunqiu shidai) (770–476 BCE) and the Warring States Period (Zhanguo shidai) (476–221 BCE). The political economy of war made a significant impact on the Yellow

BEFORE IT WAS YELLOW   77

River throughout this period. Figure 2.8 depicts the regimes that controlled the Yellow River basin and the rest of China at about 260 BCE.[41]

Erosion continued apace on the Loess Plateau and other highlands. By the Spring and Autumn and Warring States eras, there is also documentary and archaeological evidence for large-scale earthworks construction on the Yellow River floodplain itself. These are, moreover, the centuries during which writers documented relationships between people and nature in political, literary, and philosophical texts. Writers systematized older legends about the relationships between people and water in ways that reinforced the values and hierarchies of the new literate class. This classical canon set the terms of engagement between people, technology, land, and water for the entirety of the subsequent imperial era.[42]

To feed government officials, urban workers, soldiers, and ritual specialists, the contending polities of the Spring and Autumn and Warring States eras sought to exploit arable lands more intensively than ever before and to channel waters more purposefully, and they began writing documents about their activities. Some texts refer to mythic engineering activities that purportedly occurred centuries earlier, like those of Yu the Great. Others chronicle fishing and hunting, and in doing so, they incidentally document Iron Age ecology, terrain, and species diversity at a time when the river was still clear and the grasslands extensive. Still other documents describe the emergence of the state bureaucracy and its early efforts to control the river.

These centuries, when the Yellow River floodplain was contended by numerous small polities, were the eras of the first levees and canals. These earthworks and waterworks offer the earliest evidence of downstream channelization and wetlands drainage. Historical sources document sixty-one events related to Yellow River history between 773 and 223 BCE. The vast majority of the events recounted in the earliest written sources were floods and other catastrophes. However, fifteen reports referred to infrastructure construction. These occurred sporadically throughout the centuries in question.

The extant written reports of large-scale activities undoubtedly represent only a small percentage of the engineering endeavors that began during these centuries. Floodplain denizens were beginning to channel streams and wetlands into canals for drainage, irrigation, flood control, and transportation. They also began building levees that separated watercourses from floodplains, which in turn were transformed into farms and settlements. Evidence from archaeology shows that Neolithic and Bronze Age populations on the floodplain were relatively small, but "all of this changed rapidly and quite dramatically with the early Iron Age in the Eastern Zhou period, and even more markedly and more extensively with the subsequent Han imperial period."[43] Geoarchaeological excavations confirm previously unverified historical texts that collectively date the earliest construction of drainage and irrigation canals and bank and levee construction to about 2,900 to 2,700 years ago. This is around the same time that sustained interstate competition began. The levees occur in the archaeological record together with heightened rates of sediment deposit. Floodplain residents had to develop ways to mitigate the effects of long-term increases in sedimentation from the Loess Plateau. Ever-increasing sediment set in motion a cycle of further investment in flood-control works.[44]

More people, more cities, more engineering, and more sediment all appear together in the archaeological record. Numerous settlements and more extensive artifacts of material culture appeared in short order along with evidence of levee construction.[45] Not all the sediment came directly from the Loess Plateau. Rapidly alluviating streams on the floodplain itself also deposited "meters of sediment during the Zhou era."[46] Along with loess soil from disturbed land upstream, newly channeled rivers and new canals carried sediment from disturbed local slopes. Arlene Rosen explains that "the very first stages of massive soil erosion were probably due to the rapid expansion of population into these localities, deforestation for building, industry and agriculture, and erosive agricultural practices during this period."[47] For instance, deforestation of the hill slopes of the Mount Tai massif in central Shandong caused widespread floods and silty floodplain buildup, and farmers began to dig canals into the floodplain to channel the waters from the streams that drained Mount Tai in order to intensify agricultural production.[48]

The state of Qi was the dominant regime of the lower reaches of the floodplain during the Spring and Autumn Period. Duke Huan (Qi Huan Gong) (d. 643 BCE) centralized Qi administration, conquered thirty-five neighboring small kingdoms, and led a powerful interstate pact.[49] Classical sources attribute the first floodplain levees to him as well. According to later sources, Duke Huan succeeded in channeling the nine streams that the delta comprised into a single flow for the first time.[50] It is not possible to corroborate the specific events, but the story reflects that this was the dawn of floodplain management in a place that was still an active wetlands environment. The story also reveals the signal role of states that straddled the river—they could not survive without river control technology, bureaucracy, and labor management—and the symbiotic relationship between river management and expanding state power.

The *Writings of Master Guan* (*Guanzi*) is an early philosophical text attributed to the philosopher and statesman Guan Zhong (c. 720–645 BCE), who served as Duke Huan's chancellor in Qi. Guan Zhong is credited with reforms that centralized political power, created a tax code, and established state monopolies on salt and iron production. One chapter of the text focuses on destructive floods and recommends that rulers appoint water conservancy officers (*shuiguan*) to oversee waterworks construction and repairs in every county. It describes how to requisition corvée work (periodic unpaid labor imposed by the state in order to complete public works) and how to collect brush and stalks to bundle into the fascines that were the building blocks for levee construction. It specifies the annual schedule for waterworks management, the number of carts and spades needed for the jobs, and the personnel chain of command. It is, in short, the earliest evidence for the bureaucratization of a floodplain management routine.[51]

Large earthworks became widespread soon thereafter, on the North China Plain itself and in Yellow River tributaries. Embankments and drainage works were both essential parts of floodplain infrastructure. The Twelve Canals of Yinzhang (Yinzhang shier qu), also known as the Twelve Canals on the Zhang River (Zhangshui shier qu), an irrigation initiative undertaken by the state of Wei in Hebei, was completed between 403 and 221 BCE.[52] In the state of Wei, the Honggou Canal, completed sometime prior to 486 BCE, was built upon earlier waterworks. The historian Sima Qian (c. 145–c. 86 BCE) reported that by the Warring States Period, "literally millions of smaller canals led off from larger ones at numerous points."[53]

Throughout northern China, construction and consistent maintenance of levees began around this time and gave rise to the first clearly channelized courses in the lower river—the dawn of the age of imperial water management.

Upstream, the era of big state power had different ecological impacts. Much of the Loess Plateau remained under the control of nomadic herders whose suzerainty reached much further east and south than it did during later times. Early evidence from the historical record attests forty species of trees on the Loess Plateau and thirty-five species of grasses; animals including deer, elk, wolf, rabbit, and fox, and 150 kinds of fish.[54] The upper reaches of the Jing, Wei, and Luo Rivers lay outside agrarian state control for much of the era and were primarily locales of pasturage and hunting.[55] Poetry, history, and archaeology all offer glimpses of life on the Loess Plateau during the classical era. *The Classic of Poetry* (*Shijing*), comprising verses that date from the eleventh to seventh centuries BCE, depicts large grazing herds, one numbering three hundred sheep and ninety cattle, and describes the Jing River as "so clean that its bottom could be seen."[56]

The Loess Plateau remained significantly forested throughout the era, particularly in its northern and western parts. Until the turn of the Common Era, the Taihang Mountains and the Lüliang Mountains were mostly covered by forests that hunters and foresters occupied, not herders or farmers.[57] Even riparian regions remained forested. Along with *The Classic of Poetry*, *The Commentary of Zuo* (*Zuozhuan*), a work of narrative history that covers the period from 722 to 468 BCE, is full of references to forests, to trees in valley bottoms, and to forest-dwelling birds and animals. The third-century BCE *Writings of Master Xun* (*Xunzi*), a text of moral philosophy, opines on the beauty of mountain forests. *The Classic of Mountains and Oceans* (*Shanhaijing*), a fourth-century BCE compilation of terrestrial and mythic geography, references forests on the mountaintops, foothills, and riparian valleys of the Yellow River basin.[58] The records of the earliest bureaucratic states designate roles for multiple kinds of timber-removal specialists, but the forests were vast and populations were still small.[59] In short, although erosion was increasing during the centuries of the Spring and Autumn and Warring States Periods, it was coming from river-adjacent uplands in densely populated places. The ecology of the central and western expanses of the Loess Plateau and the more remote mountain slopes did not experience much anthropogenic disturbance until later times.

That is to say, even as forests began to be compromised in the core Warring States agricultural regions, ground cover remained relatively intact in most places, especially in mountains and upstream. Classical sources also depict a much less arid Loess Plateau than we know from later historical times. Many ancient and classical place-names include terms for springs, marshes, and wetlands, although the features to which they refer are nowhere to be found today. Classical texts describe places as having high groundwater tables that today require wells that are five to fifteen meters deep.[60]

The third century BCE, the final century before the first unified empires in North China, was also the end of one life span on the Yellow River. The unmanaged river—the Holocene river—had been overtaken by another one, the first human river. State-sponsored and state-controlled workshops hosted industrial-scale technology for the production of ceramics, iron, salt, fabric, and other crafts. Metal weapons had changed the terms of warfare and the size of armies as well as the possibilities of agriculture. Rammed-earth walls protected cities and borders. Charcoal to fire the forges and factories came from forests that were increasingly distant from population centers. Even though the rate of increase of sediment accumulation was gradual relative to later times, and even though the record of floodplain events is sparser than later times, this was now a managed river. The floodplain population was large, and the infrastructure to support that population was in place. So, too, was the bureaucracy that it required and the rhetorical and ideological substrate that justified expenditures of labor and revenue to manage the river.

DISCOURSES OF STREAMS AND FORESTS

The successful remodeling of landscapes, and the exploitation of human and animal labor to do so, was a complex network of ventures that rested on a robust ethical and philosophical justification. Even when classical philosophers were not writing about rivers as such, water imagery and water metaphors were central to their conceptual schemas. At the heart of the Chinese tradition, water management and moral thought consistently reinforced each other.[61]

Classical literature is suffused with commentary about the transition from wild nature to managed territory throughout the Yellow

River basin. Although few texts offer reliable evidence of specific historical events, many accounts depict early epochs of land clearance and waterworks engineering. They describe the central role that processes of landscape transformation played in creating social power and cultural meaning. They mark the Bronze Age and the Iron Age as the eras during which literate people created conceptual and technical vocabularies for discussing large-scale water control.

An important line of thinking about water management pivots around the concept of *shi*, a term that the philosopher François Jullien translates as efficacy. In the *shi* tradition, practitioners seek to evaluate the power that is inherent to a phenomenon, such as a river current, and to transform it through nonassertive activity in some way that benefits humans. The philosopher known in English as Mencius, and in Chinese as Mengzi, was born Meng Ke in a town on the slopes of Mount Tai in the state of Qi during the fourth century BCE. He served as an official in Qi three hundred years after the Duke Huan reforms and his writings exemplify the *shi* concept: "The people of Qi have a saying: 'Wisdom and discernment are not enough if you don't take advantage of efficacy, [just as] a hoe and a pickaxe are not enough if you don't wait for the right season.'"[62] Efficacy-based hydrology seeks to leverage the innate propensities of water—to flow downhill, to scour its banks, to carry sediment—in order to bring about desired results.[63]

The *Rituals of Zhou* (*Zhouli*), a work of bureaucracy and theory of statecraft that dates from about the third century BCE, also typifies the *shi* approach. The text recommends that "every canal must take advantage of the *shi* of water [just as] every embankment must take advantage of the *shi* of earth. A good canal is scoured by its own water; a good embankment is consolidated by the sediment brought against it."[64] Shen Dao (c. 350–275 BCE), an early philosopher of *shi*, puts it this way: "Those who protect and manage the dikes and channels of the nine rivers and the four lakes are the same in all ages; they did not learn their business from Yu the Great: they learned it from the waters."[65]

The Yu legend reflects an affinity for massive and technologically intensive landscape transformation in the interests of state power. However, even his story, as told in the *Tribute of Yu* treatise that dates from the sixth century BCE, emphasizes the imperative of careful study before beginning hydroengineering projects and the necessity of building drainage works that allow for floodwaters to inundate the

FIGURE 2.9

Yu the Great. This 151 CE rendition of Yu the Great is a rubbing from a relief carving at the Wu Liang Shrine in southwestern Shandong province. This public domain image of the rubbing appears on Wikimedia Commons.

plains in a controlled fashion, rather than merely confining streams behind vast levees. As the story recounts, and as the introduction to this book has summarized, Yu the Great, the legendary founder of China's mythical Xia dynasty, succeeded his father, Gun, who for nine years had tried and failed to control primordial floods through dikes and dams, which collapsed and killed many people. By contrast, Yu built drainage canals as well as levees, and he dredged the rivers' silty beds. Through these efforts, he confined the rivers to their proper channels. Only after he had done this was it possible for him to establish territorial boundaries and tribute responsibility. Yu founded the Nine Provinces, initiating the first territorial structure of the imperium and constituting a realm. Having done so, he developed the infrastructure needed to support agricultural society and set the terms of taxes and tribute for each province.[66] The Yu story is protean. It supports advocates for monumental waterworks, but people who promote less interventionist relationships to the river can use the text for precedent as well.

Written versions of the Yu story date back to the earliest bronze vessels from the dawn of the literate era. Well established by the third century BCE, when multiple versions circulated, the story had become canon by the turn of the Common Era. The "Rivers and Canals Monograph" of *The Records of the Historian*, which dates from 86 BCE, begins by detailing the accomplishments of Yu the Great, and its author, Sima Qian, explicitly situates the river management endeavors of later monarchs as accomplishments in Yu's legacy. Sima Qian references multiple versions of Yu's ventures.[67] *The Mencius* includes passages that are touch points of the *shi* tradition, but it also relates a version of the Yu story. Figure 2.9 is an image of Yu the Great from the funerary shrine of the Confucian scholar Wu Liang (interred in 151 CE); Yu is wielding a mattock, a kind of pickax for digging earth.

Although some passages in the capacious text *The Mencius* glorify Yu's transformation of the landscape, others offer an awareness of environmental degradation. One story links the primordial flood that preceded Yu's great work with deforestation and the collapse of biodiversity and also with water control. A dense network of anthropogenic transformations of the landscape collectively made agriculture possible during mythic times.[68] For example:

> In the time of Yao, when the world had not yet been perfectly reduced to order, the vast waters, flowing out of their channels, made a universal inundation. Vegetation was luxuriant, and birds and beasts swarmed. The various kinds of grain could not be grown. The birds and beasts pressed upon men. The paths marked by the feet of beasts and prints of birds crossed one another throughout the Middle Kingdom. To Yao alone this caused anxious sorrow. He raised Shun to office, and measures to regulate the disorder were set forth. Shun committed to Yi the direction of the fire to be employed, and Yi set fire to, and consumed, the forests and vegetation on the mountains and in the marshes, so that the birds and beasts fled away to hide themselves.[69]

Another passage in *The Mencius* reinforces the notion that Warring States writers who attended to the environment were thinking about deforestation and linking it to the growing power of the state: "Mengzi said: the trees of Niushan were once beautiful, but as it was situated within the borders of a large state, they were hewn down with axes and hatchets. . . . Still, through the ceaseless activity [of nature] day and night and the fertilizing influences of rain and dew, they were not without buds and sprouts springing forth, but then came goats and cattle to browse upon them. To these things is owing the bare and stripped appearance [of the mountain], and when people now see it, they think that it was never finely wooded."[70]

In *The Mencius*, deforestation is a symbol for the destruction of moral rectitude. However, this passage offers tangible information about environmental transformation as well. Mencius, a Qi state bureaucrat, was well aware of actual people taking iron tools to trees. The transformations of wood and water were not merely metaphors to Mencius and his contemporaries, who bore witness to material changes as well. The Yellow River landscape was allegorical during the Warring States centuries, but it was political and technological as well. The Warring States texts about water, water control, and the environment served as touch points for millennia to come. They propound several competing impulses, each of which justifies different sorts of waterworks intervention.

QIN AND WESTERN HAN: THE FIRST IMPERIAL RIVER

At the same time that the rulers of Qi were channeling the floodplain, the Qin regime (circa ninth century BCE to 202 BCE) was engaging in unprecedented erosion-causing activities on the Loess Plateau and building walls and other fortifications on the Ordos for the first time. Qin began as a steppe kingdom. It first appeared in historical records as a tributary state submitting horses and other grasslands animals to the Zhou court. Ancestral Qin territory was largely nomadic pastoralist land. Based in Guanzhong and encircled by mountains, Qin was fundamentally a regime of the southern Ordos, although its hinterland extended to the grasslands and deserts farther north.

By the fourth century BCE, it had become clear that none of the remaining seven Warring States regimes would be able to vanquish its rivals. Each one had a population of similar size, was equipped with the same technology, and had access to equivalent resources. Not long after the Qi court sought to centralize control over waterworks on the floodplain, members of the Qin court recruited Shang Yang (c. 390–338 BCE) to move west out of the central plains and into their peripheral domain. They requested his advice about how to transform Qin into a militarily powerful and administratively centralized kingdom. He persuaded the rulers in Guanzhong that a vigorous program of state-sponsored agricultural colonization could drive up the kingdom's population, generate an agricultural surplus, and expand the size of the military and the industrial enterprises that Qin could support.

The Qin regime sponsored a campaign of agricultural intensification. This involved several steps. First, Qin laborers built a wall across the central Ordos, and the regime staffed the fortifications with soldier-farmers. This was the first effort by an agrarian regime on the Loess Plateau to colonize land dominated by pastoralists. The act initiated a rivalry with the Xiongnu, a nomadic tribal confederation that had also moved into the Ordos during the third century BCE.

Next, Qin officials surveyed and subdivided their kingdom. They designated intended land uses throughout the domain, identifying grassland and forest land as two distinct categories of land cover, separate from agricultural land. They set aside a small part of the forests for conservation and identified the rest for exploitation and farmland conversion.[71] The reforms incentivized intensive land use. State officials allocated land to colonists and established a regime of private property in agricultural land. Farmers who exceeded quotas received rewards, and those who failed to do so were punished by enslavement. Migrants who came from outside of Qin and convicts who agreed to homestead would receive free land. Just as Qi set the terms for intensive floodplain infrastructure and intensive agriculture, Qin did the same in Guanzhong and on the Ordos. In the Chinese philosophical tradition, Mencius and Shang Yang represent the contrasting schools of Confucianism and Legalism. From the perspective of environmental history, though, they both advocated exploiting landscapes to expand state power. They simply did so from the vantage points of different regimes that occupied distinct ecological niches.

Among the range of economic, military, and administrative reforms inspired by Shang Yang's advice, the Qin government compelled farmers and workers to build irrigation canals, hew forests, and farm the Loess Plateau grasslands with mass-produced iron tools. Such implements, which first appeared in the fourth century

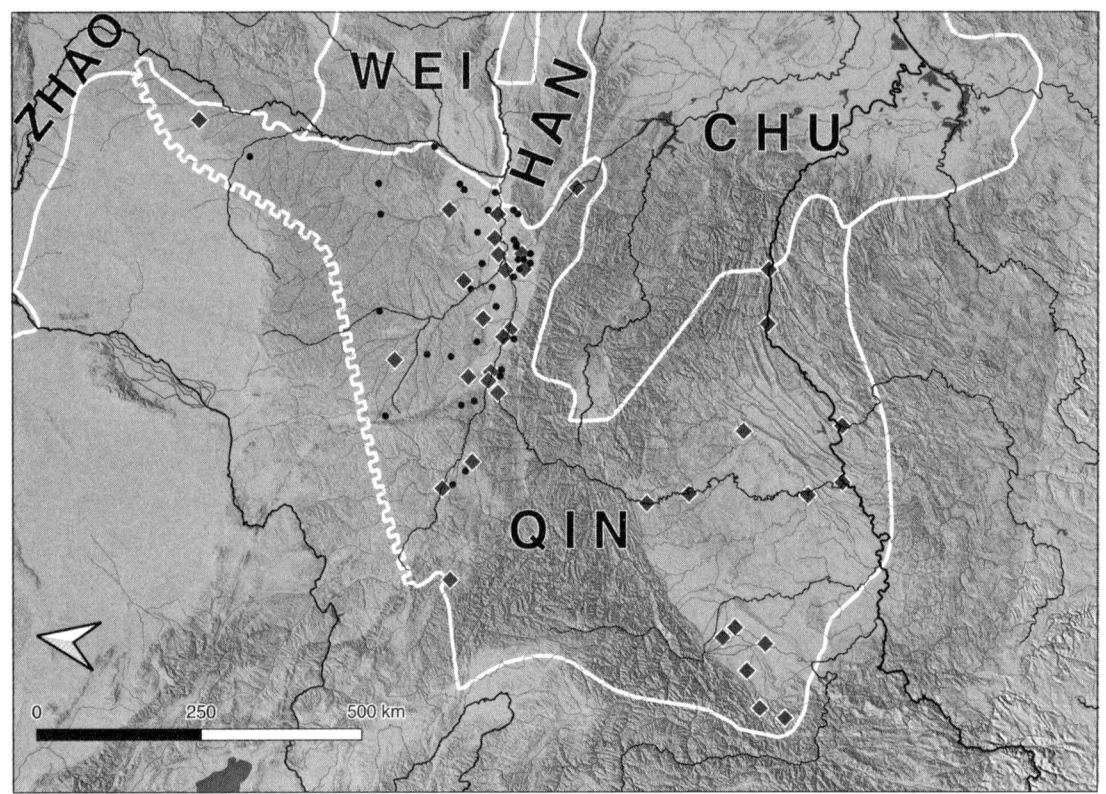

FIGURE 2.10
The Qin State before Imperial Unification. The Qin state, shown here immediately before its conquest of the other Warring States regimes, densely settled the Guanzhong region and built a wall across the Ordos. Note that the map is oriented with east at the top.

BCE, expanded rapidly under a Qin state monopoly. By the middle and late third century BCE, the large, well-fed, and well-equipped Qin army was able to carry out a series of conquests against pastoralist adversaries on the Loess Plateau and against Chinese rivals on the floodplain and in the south. Figure 2.10 is a map of the Qin state during the third century BCE, on the cusp of its campaign of conquest. It depicts the Qin wall across the Ordos, the dense settlement geography in Guanzhong, and the scattered cities and towns farther north.

In 221 BCE, Qin unified all of China to rule a population of approximately twenty million people. Later imperial regimes followed the Qin template in significant ways, and the Qi model as well, upstream and downstream alike. During its fourteen years of unified control over all of north and south China and the Loess Plateau, the Qin regime sponsored eleven forced migrations to the northwest.

Qin emphasis on agricultural colonization also necessitated an aggressive frontier policy and expropriation of hunters and pastoralists from contested land. In 214 BCE, General Meng Tian (d. 210 BCE), descended from a line of military officers and architects and known for his battles against the Xiongnu, also designated forty-four new counties on formerly Xiongnu territory, established barriers and walls including the Qin Great Wall, created a road network in the Ordos, and brought in convicts to populate the new counties.[72]

The Great Wall across the Ordos was the first piece of infrastructure built for the purpose of clearly delimiting a regime boundary across the ecological transition zone of the fragile Great Bend. The point was to create a staffed physical structure that would discourage the region's denizens from practicing multiple and interlocking modes of subsistence that shifted north and south over time according to dictates of weather, climate, geopolitics, and political economy. The wall demarcated a zone of agricultural intensification throughout the region south of the wall. Three hundred thousand troops and an enormous number of wall-building conscripted laborers lived on the Loess Plateau during the Qin.[73]

The Qin was the first Chinese regime to systematically organize policy around the idea that big infrastructure paid for itself in agricultural intensification and tax revenue. Other than the wall across the Ordos, the most notable example of this was the Zhengguo Canal in western Guanzhong, which was built in 246 BCE parallel to the Wei River near its confluence with the Jing River. The canal irrigated more than forty thousand *qing* of land, about two hundred thousand acres, diverting water from the Jing to the north bank of the Wei River. It supported vast increases in productivity around the new imperial capital at Chang'an at the dawn of the imperial era. However, its efficiency soon declined because of silt accumulation. The challenge of eroded sediment washing onto floodplains was already clear to commentators at the time.[74]

The Jing River travels southeast from its headwaters, traversing the part of the Loess Plateau where the sediment is thickest, before disgorging into the Wei. As the Jing and its tributaries incised the soil layers in a region that was coming under cultivation for the first time, they formed gullies and canyons that lay below the level of the cultivated land. Irrigation works needed to start out at higher and higher elevations to avoid the gullies, and new feeder mouths had to be built again and again near the Jing and Wei confluence as older ones became useless. Infrastructure like this "fitted less and less easily in its environment even from an early date."[75]

Various regimes abandoned and revived the Zhengguo Canal repeatedly over the following millennium, but after its early decades, it never again irrigated more than ten thousand *qing* of farmland.[76] A poem in *The History of the Han Dynasty* (*Han shu*) extols the Zhengguo Canal. The author acknowledges that the canal water was full of sediment but frames this as an opportunity as well as a challenge. At that time, sediment would have still included fertile soil, not the sand and gravel that predominated after centuries of persistent erosion. The poem clearly reveals that the waterworks system hedged against unpredictable weather and allowed for intensive agriculture in Chang'an's foodshed in the Wei Valley, even as it transformed the ecology of the region:

> Picks and shovels take the place of clouds,
> And an open canal is like the rain.
> In one *shi* [almost thirty kilograms, or about sixty pounds] of Jing water,
> Several liters [*dou*] are silt.
> How it irrigates and fertilizes,
> And makes our crops grow high.
> It feeds the officials at the capital,
> And millions of mouths [besides].[77]

The Qin regime that unified the realm did not survive the death of its founder in 210 BCE. Following a few years of fighting after his demise, a rebel from the southeast destroyed the Qin capital and declared himself the founding emperor of a new regime called the Han. At the scale of years and decades, the unification of China under the Qin and Han regimes did not spell a major change for the Yellow River or its denizens. The East Asian monsoon was generally weak during the last two centuries before the Common Era, a relatively arid period during which climate variability was high and droughts, often followed by devastating floods, occurred regularly.[78] The population continued to grow apace and to urbanize, facilitated by steady agricultural and technological intensification, both on the floodplain

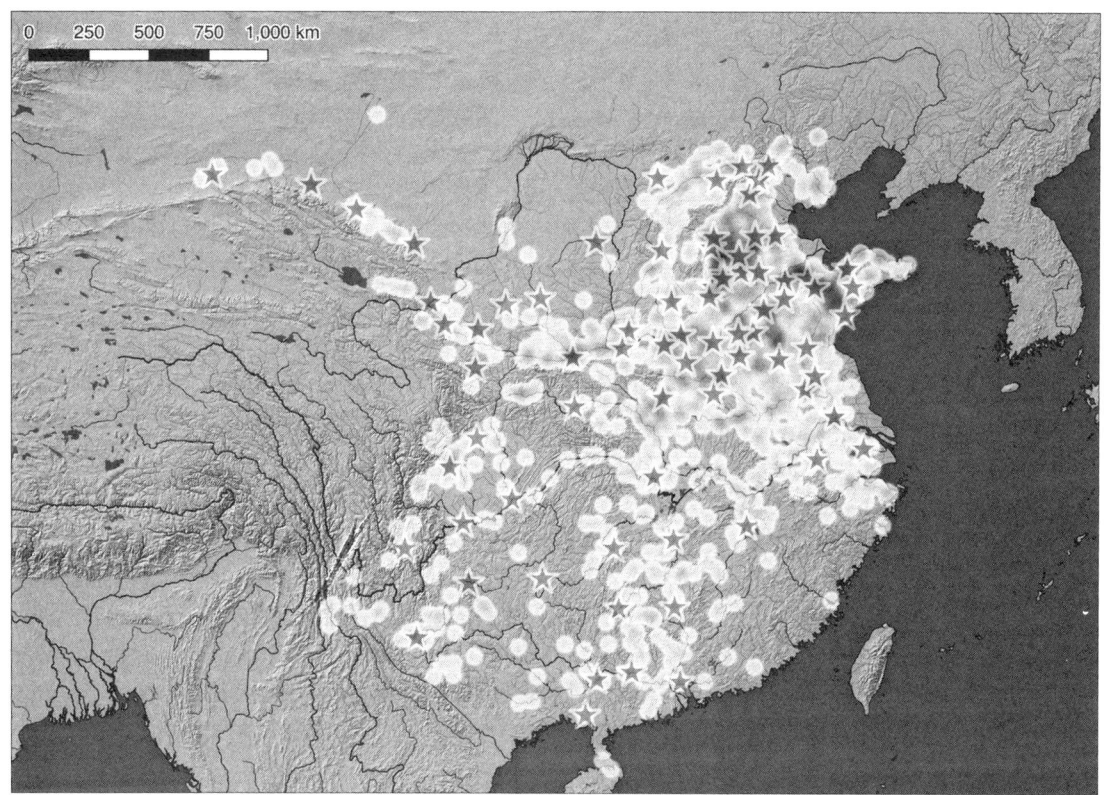

and on the Loess Plateau, resulting in significantly larger demands on natural resources.

Figure 2.11 depicts the distribution of population and prefectures at the time of the census of 2 CE. At that time, nearly forty million of China's fifty-seven million people lived either on the Yellow River floodplain or along the Wei, Fen, and Luo valleys. A line of prefectures on the banks of the Jing and Wuding Rivers spanned the central Loess Plateau. The floodplain counties, the districts on the path of the Yellow River floods, had an average population density of 112.74 people per square kilometer. Farmers converted grasslands, wetlands, forests, and mountain slopes into agricultural and grazing land both upstream and downstream. Workers channelized the lower course and settled alongside it. State officials chronicled every one of its floods. Woodcutters deforested the hills of the southern Ordos to build and heat the capital at Chang'an and to power its forges.

FIGURE 2.11
The Settlement Geography of North China in 2 CE. The floodplain reached its highest population density in 2 CE. The population of Guanzhong at that time was dense as well. North of Guanzhong there were prefectures—large outposts that oversaw military activity—but there were few counties.

FIGURE 2.12
Han Dynasty Farming and Blacksmithing. This paper rubbing of a stone carving depicts the political and cosmic context for iron working around the turn of the Common Era. On the top panel, artisans are manufacturing swords under the observation of auspicious dragons and civil officials. The bottom panel documents scenes of agriculture with iron plows and hoes. Courtesy of the Institute of History and Philology, Academia Sinica.

Iron technology improved dramatically, and iron tools became widespread.[79] Cast-iron plowshares, rakes, and other implements allowed farmers to break sod and roots in order to bring more land and different land types under cultivation, to more efficiently cut timber, and to more easily dig canals and build levees. With effective tools, farmers could plow fields more deeply: clearing ground cover, loosening soil, and removing boulders, thereby increasing the rate of erosion on the loess lands. Figure 2.12 is a rubbing of a stone carving about the Han iron industry. Figure 2.13 depicts this era from the perspective of floodplain event history. Historical sources record a total of sixty-two events between 282 BCE and 11 CE. These were centuries of relatively active water infrastructure development and maintenance. The pace at which sources attested both management and catastrophe increased in the final century before the Common Era, particularly around the turn of the era.

A major campaign of conquest in 121 BCE penetrated deep into Central Asia. Until then, the northern and western perimeter of the Great Bend of the Yellow River marked the border between Han and Xiongnu territory. The Qin wall had been abandoned. There had been heavy Han defenses and a large number of Chinese forts along the northern and western perimeter of the Ordos. This was a generally arid century, and climate conditions surely exacerbated political and military conflict. People built irrigation canals to reroute water, they consumed wood and topsoil, and they began to experience the formation of sand dunes in their settlements. Some lakes completely dried up because their water supply had disappeared.[80]

Inside the Hetao bend, a mix of farmers, Xiongnu colonists, and allied indigenous pastoralist residents occupied the Ordos. The bor-

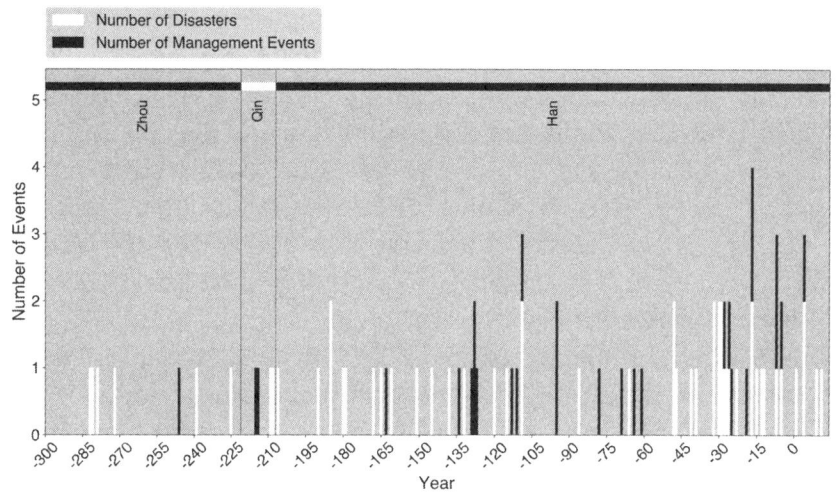

FIGURE 2.13

Event Timeline, 300 BCE to 14 CE. Management events preceded the disaster era as people built levees, drainage works, and transportation works to permit dense settlement on the floodplain.

der was heavily fortified, and the Han and Xiongnu regimes struggled for suzerainty. The Xiongnu launched frequent raids on the region. In the 1980s, the historical geographer Hou Ren-zhi argued strongly for an almost purely anthropogenic origin to desertification in the Ordos, beginning with the Han-Xiongnu conflict. His position, based on textual evidence, has been tempered by more recent scientific surveys, but the temporal correlation of war, colonization, erosion, and alluvial activity on the floodplain is strong, even though causality is not entirely clear.[81] More recent environmental and archaeological studies in and around the Mu Us Desert reveal that desertification spread under the influences of both arid climate conditions and intensive agriculture in ecologically marginal regions.

Han government policy in the final century before the Common Era continued to encourage migration north and west onto the middle and upper reaches of the Yellow River and its tributaries. The goal was twofold. One goal was to relieve population pressure on the floodplain. The other was to settle colonists in the Ordos to make it into Han agrarian space as part of a strategy to suppress the Xiongnu.

The colonists were expected to be self-sufficient in short order.[82] The Ordos population tripled, rising to 7.9 million by the turn of the Common Era. Large-scale deforestation occurred in the southern valleys of the Ordos. One frontier settlement program in 127 BCE moved six hundred thousand to eight hundred thousand new residents to the region. Another 725,000 settlers came to the Ordos during a colonization campaign in 119 BCE.[83] Nearly 8.3 million hectares of land came under cultivation, an area that is approximately equivalent to the size of Maine or Indiana. The population of the loess hills rose to 2.8 million people, who subsisted to a significant extent by irrigated farming.[84] After the colonization of the far west, surveyors traveled to the headwaters of the Yellow River and described its origins with some accuracy. The whole watershed had become Chinese territory.[85]

Following the military campaign of 121 BCE, "the greatest expansion movement in Chinese history," the northern border of the Western Han lay far north of the Ordos Loop. The regime no longer maintained the Qin walls across the Loess Plateau.[86] Immediately to the northwest of the Yellow River basin and the Loess Plateau, "tree felling and cultivation broke surface layers of alluvium and clay, water and wind began to erode and expose underlying beds of sand. Blowing sand encroach[ed] upon the cultivated area and eventually drifted across the ruins of former cities."[87] Just west of the Yellow River in Inner Mongolia lay a district of lakes and wetlands that supported a Han-ruled population of one hundred thousand in 127 BCE. It subsequently transformed into desert.[88]

The period after 121 BCE marked the beginning of a civilian phase of colonization on the Ordos. The region continued to see large-scale, state-sponsored settler colonialism: migration, occupation, and intensive farming, particularly near the capital at Chang'an. This marked a systematic peopling of the territory with Chinese farmers who wielded iron tools. The process followed the blueprint of a 169 BCE document by Chao Cuo (ca. 200–154 BCE), a political adviser who called for expropriating Xiongnu and allied pastoralists from the Ordos and replacing them with settler colonists. He advocated for establishing walled cities with moats wherever locations had water and arable land. According to Chao Cuo, each settlement should have one thousand households, and settlers should train as militia units to suppress Xiongnu uprisings. As he also put it: "It is necessary to settle permanent residents in border regions since expeditionary soldiers from other parts of the empire do not understand the character and capacities of the Xiongnu." He called for "the government to build frontier communities that are rooted locally, tightly connected, mutually assisted, and militarily united." Entire families moved in,

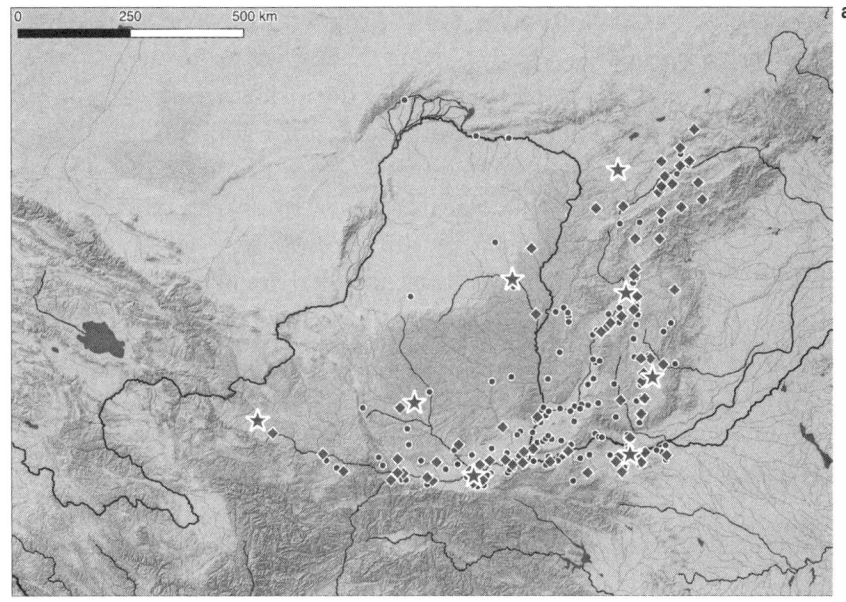

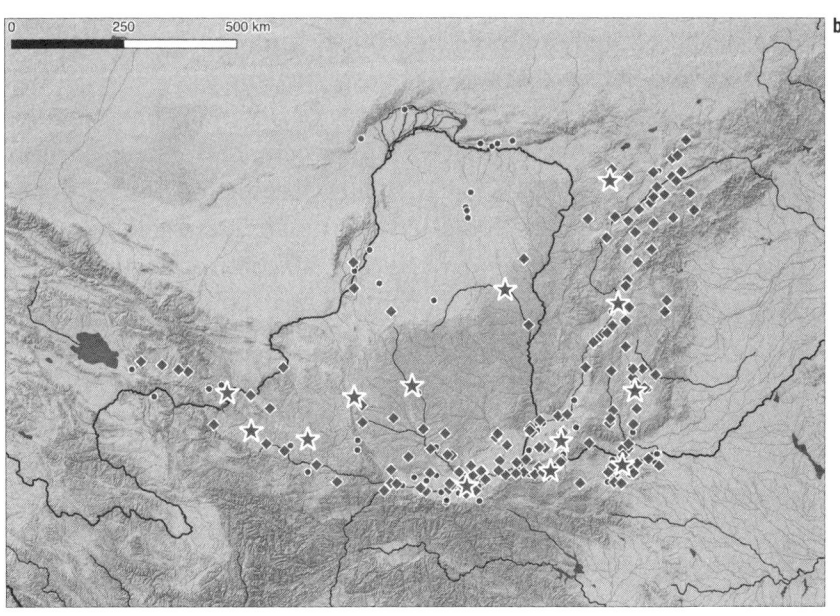

FIGURE 2.14
Qin and Western Han Sub-County Units on the Loess Plateau. Colonization on the Loess Plateau and the Yellow River tributaries to the east of the Ordos expanded between Qin times (a) and Han times (b).

exempted from tax and labor service, drawn from convicts, slaves, and free commoners, and supporting themselves as farmers.[89]

Figure 2.14 maps the colonization of the Ordos and the expansion of population and state power on the eastern and southern Loess Plateau during the last two centuries before the Common Era. In 221 BCE, there were eight prefectures, seventy-seven counties, and 147 other attested settlements on the Loess Plateau, all situated primarily in a southwest-to-northeast arc that included Guanzhong, the Luoyi valley, and the Fen River valley. Three prefectures marked a relatively southerly frontier across the central Ordos. About two centuries later, in 9 CE, there were fourteen prefectures, 154 counties, and eighty-two other attested settlements. The settlement arc had stretched into the Liupan mountains west of the Ordos, and its core had added numerous administrative seats. The Yinchuan Plain, the rain-shadow valley at the western edge of the Hetao, or Great Bend, was densely populated as well. Counties and other settlements dotted the central and northern Ordos, including in places that are now desert.

Chao Cuo's model followed Qin policy and built on precedents for settler colonialism that had been evolving since the Bronze Age. The framework included violent dispossession of indigenous populations, forced and voluntary migration to settlement frontiers, government support for new immigrants, military protection of regions being colonized, land cultivation by civilian and military colonists, free land for immigrants, administrative organization of immigrants—first into military districts and then into civilian counties and prefectures, military control of the frontier region, and interdiction against outmigration.[90] In this way, the early Chinese empires populated the semiarid and multicultural Ordos grasslands and forests with a large farming population.

Archaeological evidence actually depicts the stages of colonization. There were a range of cultures on the Ordos prior to the second century BCE. These were replaced first by uniform Xiongnu-style cultural assemblages. Han Chinese sites followed beginning in the first century BCE. Han conquerors established administrative seats, military fortifications, and farms, especially in the northern Ordos: along the northern part of the Hetao bend, and in what is now the Mu Us Desert but was then arable land. From archaeological and

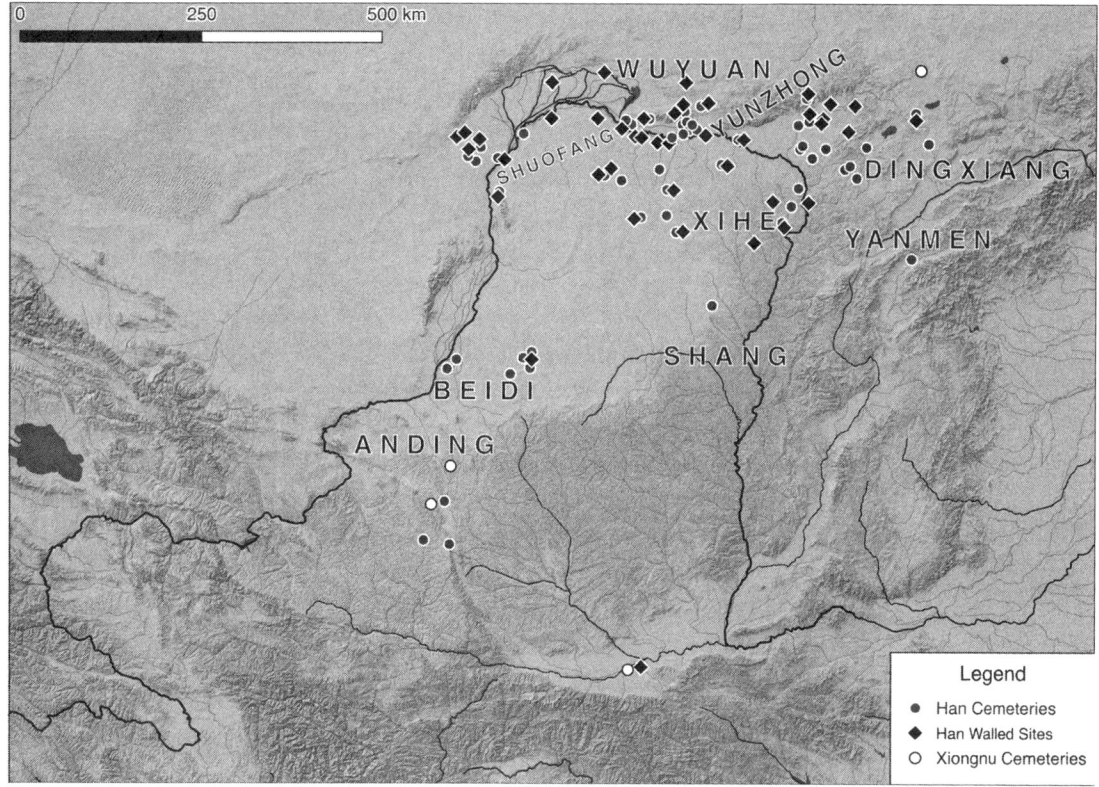

FIGURE 2.15
Archaeological Evidence of Western Han Settlements in the Ordos. Former Han provinces and settlements occupied locations in the Mu Us Desert (many of them in places that are no longer habitable) and north of the river as well. Based on Miller, "The Southern Xiongnu."

textual evidence, we know that troops and civilian colonists succeeded in defending and provisioning garrison colonies and controlling trade and economic networks, and natural resources such as salt deposits besides. Many non-Han names persist in the Ordos in historical sources and artifacts from tombs. The indigenous populations never departed: they were simply colonized and displaced, even as intensive agriculture became the dominant mode of subsistence.[91] Figure 2.15, a map of Han archaeological sites in the Ordos, enriches the story told in figure 2.14. It reinforces the fact that the northern Ordos, now anchored by the dunes of the Mu Us Desert, was then settled territory. There was also a dense population north of the Hetao bend in lands that were only sparsely occupied in later times.[92]

The Ordos settlements, backed by state-supported infrastructure projects, did not always account for the behavior of the restless and

BEFORE IT WAS YELLOW 97

alluviating river system in which they occurred. A major land reclamation project in the middle of the second century BCE built new irrigation canals. The canals brought five thousand *qing* of new land under cultivation, but the river shifted and silted so much that the canals became useless within a few decades.[93]

Massive timber operations built and fueled the capital at Chang'an, the expanding population of towns and cities throughout northern China, and the network of iron forges. Forests were lost to firewood collection, charcoal making, land reclamation, metallurgy, brick making, and construction. By the Western Han, trees for construction at the capital no longer came from Guanzhong itself.[94] Woodcutters traveled as far afield as the Wei headwaters in search of trees.[95] Parts of the Loess Plateau that were formerly covered by grasses and trees "turned into barren land after long term vegetation destruction."[96] Water tables began to drop.[97]

The Writings of the Huainan Masters (*Huainanzi*), a text of political philosophy written sometime before 139 BCE, refers to "whole forests burned for the hunt and great tree trunks scorched and charred." Referring to fuel for bronze and iron forges, the text comments: "No tall trees were left on the mountains, and the silkwood, oaks and lindera trees disappeared from the groves. [Untold amounts of] wood was burnt to make charcoal and [great quantities of] plants turned to white ash in bonfires [for potash], so that the anise and jasmine could never meet their perfection. Above [the smoke] obscured the very light of heaven, and below the riches of earth were utterly exhausted. All this was due to the use of fire."[98] This text is surely hyperbole, but it provides a sense that industrial-scale timber consumption was a new and noteworthy phenomenon during the second century BCE, and that it was alarming to the people who witnessed it.

Nevertheless, compared with later times, plenty of ground cover remained intact in the middle reaches of the Yellow River. Some mountains and riparian valleys were still forested. *The History of the Han* references other forests in the upper reaches of the rivers of the Ordos, in the Liupan Mountains west of the Ordos, and in the Lüliang Mountains that drain the Fen River. It appears that Han exploitation of the loess region was limited primarily to lowland regions and the lower courses of the major rivers.[99] On plateau grasslands, horse pasturage was intermixed with agriculture.[100]

All these activities had an impact downstream. By the turn of the Common Era, populations were large enough and technologies sufficiently ecologically impactful that erosion had begun to affect the entire alluvial plain. Erosion remained low by the standards of later centuries: average sediment deposition was still less than one centimeter per year. However, it had already increased by several orders of magnitude relative to earliest readings. Large-scale farming, particularly wheat agriculture, had made society increasingly reliant on the Yellow River and its tributaries, which provided rich soil and irrigation water. Floodplain residents moved close to major rivers to gain access to silt and water. The slopes of Mount Tai had begun losing their soil to anthropogenic erosion long ago. By the turn of the Common Era, the sediment from the uplands onto the plain during heavy rainstorms had become almost devoid of silt and clay. What remained was primarily sand, which suffocated fields without contributing to their soil fertility.[101] Farmers came to depend on upstream sediment to fertilize their crops.

Large landowners, primarily wealthy aristocrats, mobilized labor to construct canals, reservoirs, and irrigation facilities, and to organize water for their own interests. They also drained ponds and wetlands and transformed them into farmland. No longer available to contribute to the floodwater drainage ecosystem, these low-lying, swampy fields routinely flooded when waters were high. Most impactfully, landowners and officials constructed levees to create a domain for intensive agriculture that was separate from water and its entrained sediment, changing local hydrology and extinguishing small creeks or even reversing their direction of flow. T. R. Kidder's excavation site on the Hebei plain at Sanyangzhuang reveals evidence of floods, levees, and other landscape modifications, as well as pollen evidence of large-scale land clearance for wheat, millet, and mulberry trees (to feed silkworms). Arboreal pollen diminished, and herbaceous pollen rose.[102] The archaeological record, like the event database that anchors this book, links the commencement of levee building and the earliest documented flood disasters to one another during the early imperial period of agricultural intensification that transformed the Loess Plateau, the floodplain, and their relationship to one another.[103]

By confining water to a narrow channel and extinguishing adjacent drainage systems, levees inherently increased both riverbank erosion and sediment accumulation in the riverbed. The new interlocking network of natural and constructed hydrosystems exposed residents to extreme flood risk. Although protective earthworks decreased the frequency of the routine inundation that had previously characterized the floodplain, they set the stage for disastrous high-amplitude deluges on an alluvial plain that was densely populated for the first time.[104]

These transformations supported a dense population and a prosperous society. Agricultural surplus on the plain helped feed the armies and settlers who expropriated the indigenous residents of the grasslands and other frontiers and who colonized the Loess Plateau. The more that alluviating sediment from the middle course and the Shandong hills became a part of life for floodplain residents and farmers, the less willing residents and the regime were to permit the river to meander freely. In addition to locally sponsored projects, the court undertook forty major irrigation, transportation, and flood-control projects on the lower and middle Yellow River and its tributaries. These are the initiatives that appear in the event database. The first large-scale floodplain construction projects, including levees, canals, and wetlands drainage, date to the third century BCE. Irrigation projects extended even to the upper course of the river. Coding and analysis of the infrastructure activities from this period in the event database reveals that 30 percent of the attested management activity in the final two centuries before the Common Era consisted of events of new construction, 27 percent of activities involved proposal and discussion, and 10 percent of the events consist of general water management. Only two events—5 percent of the total corpus of attestation—are repairs of existing structures. This was the era in which the built environment of the hydrosocial floodplain came into place for the first time. The maintenance implications of the engineered system were not yet in focus.

Collectively, the Han waterworks regulated the river, but they complicated the problem of its subsequent management and constrained future choices. They compelled all future regimes to expend vast resources to maintain existing waterworks and the precarious populations that depended upon them. By the time of the great floods at the beginning of the Common Era, the river was perched above the floodplain.[105] This was the beginning of the high-disaster and high-management era in Yellow River history: intensive levee construction and interventionist flood management. Engineers were experimenting with various techniques and writing about multiple river management philosophies. All this was transpiring amid the rapid colonization of the Ordos and rapidly rising erosion there.

The Yellow River broke its banks and flooded in 168 BCE, requiring a large force of laborers to repair the levees and return it to its course. It flooded again in 132 BCE, breaching toward the south, flowing into the marshes and lakes of the floodplain and from there into the Huai River.[106] Laborers repaired the breach quickly, but it soon failed again. The emperor's diviners and numerologists urged him to accede to the will of the cosmos and not to repair the breach, particularly because he also had to respond to a Xiongnu invasion around the same time.[107] Twenty years of damage and failed harvests passed. Figure 2.16 depicts the site of the breach and all the recorded events between then and 108 BCE. The historical record attests five distinct disaster events and nine events of infrastructure management. As the map reveals, all the disaster events clustered around the breach site, but the works of civil engineering spread out broadly across the floodplain. Much of the effort turned toward creating new transportation canals and realigning existing canals to reflect the new river course. The few substantive events to repair the breach itself were unsuccessful.[108]

In 110 BCE, after twenty years of flooding, the emperor performed major rituals to propitiate spirits of mountains and rivers, but still the breach persisted. The following year, a season of drought, the emperor prayed for rain. He also took advantage of low water levels to raise a force of between twenty thousand and thirty thousand laborers to repair the breach. He inspected the operation in person and cast offerings into the rift. He ordered all the members of his retinue to carry fascines, massive bundles of sorghum stalks that were a readily available, cheap, lightweight agricultural waste that served as building material throughout the imperial era on the timber-poor floodplain. This was intended to reflect the court's commitment to the cause. However, the imperial entourage discovered that the farmers had already burned off the brush in their fields, so

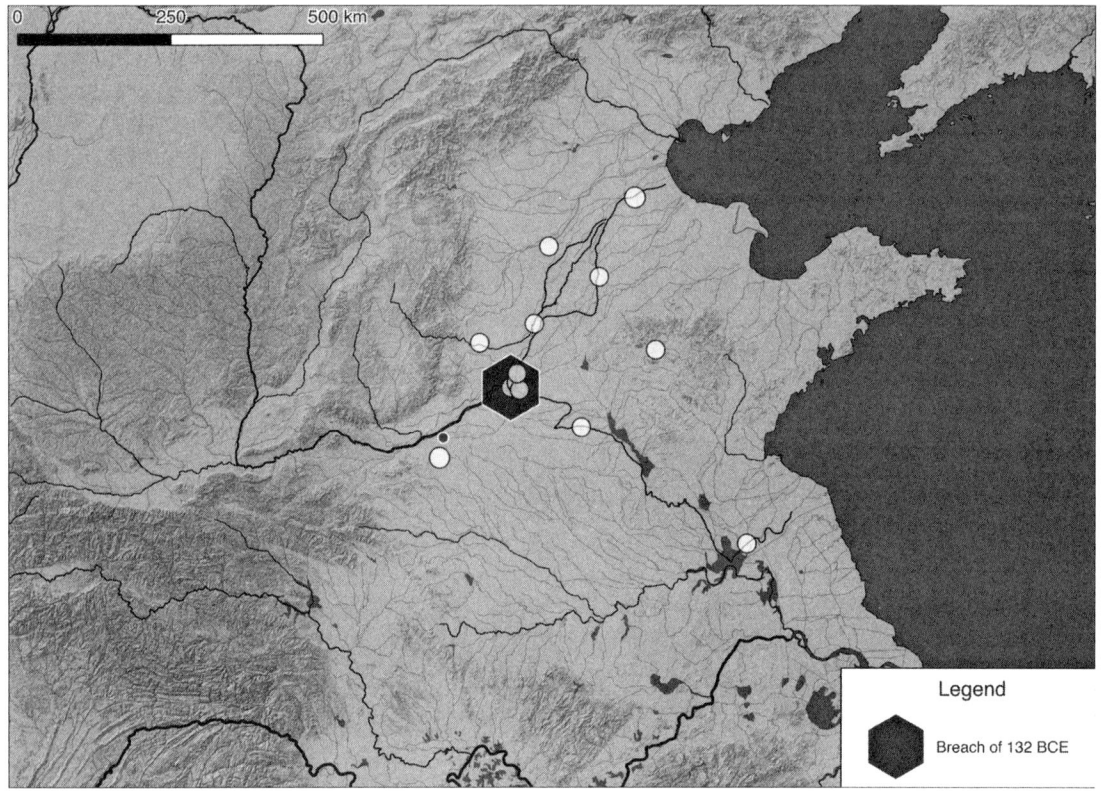

FIGURE 2.16
The 132 BCE Breach and Its Aftermath. The breach in 132 BCE occurred in the heart of the avulsion zone and in the most densely settled part of the empire. There were few management events in the years immediately following the breach.

instead they drove bamboo poles into the riverbed to make a low dam.

Sima Qian, author of *The Records of the Historian* (*Shiji*), and also a palace attendant at the time, was one of the wood carriers and a direct witness to the event. "How tremendous are the advantages of these bodies of water, and how terrible the damages!" he exclaimed, summarizing in one sentence the paradoxes of life on the floodplain. The emperor himself wrote a poem commemorating the activities (box 2.1). The poem attributes the flood to slope erosion and vividly describes its effects and the efforts to ameliorate it, using both practical and supernatural techniques. It depicts a hydrosocial world of people, animals, and spirits in which the court sought to balance competing requirements: to maintain wetlands for drainage and fish habitat, to burn stubble to fertilize farms, and to transport building materials from the wilds to the managed river.[109]

Box 2.1
Emperor Han Wudi's Poem at the Site of the Huzi Breach

In the "Rivers and Canals Monograph" (*hequshu*) of the *Records of the Historian* (*Shiji*), Sima Qian reports that in 132 BCE, the Yellow River breached its dikes at a spot known as Huzi (Bottle Gourd) and avulsed southeast into the Juye Marsh to take the course of the Huai River.[1] After the river changed course, its water no longer filled the irrigation canals of the floodplain or the transportation canals that carried grain to the capital. For twenty years, the river in spate produced repeated floods. These disasters were interspersed with droughts, and harvests were poor.

In 110 BCE, Emperor Han Wudi (157 BCE–87 BCE) decided to perform the colossal Feng and Shan (*fengshan*) rituals, sacrifices intended to reinforce the emperor's cosmic mandate though homage to the deities of heaven and earth. The Feng ritual involved building an altar at the summit of Mount Tai, and the Shan ritual entailed clearing away earth at its foot. Wudi traveled throughout the realm propitiating various mountains and rivers, and he cast jade and a white horse into the river at the site of the Huzi breach. Filled with despair at the challenges of river management, he composed the following verse:

The Huzi flood! Oh, what will we do?
The turbulence! Oh, the villages have all become rivers.
The villages are rivers! Oh, the land cannot find peace.
We labor without end! Oh, the mountains have become the plains.
The mountains are the plains! Oh, and in the Juye Marsh,
[Even] the fish are despondent! Oh, the cypress heralds winter,
The river has abandoned its course! Oh, it has left its constant channel.
The charging dragons! Oh, loosed to travel afar.
The former stream! Oh, we beseech the deities [for its return].
If not for the Feng and Shan rituals! Oh, how would I know all this?
Ask Uncle River for me! Oh, how could you be so inhumane?
No end to the raging flood! Oh, the worry of my people.
[The city of] Niesang is afloat! Oh, the Huai and Si [Rivers] are full.
For so long [the river] has not returned! Oh, its time is tardy!
The river in tumult! Oh, its turbulent current!
It contaminates the north! Oh, it cannot be dredged!

> Box 2.1 (continued)
>
> Take up the long poles! Oh, sink the beautiful jade!
> Praise Uncle River! Oh, the brushwood is not enough!
> The brushwood is not enough! Oh, the people of Wei are at fault!
> They have burned every stalk! Oh, how can we resist the waters!
> Strip the forests of bamboo! Oh, repair the levees with stones!
> The breach at Xuanfang! Oh, ten thousand blessings will come![2]
>
> The poem reflects several of the themes of this book: the destructiveness of the floods, the presumptive role that great monarchs were expected to play in water management, and the ingenious techniques and immense labor that water management demanded. According to Sima Qian, laborers finally succeeded in closing the breach soon after this performance. They returned the river to its original northerly course and followed that achievement with another round of canal building.
>
> ---
>
> 1. The incident appears in Sima, *Shiji* 29.1409–1415; the poem itself is on page 1413 (Beijing: Zhonghua shuju), digital version available at Scripta Sinica. The author's translation of the poem references Watson, *Records of the Grand Historian*, 236–37.
> 2. Sima, *Shiji*, 29.1413. The translation of the poem is my own, with reference to Watson, *Records of the Grand Historian*, 236–37.

The efforts at breach repair, which mobilized tens of thousands of laborers to move earth for multiple canal and levee projects that each took two or three years to complete, set the stage for such activities throughout the remainder of the imperial era.[110] This was the first time that the state determined how to mobilize expertise, labor, material, and political capital to reroute the sediment-clogged river into a preferred bed they had chosen for it, and how to engineer it into a single course across its floodplain. After the breach repair and the canal-building initiatives that followed soon thereafter, floodplain population rose rapidly, and waterworks construction continued apace.[111]

It was only in the eighteenth century that on the floodplain the river would be entirely confined within levees. However, in some locations, it had begun to perch at an elevation higher than the surrounding floodplain by the beginning of the Common Era.[112] Even after large-scale water management commenced, the physics of gravity and sediment deposit meant that the river was always liable to reoccupy the primordial watercourses, lakes, and wetlands of the floodplain. When it overtopped or burst the levees, the results were catastrophic. Moreover, as the levees lifted the riverbed above the plain, gravity caused moisture to seep into the ground alongside it. It caused salinization as it evaporated, killing trees and crops.[113]

By the turn of the Common Era, commentators were clear about the dynamics of the system that they inhabited. A class of experts explained the causes of floods and described the techniques best suited to managing them. Around 30 BCE, a general named Guo Chang, who had fought in the Xiongnu wars, was tasked with building drainage canals on the floodplain to relieve pressure on the levees. His associate Feng Qun, a regional commander (*duwei*), pointed out that the ultimate problem was not the high water, but the sediment that settled into the river and blocked its course in low-lying terrain. Feng Qun recommended building more drainage canals rather than more embankments.[114] The debate echoed the discourse of Yu the Great and his father, Gun, and it is one that would recur again and again throughout imperial history.

This was also an era of engineering innovation. In 28 BCE, responding to a massive flood that covered an area of about 10,000 square kilometers and affected four prefectures, a man named Wang Yanshi, a local expert who did not hold an official position at court, perfected the technique of using gabions, long and skinny bamboo baskets filled with rocks, as building blocks for breach repair.[115] Another flood two years later was also said to have come under control only after water management experts deployed new and specialized techniques.[116]

The first document about systematic Yellow River management, Jia Rang's treatise "Three Theses on River Regulation" (*Zhihe sance*), dates to 7 BCE, an era of catastrophe and instability both along the river and in political culture. Like Wang Yanshi before him, Jia Rang was a river expert who did not hold an official position at court. He studied the history of writing about the river and conducted fieldwork on the floodplain.

The strategies that Jia Rang propounded began by acknowledging the fundamental fact of silt accumulation. Like Feng Qun, he believed that high embankments could never fully control the sediment-clogged water. Jia Rang's plan to ameliorate these problems was for the state to decree and enforce a partial depopulation of the floodplain. In short, he advocated for building dikes far apart rather than close set, permitting the river to meander between low protective barriers. He argued for this over the high-intervention and high-population alternative: straightening the course and speeding the river's current to push sediment toward the sea. He also recommended building a vast network of drainage canals with sluice gates and stone retaining walls to draw floodwater out of the riverbed and onto the plain during times of high water. This would reduce the risk of catastrophic disasters but would also take farmland out of commission. This was not an option that the court could countenance, and the advice in "The Three Theses" was never implemented.[117]

"The Three Theses" also identified how levee construction initiated slow processes of environmental change and made floods less frequent but more catastrophic. Jia Rang correctly observed that frequent inundation on the plain had prevented ancient people from settling there, and he admitted that levee construction was the prerequisite for regional development on the floodplain. However, he also knew that the soil around the river was becoming saline and killing soil and crops in the vicinity, and he understood that this was happening because the confined and perched river caused moisture to accumulate alongside the levees rather than draining into the river.[118]

Jia Rang was not the only commentator at the time to critique the system that had come into place during the preceding decades and to recommend a new path forward. Other water conservancy experts of the era turned their analysis to the chain of causality that led from settlement on the Loess Plateau to catastrophic flooding on the alluvial plain. In 4 CE, Zhang Rong, who served as commander in chief (*da sima*) at the imperial court, connected every link. State sponsorship of colonization led to cultivation on loess hill slopes. That caused erosion, which was exacerbated by annual rainfall variation, which led to sediment deposition in the riverbed. The poorly conceived levees that people built in response to sediment accumulation could not prevent the floods that inevitably resulted. Here are his words:

> The nature of water is to flow downward. When it flows swiftly it scours its own bed and scoops out hollows and rather deep places. Now the water of the [Yellow] River is heavy with sediment. Every sixty pounds [*shi*] of water contains as much as five and a half liters (six quarts) [*dou*] of silt. In every county, both east and west of the capital, the people all direct the water coming from the mountains, which flows into the Yellow River and the Wei River, to irrigate their fields. In the spring and summer, the streams dry up, and this is a time of little water, therefore the water flows slowly and [deposits more silt which] blocks all openings, and it is shallower. But when the rains come the water rushes violently down so that there are floods and breaks in the dikes. And then the government and the people go on building dikes until the level [of the river] becomes slightly higher than the surrounding land. It is like building [ever higher] dams to store up water. It would be better to follow the nature of water and not [permit so much] irrigation; then the hundred rivers would flow freely and the waterways would regulate themselves [*shuidao zili*], and there would be much less danger of floods breaking through, with all the harm they bring about.[119]

In short, two thousand years ago, the relationship between erosion, sediment deposit, and flooding was clear to close observers of the Yellow River. The millennium before the Common Era ended with experts making scientifically trenchant but politically impossible suggestions about large-scale changes in policies of landscape transformation, frontier colonization, and population management. Their warnings would soon prove prescient.

A NEW MILLENNIUM: DISASTROUS FLOODS AND THEIR AFTERMATH

With the Yellow River constrained behind levees on the populous floodplain and with erosion rising on the colonized Loess Plateau, commentators attested immense floods during the first two decades of the Common Era. This was a time of considerable climate variability, when periods of drought punctuated catastrophic rains.[120]

FIGURE 2.17
Management and Disaster Events, 2–220 CE. After the floods early in the first century CE, management events and disaster events were both attested less frequently.

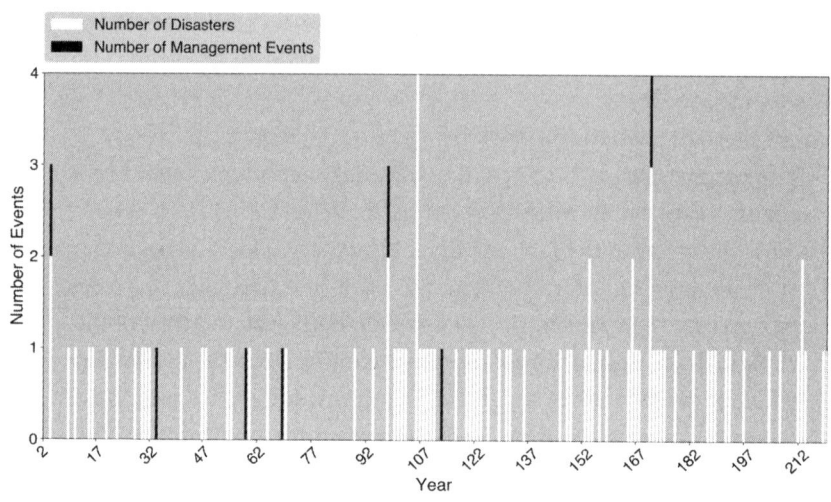

Major floods in 1 and 3 CE presaged a massive course change in the river that would shake the empire between 14 and 17 CE. All this transpired during a time of profound political unrest. During a time of factional strife and court infighting, in 9 CE, Wang Mang (c. 45 BCE–23 CE), an aristocrat who was regent to a puppet boy-emperor and kin to a Han empress, usurped the throne and declared a new dynasty. During a fourteen-year reign, which ended with his death in combat in 23 CE, he and his supporters battled Han restorationists, peasant rebels, and foreign invaders while also attempting to launch major administrative reforms. The beleaguered Wang Mang regime did not make it a priority to ameliorate the disaster on the floodplain, nor did the restorationist Han court in its early decades. The historical sources report a total of only four infrastructure repair and management events in the first seven decades of the new millennium, although they attest disasters in all but three of those years. Figure 2.17 depicts all the events attested in historical sources between 2 CE and 220 CE. Only a few of them involved earthworks management. Moreover, only three of the disasters are described in the sources as levee breaches—a clear sign that few maintained levees existed during these centuries.

The massive floods caused calamitous levee failures in 3 and 11 CE. The breaches followed a series of earthquakes. The breached river split the lower course into two branches spreading out across much of the historical floodplain. The old course continued to flow to the north of Mount Tai, whereas the other flowed south to join the Huai. In spite of substantial levee building during the preceding century, the earthworks system was still a patchwork, and much of the floodplain was still unprotected by barriers.[121]

In 14 CE, a "catastrophically disruptive" flood breached the levees not long before harvest time, forming a shallow lake that covered the west-central plain, inundating much of the central plain, and deluging as much as 1,800 square kilometers, almost 700 square miles, a region crisscrossed with canals and densely populated, home to more than 40 percent of China's population.[122] The effects of the flood of 14 CE continued without respite for three years. The inundation triggered popular unrest as well as organized resistance on the part of Han loyalist aristocrats. The starving farmers who lived west of Mount Tai, caught between the two untrammeled branches of the river, formed an army that marched west toward the metropolis of Luoyang, where they defeated Wang Mang's troops in 18 and 22 CE. Ultimately, Han loyalists vanquished both the peasant army and the Wang Mang forces and reestablished the Han ruling house in 23 CE.[123]

The reconstructed regime, known as the Eastern Han, faced almost a century of reconstruction. The Guanzhong Plain, eroded and deforested, had become unable to sustain the imperial capital at Chang'an. It would be five hundred years before Chang'an was the headquarters of a regime again. The revived Han ruling house instead established its capital east of the Ordos at Luoyang. The capital region in the Yiluo River valley became the site of massive waterworks investment that is not reflected in the events database. An engineered system of canals and other waterworks connected the city directly east to the floodplain. Engineers rerouted the Luo River north so that goods shipped up the Yellow River could be conveyed directly to the capital. The new waterworks also allowed the city to have a cosmologically favorable orientation with the Mang Mountains to the north and the Luo River to the south.[124] Archaeological evidence reveals that floods in the vicinity of Luoyang began to entrain gravel along with fine sediment soon after the capital moved there. This signifies that the Mang Mountains lost their soil, trees, and other vegetation almost as soon as the city wall was constructed.[125]

East of Luoyang, the vast floods on the alluvial plain lasted for over sixty years. For decades, the river spread out across much of its ancestral floodplain, creating wetlands and shallow lakes during each wet season. This occurred on lands that had long since been claimed for agriculture and intensive residence and had become the most densely populated, intensively farmed part of the Han empire. Many people were killed outright by floodwaters or became refugees in the immediate aftermath. Famine, banditry, and disease plagued flood survivors. The southern branch was not stanched until 70 CE. All of this launched a great migration to the south.[126]

Between the census counts of 2 and 140 CE, the recorded population declined from 57.7 million individuals to 48 million individuals. The population of North China—essentially the Yellow River basin—declined by nearly half, but the population of the rest of the empire grew substantially, until it became almost equally distributed between north and south. Within the Yellow River basin, some people moved west, out of the floodplain and toward the edge of the Loess Plateau around the capital at Luoyang; others moved east, out of Guanzhong and the Ordos, arriving at the same destination. In 2 CE, 44 million people lived in north China and 14 million in south China. In 140 CE the corresponding figures are 26 million and 22 million (fig. 2.18). In the west, with the retreat of the capital from Guanzhong and the Ordos relinquished to the Southern Xiongnu and the Tibetans, the population decreased by 6.5 million people.[127]

The move to Luoyang was one element of a large-scale contraction of state power and a retreat from the colonization of the Ordos. The regime did establish new fortifications on the Loess Plateau, but almost all of them were situated east of the Hetao bend. They spanned the eastern part of the Loess Plateau, north of the capital. Although there was some skirmishing in the Ordos right after the Han restoration, the regime ultimately abandoned that territory to Xiongnu pastoralists, who moved in by the 40s. Once Chang'an was no longer a capital, the strategic importance of the Ordos diminished. The Later Han regime, struggling to collect adequate tax revenue from powerful aristocrats and restive peasants, could no longer fund an army to defend its colonized frontier. A treaty of 50 CE allocated the Ordos, northern Shanxi, and Gansu to the Southern Xiongnu, a nominally loyal, semi-independent state allied to the

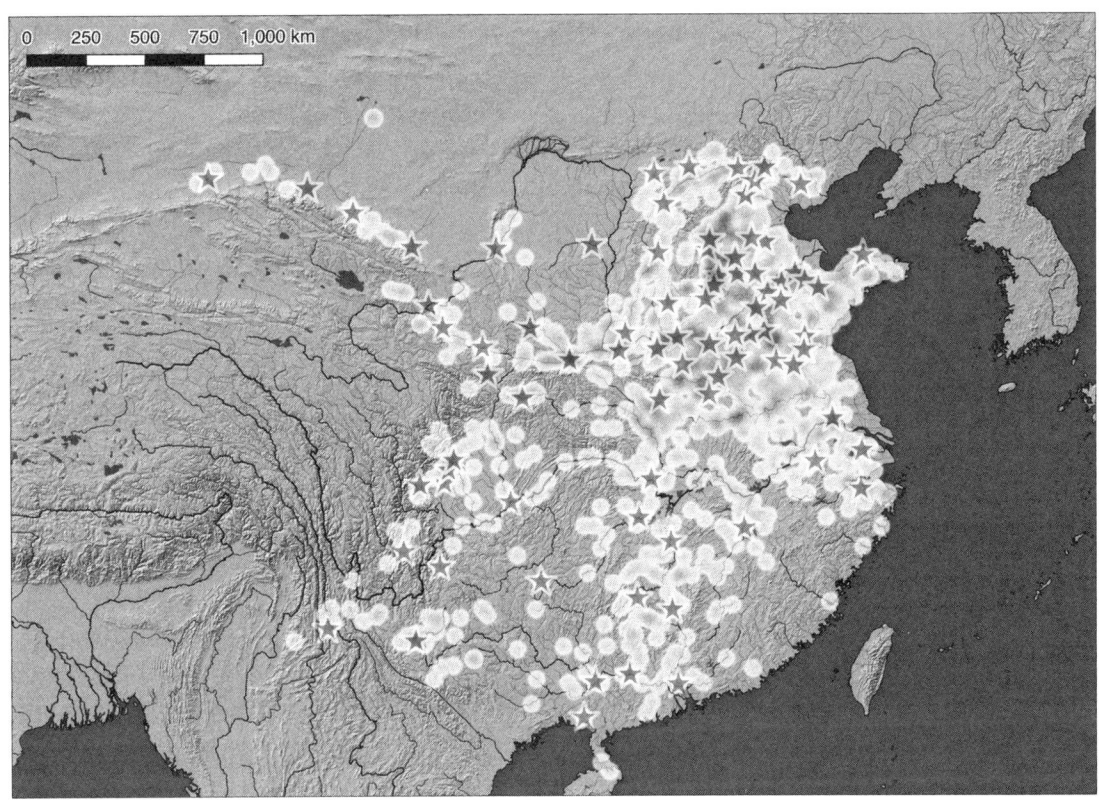

FIGURE 2.18
Population Distribution in 140 CE. By 140 CE, even though the distribution of prefectures remained oriented to the north, the density of population was beginning to shift out of the floodplain and toward the south.

Han. In the first millennium, the Ordos was ethnically diverse and "persistently multicultural."[128] The Southern Xiongnu reasserted control over a region of grasslands and riparian forests that they and members of their political network had long occupied, primarily as pastoralists, prior to the aggressive settler colonialism of the previous two centuries. The Xiongnu and their pastoralist allies occupied the whole Ordos with their herds. Tombs there from the first century CE are all culturally Xiongnu.[129] From their base in the Ordos, the Xiongnu skirmished periodically with Turkic, Mongol, and Tibetan rivals to the north and west. However, there was little fortification or sedentary agriculture there for hundreds of years. Intensive agriculture on the Ordos resumed only sometime after the unification of China under Sui rule in 589 CE. Figure 2.19 depicts the history of Loess Plateau settlement geography between 220 BCE and 800 CE. The expansion of county settlements during the first half of the Han is vividly clear, as is its collapse along with

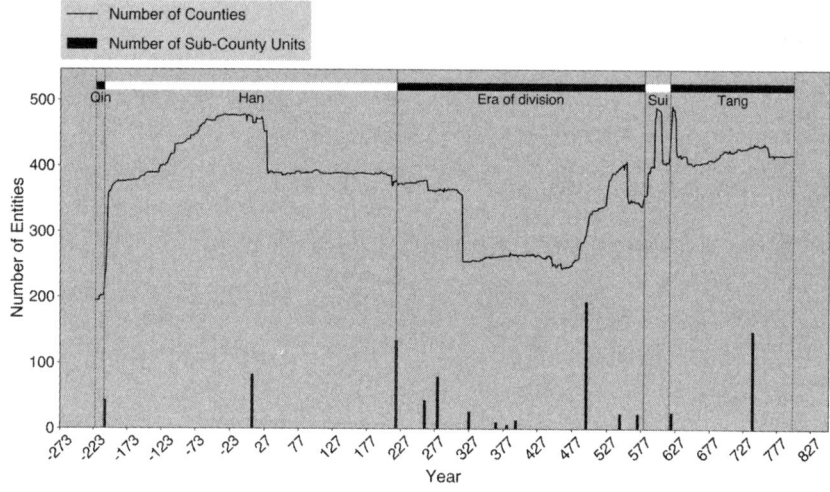

FIGURE 2.19
Loess Plateau Political Geography, 220 BCE to 800 CE. The density of settlement on the Loess Plateau fell dramatically after the end of the Western Han, and again after the end of the Eastern Han. It remained quite low during the 300s and 400s before rising again.

the Former Han regime. Over the following half millennium, the number of county seats continued to contract, although some post-Han regimes recorded a significant number of settlements without administrative rank. The reestablishment of county seats in the sixth century set the stage for the reassertion of imperial power in the decades to come.

Sporadic Eastern Han attempts to resettle Chinese farmers in the region failed. Indeed, Chinese inhabitants responded to mounting pressure from the Southern Xiongnu and their allies by abandoning their villages and farms. Some were evacuated by the government, and others withdrew independently. In 30 CE, the Eastern Han regime abolished more than four hundred counties in the northwest, more than one-quarter of all the counties that had existed at the time of the 2 CE census. The most populous jurisdiction, Xihe Commandery, which spanned the heart of the Ordos pastures, experienced the greatest declines. In 2 CE, a population of 698,836 lived in thirty-six counties. In 140 CE, 20,838 people lived in thirteen counties.[130] The small number of remaining Chinese settlements were limited to the southern part of the Ordos region.[131] Immigrants and refugees moved south, crossing the Qinling Mountains to settle in Sichuan.[132] This marked the beginning of a centuries-long renewal of forests, grasslands, and hydrology in the Ordos.

Correlation is not causation, and complex hydrosocial systems resist simple and reductive explanation. That said, there are few attestations of flooding between the first century and the eighth century, despite the fact that there are also few records of floodplain management. These eight centuries of low avulsion and few records are also an important part of the lifespan of the Yellow River. However, there was one major management undertaking at the very beginning of that time span. Soon after taking power, the restoration ruling house hired thinkers who understood hydrology and river engineering to recommend ways to channelize the lower course of the river in a stable and practical fashion.[133]

In 69 CE, the court assigned Wang Jing (30–85 CE), a Shandong native with a background in astronomy and divination as well as hydraulic engineering, to develop and oversee a plan to regulate the river and restore it to a single course. Armed with texts that included *The Classic of Mountains and Oceans* (*Shanhaijing*), *The River and Canal Monograph* (*Hequshu*), and *The Diagrams of the Tribute of Yu* (*Yugongtu*), he developed the first comprehensive effort to plan a systematic waterworks program at the scale of the whole floodplain.[134] His goal was to devise a landscape that focused on flood diversion rather than earthworks barriers. Wang Jing and his collaborators identified the tributaries, lakes, and swamps that could contribute to flood diversion and sediment discharge.[135] In the wake of population collapse, political circumstances enabled Wang Jing and his colleagues to put some of the ideas from Jia Rang's "Three Theses" into practice. They commandeered thousands of migrant workers to build a vast network of canals and reservoirs throughout the floodplain, taking advantage of natural drainage systems, terrain, and wetlands distribution. The new waterways were designed to protect transportation routes and settlements, and to reduce pressure on levees during times of high water. I have no choice but to follow the historical record in ascribing all the success of these measures entirely to literate and titled individuals. However, the sources also praise them for their inspection tours and field trips, and agency was surely distributed widely among the residents and artisans who were their anonymous collaborators.[136]

On the main channel of the river, Wang Jing and his collaborators identified locations where the river could be straightened, using close-set levees to accelerate the current and to push sediment downstream. They supervised the construction of a great many sluice gates on rivers and throughout the canal system. These could be closed to entrap water for irrigation and transportation during dry seasons, and opened to drain water during wet times so as to avoid catastrophic levee breaks. They also established a system for routine inspection to ensure regular repairs, appropriate schedules for opening and closing sluices, and calendars for county magistrates to commandeer corvée labor. They oversaw the reconstruction of the Bian Canal, the main transportation route from the rice-growing Yangtze delta to the capital at Luoyang.[137]

Wang Jing and his collaborators recognized that silt could easily overwhelm the shallow canals and reservoirs at the heart of the system he designed. Over the resistance of his critics, he insisted on installing structures for separating clear water from silty water: "one water gate every ten *li* [approximately five kilometers]" (*mei shili li yi shuimen*).[138] The water gates were a preindustrial version of the silt-washing structures of the contemporary Xiaolangdi Reservoir (plate 21). They were a type of low dam known as a weir in English. They created a barrier just above the water level over which clear water could flow easily while causing heavier sediment to precipitate to the riverbed, where it could be dredged out during the dry season.[139] The Wang Jing construction privileged the creation of a large number of simple structures that could be locally maintained rather than a smaller number of monumental earthworks. The hydrosocial landscape he devised took silt management as a central consideration. The new river course was direct to the sea, the current was rapid, and silt deposition was limited.

By the time of the 140 CE census, two generations after the Wang Jing initiatives, the Yellow River floodplain was populated once again, though not nearly as densely as it had been in 2 CE. Farmers and townspeople were protected by a canal, dam, reservoir, and levee system that managed silt, drained wetlands, channeled rivers, and carried water to fields. Of the forty-five disaster events that appear in the historical record between 65 and 140 CE, only one was a levee breach. Levees were simply not a significant aspect of floodplain engineering at this time.

The Wang Jing infrastructure certainly played a role in ameliorating flooding during subsequent decades and centuries and in redefining and reducing the kinds of high-water events that counted as disasters. However, as extraordinary an accomplishment as they may have been, these kinds of preindustrial waterworks, built primarily from earth, wood, bamboo, and sorghum stalks, would have required massive annual upkeep to remain in good working order for long. Canals, reservoirs, and riverbeds needed regular dredging, even in an era of sedimentation that was lower than it had been during the Western Han. Managers and laborers needed supervision and payment. There is little evidence of any of that in the historical record. If those activities were occurring, they were transpiring through innumerable local and vernacular efforts that are not visible in the historical record. It is impossible to assign percentages of causation in complex historical systems, but it is indubitable that diminished rates of erosion from the Loess Plateau were one factor contributing to attested stability on the medieval floodplain. The history of the Loess Plateau correlates better to flood history than either climate trends or the history of water infrastructure.

The floodplain remained relatively stable for eight hundred years after the early first century of the Common Era. Between the third and seventh centuries, the North China Plain was under the control of numerous short-lived Chinese, Turkic, and other regimes. The wars of this era sometimes included intentional waterworks destruction. Waterworks were sometimes neglected and sometimes maintained. *The Book of Jin* (*Jinshu*), which was published in 648 and covers the history of the Jin dynasty between 265 and 420, attests some dam and flood diversion construction and some commentary about the legacy of Yu the Great. The Jin and other post-Han regimes maintained and extended transportation and irrigation canals.[140] However, even on the floodplain, the regimes of this era valorized martial clans and prioritized hunting over intensive farming. Although there were sporadic water management events, much of the floodplain probably reverted to a state of multiple mobile courses and active wetlands. A population snapshot from the late third century (fig. 2.20) shows the Ordos outside of Chinese control, relatively dense populations along the then-northerly course of the Yellow River and along the major tributaries, and a growing population south of the North China Plain.

Few political documents attested floods and breaches during the post-Han centuries, the records from geoarchaeology and environmental science do not reflect high sediment deposition, and the river never changed course. This further reinforces the logic that the

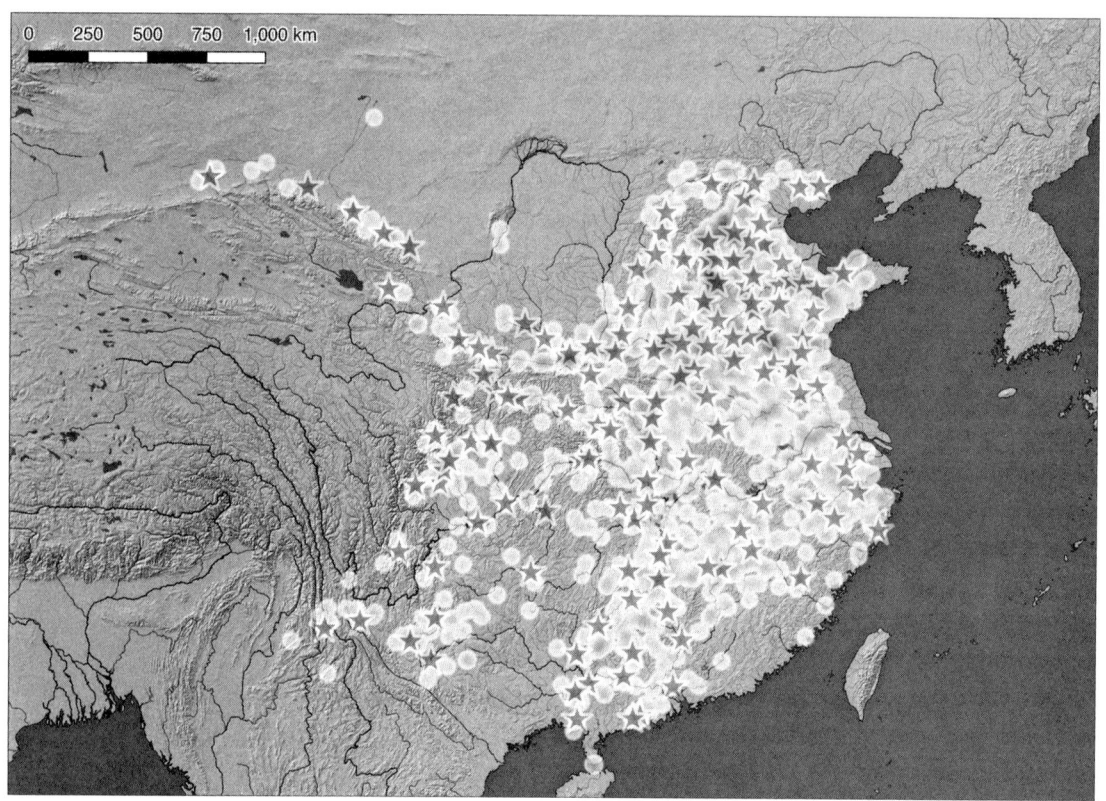

FIGURE 2.20
Population Distribution, circa 280–289 CE. In the third century, the Ordos was almost entirely depopulated, and population continued to spread toward the south.

stable era resulted from low erosion as well as fewer high-barrier waterworks and less-intensive floodplain agriculture. The system remained stable until the tenth century.[141] It is not that chroniclers were not paying attention: if there had been floods during this period, sources would have attested them. Dynastic chronicles like *The Book of Jin* continued to record historical events. In addition, Li Daoyuan (c. 466–527), a geographer, writer, and court official, compiled *Commentary on the Water Classic* (*Shuijingzhu*) around the turn of the sixth century. The forty-chapter book is a geography of 1,252 watercourses and canals and the history and culture surrounding them. It describes the source, course, and major tributaries of each river, and it includes transcriptions of stone inscriptions and fishermen's folk songs. The text built on the now-lost third-century *Water Classic* and on Li's own fieldwork around Henan, Shandong, Shanxi, and Jiangsu. It does not describe the Yellow River or the rest of the floodplain as a region prone to disaster. *The Records of Counties and Prefectures*

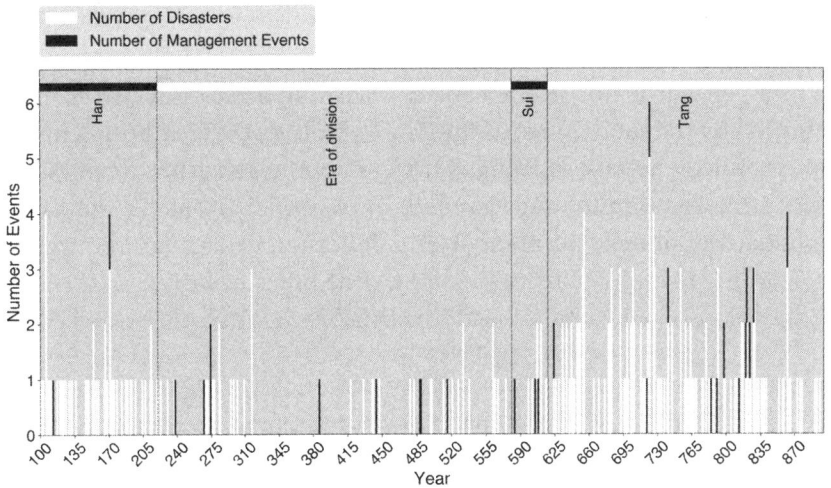

FIGURE 2.21
Event History, 100–900 CE. In the centuries between the 200s and the 600s the number of events attested on the Yellow River was extremely low.

from the Yuanhe Era (*Yuanhe junxian zhi*), an early ninth-century geographical encyclopedia that describes the spatial history of the realm, collates many medieval records about locales on the alluvial plain, likewise has little commentary about Yellow River floods. Figure 2.21 reveals just how few disasters appear in the historical record between the early 200s and the early 600s.

On the medieval Loess Plateau, where pastoralism predominated over agriculture, trees and grasses rebounded. The population contracted there under the influence of both political instability and hot and dry climate conditions.[142] The number of settlements diminished, and so did the number of people who practiced agriculture, particularly in regions where farming was marginal and unpredictable. Agriculture concentrated in the Guanzhong Plain and in the Fen valley.[143] No regime maintained a capital in Guanzhong during the period of division. Forests and grasslands rebounded. The medieval author Guo Zhongyan reported that, from summits in the Lüliang Mountains east of the Ordos, he could see nothing but forests for hundreds of kilometers in all directions, with no signs of human habitation.[144] The Jing River ran clear, and the Muddy River, the Nihe, became known as the White Horse, or Baima.[145] Timbering occurred primarily around fortified sites and did not reach deep into the mountains except in those locations.[146] *The Commentary on the Water Classic* attests twenty-seven lakes in the Ordos.[147] Some of the regimes that occupied the Loess Plateau during these centuries are known to have dug irrigation canals. Farming remained a feature of subsistence in the Ordos, but its intensity waned relative to earlier times.[148]

On verdant plains in the central Ordos, in a location that is now abutted by the sand dunes of the Mu Us Desert, the Southern Xiongnu founded the city of Tongwancheng as their southern capital in 419 CE. The capital, surrounded by seasonal nomadic encampments, supported a permanent population of around ten thousand. According to a report from the time, "The hill is beautiful, in front of it the plain is wide, and around this there is a lake of pure water."[149]

CONCLUSION

As table 2.1 shows, the early imperial population of Shaanxi (the province roughly coterminous with Guanzhong and the Ordos) peaked at just over three million people in 2 CE, before falling below one million two centuries later. By the early seventh century, the population exceeded its previous apex. The two centuries that commenced with imperial reunification in 581 and extended through the mid-eighth century featured ambitious floodplain reconstruction, with a focus on large-scale transportation and irrigation canals and on levee and reservoir construction. Upstream, there was a capital at Chang'an for the first time in more than half a millennium, in a region that was once again surrounded by clear water and ample forests. The Sui and Tang regimes renewed attention to settler colonialism and militarization on the Loess Plateau. These activities, and their consequences, are the subject of the next chapter.

In one sense, the floods at the beginning of the first millennium are one of the most significant events in Yellow River history. Commentators at that time and in the two thousand years since have correctly attributed them to catastrophically rising rates of sediment accumulation caused by erosion that resulted from colonial agriculture on the Loess Plateau. That process was, in turn, the culmination of a progression of agricultural intensification and rising erosion that had been underway since the late Neolithic era.

From another perspective—from the vantage point of attested Yellow River events in the historical record in later centuries, and from the perspective of subsequent high rates of sedimentation—the entire era of river history that extends from earliest times to the ninth century is a single long era. During that whole time, in spite of sediment accumulation in particular tributary valleys, the river traversed

TABLE 2.1

Ancient and Medieval Population in Shaanxi

Year	Dynasty	Population	Percentage of total imperial population
1025 BCE	Western Zhou	1,000,000	7.29
210 BCE	Qin	3,000,000	15.00
2 CE	Western Han	3,191,624	5.33
140 CE	Eastern Han	972,134	1.98
280 CE	Jin	722,595	4.49
609 CE	Sui	3,737,630	8.00

Source: Cao, *Shaanxi shengzhi renkou zhi*, 331.
Note: These population figures should be taken as a heuristic rather than a precise count given the problems with conducting and comparing ancient censuses. Population tripled between earliest estimates and the Western Han and then contracted, regaining and exceeding its historical peak in the Sui dynasty.

a largely unengineered floodplain in a relatively stable fashion, and floods were only sporadic. Forest and ground cover remained reasonably intact on the Loess Plateau, where population remained limited. From this point of view, the period of the late Western Han and early Eastern Han was an anomaly in an otherwise generally tranquil eon of river history.

By suggesting this perspective, based on the long-term event record and sediment record, this book breaks from conventional historiography. To be sure, the historical and archaeological records of the classical era reveal that human activity could easily disturb loess soil and bring it into motion, with visible effects downstream. However, these dynamics were of relatively low magnitude compared to later eras. The events of the early imperial era were significant indeed. This is partly because they were a turning point on the river, and partly because they set the terms of discourse and policy that would govern the relations between people and the river for the two millennia that followed.

By the end of the period covered in this chapter, the floodplain and the Loess Plateau had divergent forms of political economy. The Ordos was established a frontier of conflict between agrarian and pastoralist lifeways and Chang'an was recognized as a potentially suitable place for a capital. The terms of debate over floodplain policy were clear and well documented. The fact that the Loess Plateau was the origin of the sediment of the floodplain was evident to careful observers. Rulers and engineers could point to precedent for their respective positions: whether to channel the river within high levees and force sediment out to sea, or whether to disperse water and silt into drainage canals and reservoirs.

When a centralized state reasserted authority over the historical Han territory in the late sixth century, there was no necessary reason to believe that it would retrace the steps of the earlier period. The reason that came to pass, under conditions of unprecedented rates of erosion and sedimentation, is the subject of the next two chapters.

CHAPTER 3

Loess Is More: The Middle Period Tipping Point, 750–1350 CE

During the 900s and 1000s, sediment flowed into the Yellow River at a pace that was an order of magnitude more rapid than its highest prior rates. As one team of ecologists explained, "The change in the sediment supply to the coastal areas occurred suddenly, probably indicating rapid deforestation at about 1,000 years [before the present]."[1] Another team used evidence from loess paleosoil sequences, pollen, historical geography, and coastal sediments to arrive at the same conclusion: soil erosion on the Loess Plateau rose to around its current rate about a thousand years ago. The cause was primarily human destruction of natural vegetation, along with other changes in land use.[2] It is possible to pinpoint the ecological decline of the Loess Plateau with some precision. Until the tenth century, loess soil, amply covered with vegetation, was able to absorb rainfall, but thereafter, as deforestation and grassland disruption increased, the Ordos basin commenced a phase of severe and rapid erosion. The era of frequent flooding on the floodplain began at the same time.

Erosion rates accelerated over the course of three phases of human activity, each with a clear start and end date tied to the narrative of

FIGURE 3.1

Moisture Index, 700–1400. The eighth and ninth centuries were a time of protracted drought on the Loess Plateau and throughout the Yellow River basin, with the exception of a moist period around the turn of the ninth century. After an unusually moist tenth century—the turn of the disaster era on the Yellow River—moisture levels fluctuated up and down in the ensuing centuries.

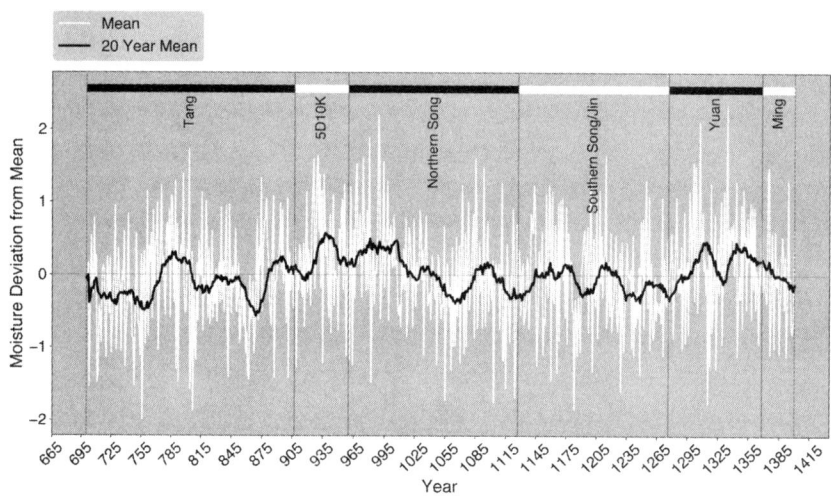

political history. In the year 589, the Sui regime (589–618) unified the historical Han territory and reestablished Chang'an as an imperial capital. Their successors, the Tang (618–907), maintained that designation of the metropolis. Guanzhong and the steep loess hills that surrounded it once again served as foodsheds and timbersheds for a cosmopolitan city. However, erosion rates did not increase very much. The Ordos remained under the control of a single regime, so there were no pressures for settler colonialism or wall building. It was sparsely populated and not contended or fortified. As figure 3.1 reveals, Tang rule transpired during an extremely dry climate period. The Ordos would have been more suitable at the time for grazing than for farming.

Drought would have had a direct impact upon the growth rate of Loess Plateau vegetation, the extent of desertification, and the annual volume of river water. Indirectly, it would also have had a profound effect on the size and distribution of human and animal populations; the plateau's steep north-south moisture gradient meant that small changes in precipitation could transform survival strategies in any given place. During moist years, populations could grow and spread out and landscapes could support settlers and soldiers alike. During drought, farmers, herders, hunters, and animals crowded into smaller habitable territories. This could cause internal social friction but also leave antagonists isolated, at a distance from one another.

In 755, a massive uprising known as the An Lushan Rebellion (An Lushan zhi luan) triggered a series of events that caused sovereignty in northern China to splinter between competing warlords, and ultimately between rival claimants for the imperial throne. Figure 3.1 depicts a moist period early in this process, around the turn of the ninth century. Disarray in the Tang empire also allowed various states of the steppe and Tibet to occupy parts of the Ordos and the rest of the Loess Plateau. The middle and upper courses of the Yellow River became sites of frequent battles, repeated fortification initiatives, refugee migrations, and competing waves of military and civilian colonization. Erosion began to rise rapidly.

This confluence of events marked the turning point that caused the Yellow River to convert from its low-disaster early period of slow environmental transformation to a frequently calamitous final imperial millennium of high-velocity ecological change. Even the name of the river changed. Most commonly known as the Great River (Da he) until the mid-Tang, the Yellow River (Huang he) moniker came into common usage by the early ninth century in reference to the hue of the loess sediment that it had begun to carry.[3] A 798 map and survey of the upper reaches of the river is the first document to use the term *Yellow River* in a title.[4]

In 960, near the beginning of a lengthy moist period, a new regime known as the Song (960–1276) unified much of the former Tang territory. However, the Song rulers did not succeed in recolonizing the northern Ordos. That region remained in the hands of a rival people, the Tangut, who established a regime known as the Xi Xia (1038–1227) and founded a capital on the banks of the Yellow River on the Yinchuan Plain. Hundreds of thousands of troops and allied civilians on both sides of the border confronted one another across a heavily fortified frontier that arced across the Ordos region south of the Mu Us Desert and through the Liupan Mountains at the southwestern perimeter of the Loess Plateau. A commercial timber market arose around the same time. Erosion rose exponentially.

The Song regime lost control of northern China in 1127, a time of rapidly fluctuating moist and arid cycles. For more than two centuries thereafter, the entire Yellow River basin was in the hands of sovereigns who did not fortify a border on the Loess Plateau or encourage intensive agriculture there. That transition marks the end of the second life span of the Yellow River.

FIGURE 3.2

Disaster to Management Ratio, 750 BCE to 750 CE. The Tang regime devoted little attention to floodplain management even as the frequency of reported disasters began to rise in the seventh and eighth centuries.

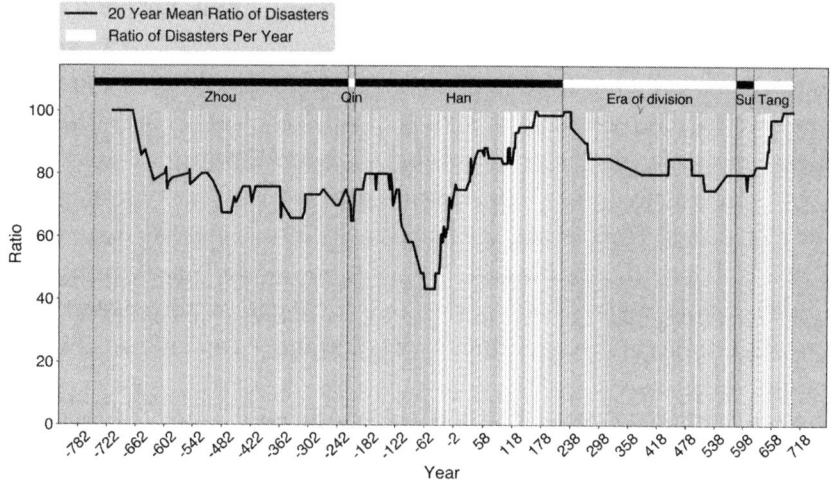

On the floodplain, the Yellow River's uneventful era ended abruptly during the very wet decade of the 920s. Figure 3.3 reveals that a remarkable correlation began in the second half of the tenth century. Before then, the Loess Plateau was able to absorb moisture, and rainfall was not necessarily connected to flooding. Thereafter, deforestation and ground-cover degradation meant that heavy rains caused sediment to wash through loess hills and gullies and into the Yellow River and its tributaries. From then onward, disasters became more likely to occur during rainy years.

According to the historical record, at least one flood or breach occurred every year between 924 and 954. Beginning in 958, historical sources attest disasters on the alluvial plain almost annually, and almost always in multiple. During an entire generation of abnormally high rainfall that transpired at the same time as the rapid rise in the erosion rate, there were only seven calamity-free years between 958 and 1029. There were three years—983, 984, and 991—when the attested number of floods and breaches rose into the double digits. None of these figures has any precedent in recorded history.

Hydrocrats around the turn of the millennium understood that they were seeing unprecedented amounts of silt and that they had to adjust their expectations away from the historical sources that they consulted for precedent.[5] The pace of calamity ebbed for a few decades during the first half of the eleventh century as people began rapidly building and maintaining levees and drainage systems. The

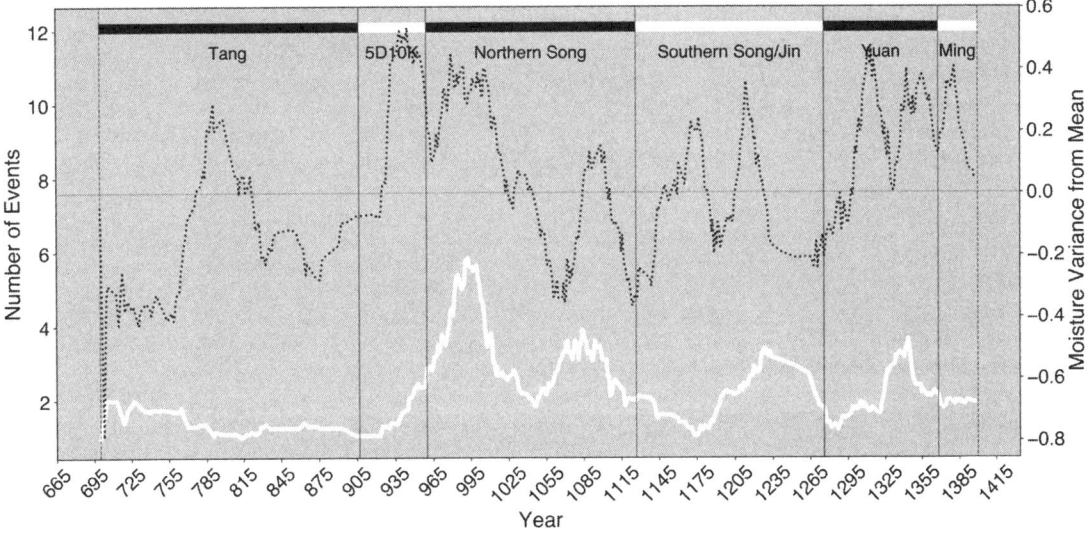

FIGURE 3.3
Disaster and Moisture Correlation, 700–1400. This extraordinary image reflects the only period in Yellow River history when the rate of attested disasters correlates perfectly with the moisture rate. Before the tenth century, when the Loess Plateau was intact, and floodplain architecture was sparse, high rainfall did not presage flooding. After the fourteenth century, when floodplain management became more aggressive, the rainfall and flood records diverge once again.

urgent need for river management may be one of the factors that prompted the emergence of managerial statecraft related to the river in the tenth and eleventh centuries.[6] Figure 3.2 reveals that the ratio of attested disaster events fell relative to attested management events beginning in the ninth century, even as the total number of flood catastrophes rose.

At no time after the 900s did the rate of recorded calamities recede to its earlier rate. From then onward, although there were fluctuations, the river continued to flood during both wet times and dry and during times of greater and lesser attention to maintaining waterworks. Figure 3.4 depicts the history of the floodplain during the centuries between 700 and 1400. Floodplain residents—farmers, hydrocrats, laborers, city dwellers, and emperors alike—learned to live in a world of river turbulence that had no historical precedent. Floods and droughts exacerbated diplomatic and political conflict and also the misery and exploitation of vulnerable people.

The transition to the new epoch of river upheaval largely coincided with Northern Song suzerainty over the North China Plain (960–1127). The Northern Song presided over China's smallest nominally unified state and its most unstable frontier. Military and environmental peril helped to drive technological innovation in this

FIGURE 3.4

The Geography of Floodplain Events, 700–1400. The number and location of attested disasters and infrastructure management events changed dramatically between (a) 700 and 920, (b) 920 and 1029, (c) 1029 and 1090, (d) 1090 and 1165, (e) 1165 and 1220, and (f) 1220 and 1350, along with the course or courses of the Yellow River during each date range. The date ranges are based on natural breaks in the data and are not of equal duration. These maps reveal that there was little management until 920. Thereafter, there was an unprecedented rate of disaster and a great deal of management as well. In the eleventh century, the river turned north, sending human attention to a new part of the floodplain, leading, by the date range depicted in 3.4d and 3.4e, to dramatic attention to management activities. By the end of the era depicted, attention to the river was spread throughout the floodplain.

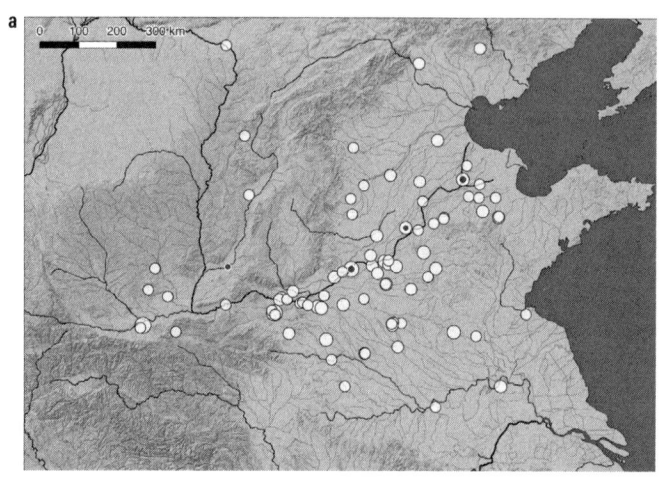

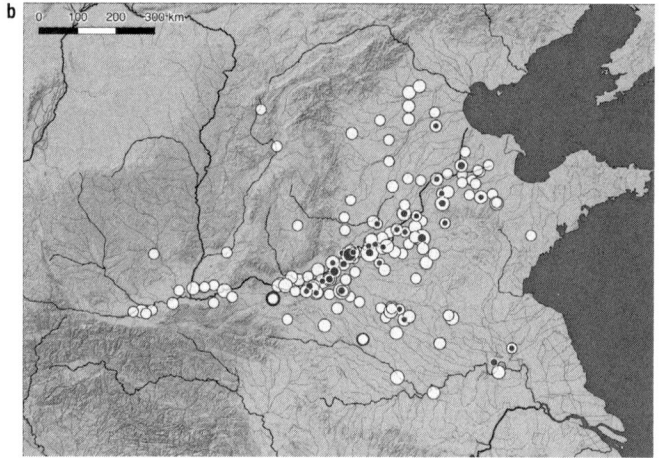

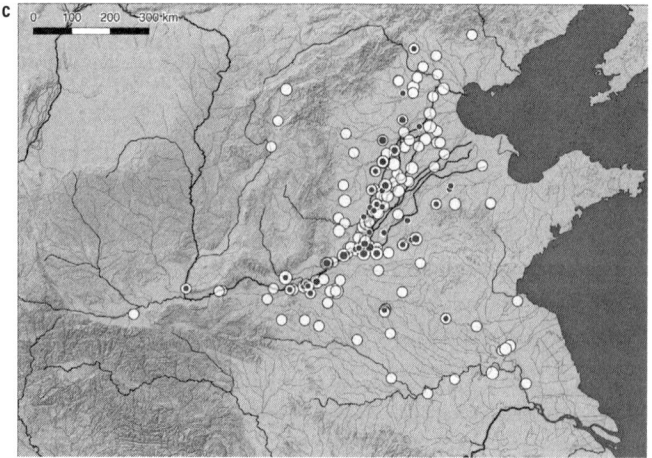

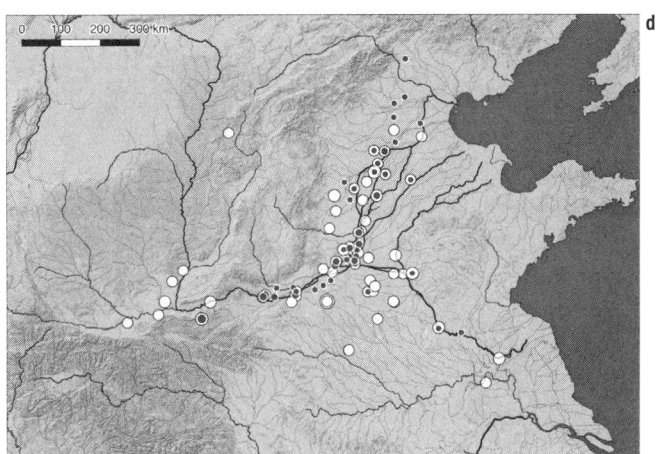

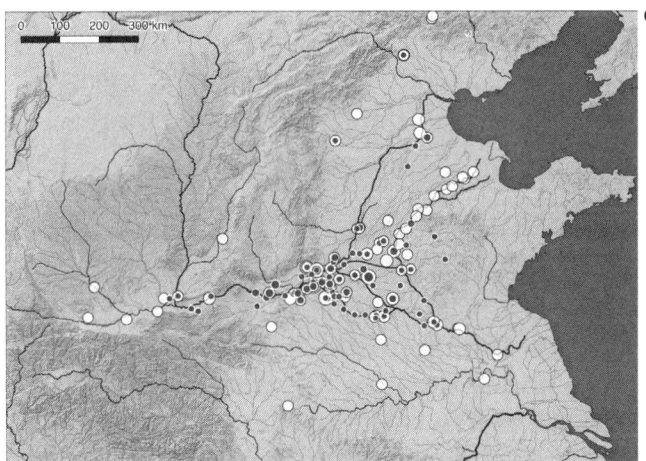

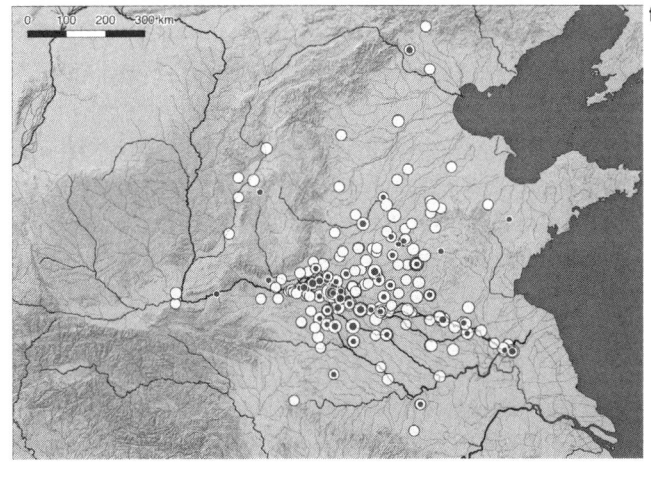

period. In river management, as in so many other domains, Song inventions set the tempo for the imperial centuries that followed. Since the 1900s historians of China have identified the transition from Tang to Song rule as a major watershed in imperial history. Political and elite power took novel forms. With access to new strains of fast-ripening rice, the population doubled, became more urban, and became more oriented to the Yangtze River watershed in the south than to the historical Yellow River heartland in the north.[7] Nobody has yet written an environmental history of the Tang-Song transition, nor have historians identified ecological upheaval in the Yellow River as a cause, an effect, or a component of the much-scrutinized transformations characterizing this era. It is my hope that this chapter begins to rectify that oversight.

BEFORE THE EVENTFUL ERA

Beginning in the late sixth century, the reassertion of central state power, first under Sui control and then during the Tang, fueled population growth. The census of 280 CE attested a Shaanxi population of only 722,595 households, which constituted 5 percent of the recorded population of the regime at that time. The next census, from 609, recorded a Shaanxi population of 3,747,630, representing 8 percent of the whole Sui population and a fivefold increase over the 280 census. Likely because of exigencies of imperial transition and different census methodologies as well, the population in 618 had declined to 1,698,143 (6 percent of total), but by 742, the population had risen to 4,318,613 (10 percent of total), a number that would not be surpassed until the eighteenth century. Most of this population was centered on the Guanzhong Plain around the capital at Chang'an.[8] The sediment record does not appear to indicate that intensive settlement in Guanzhong alone had a significant impact on erosion history. Figure 3.5 shows the population distribution throughout the Tang empire at the time of the 742 census. Chang'an was at the center of the populous Guanzhong Plain, but the Ordos was only sparsely inhabited.

Chang'an was the world's largest city, a cosmopolitan urban center with a population of over one million Chinese and foreign residents. Its outer walls, a vast rectangle of 8 by 10 kilometers, or 5 by 6 miles, and the palaces and boulevards enclosed within, were oriented according to the compass and the cosmos. Gongs beat hundreds of times every morning and evening to mark the start and end

PLATE 1
The Yellow River Study Area.
This image depicts the entire region described in this book as it is situated in the context of the whole of East Asia. The Loess Plateau is the source of the sediment deposited on the alluvial plain. The colored lines depict the many courses that the Yellow River has taken across the plain during historical times.

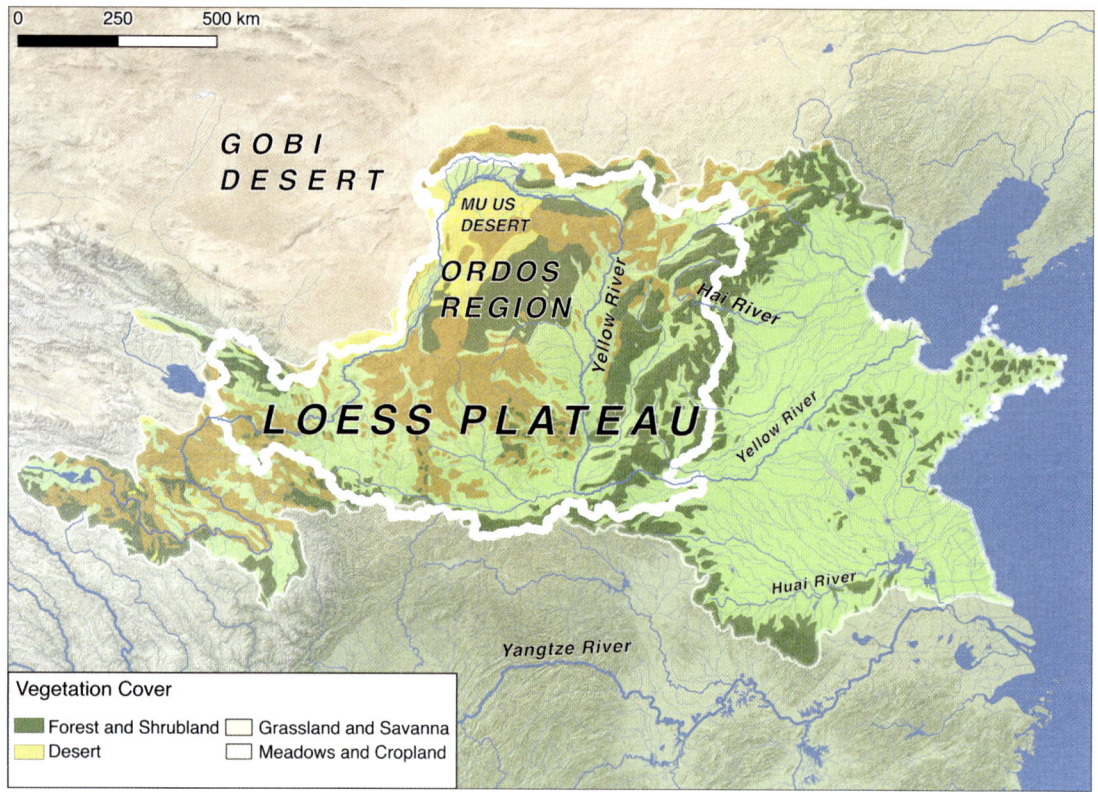

PLATE 2
The Yellow River Watershed and Its Ecological Regions.
The Yellow River floodplain is primarily arable land. The Loess Plateau is a diverse network of ecosystems that includes desert, scrub, forests, and grasslands.

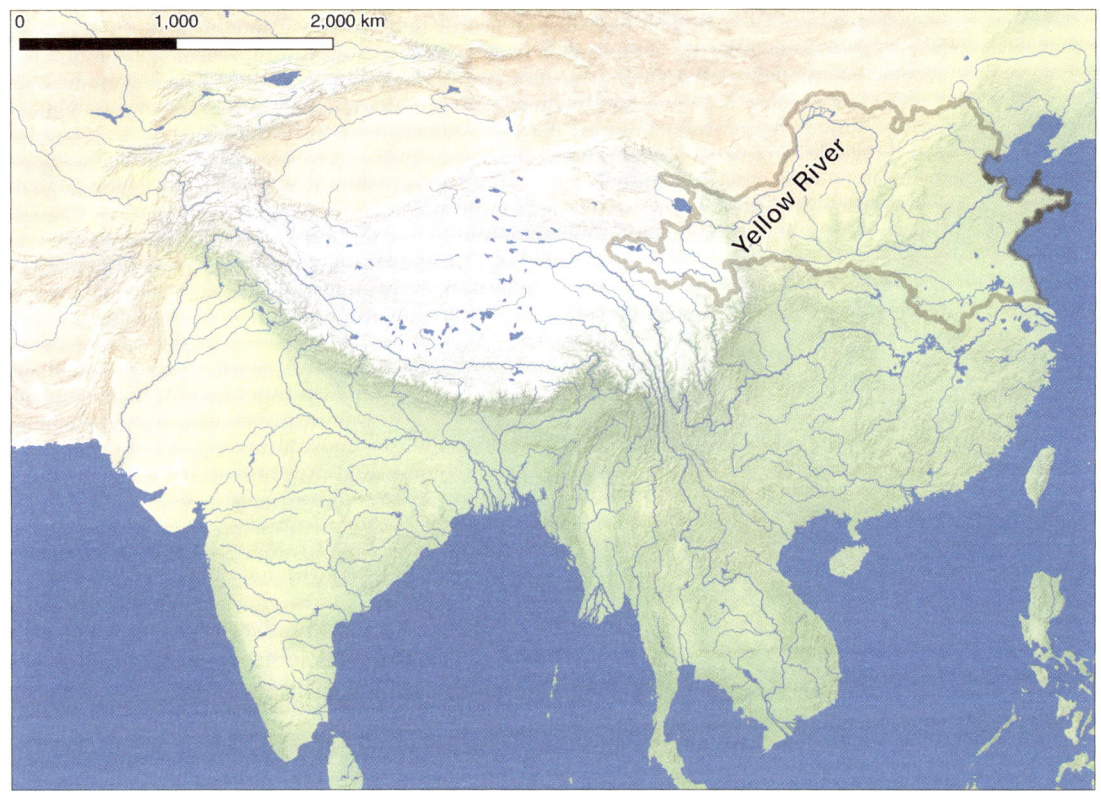

PLATE 3

Eastern Asia Elevation and Hydrology. The Yellow River is one of the many Asian rivers that descend from the Tibetan Plateau. The high-altitude position of Tibet in relation to the rest of China means that almost all of China's major rivers flow from west to east. Canals have been the means for connecting the east-flowing rivers to one another.

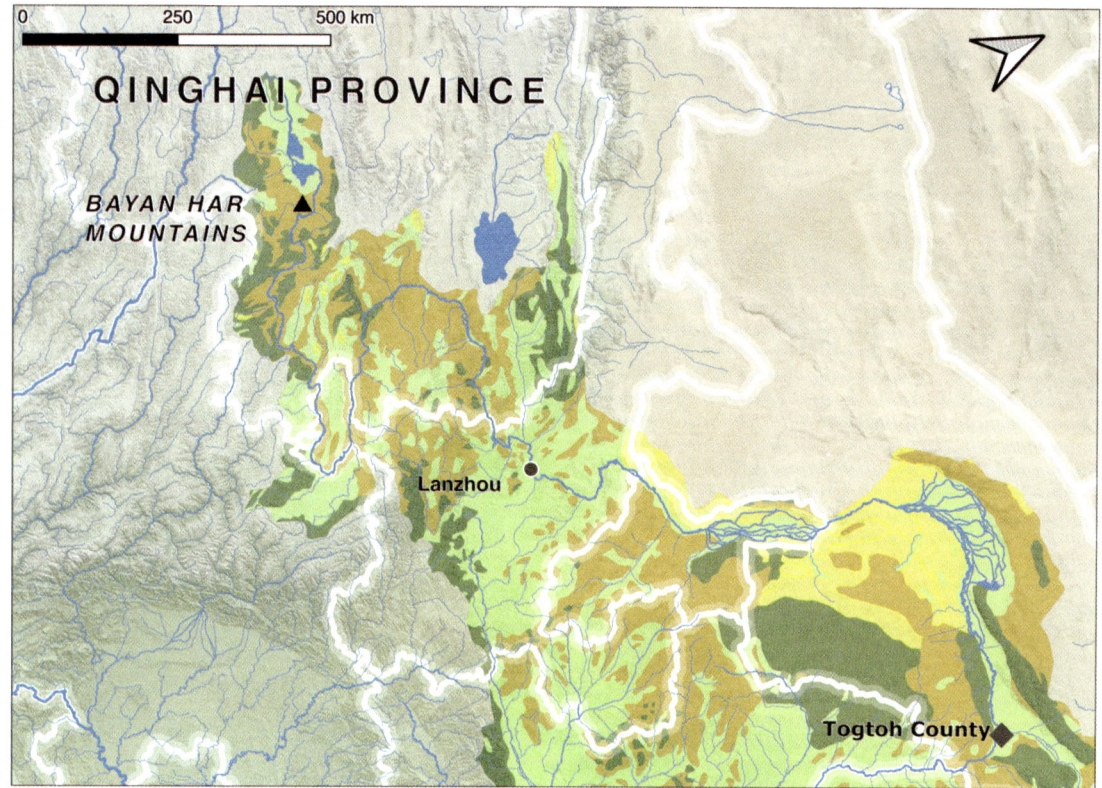

PLATE 4

Upper Reaches Overview. The upper reaches of the Yellow River, from the Bayan Har Mountains to Togtoh County, traverse a high-altitude, ecologically diverse region. This map is oriented with the west at the top. It depicts modern provincial boundaries as well as river ecology.

PLATE 5
Ngoring Lake, Sanjiangyuan National Nature Reserve. This photograph depicts the Yellow River at its headwaters on the Tibetan Plateau. Photograph © Ian Teh.

PLATE 6
Gyaring Lake. Erosion is clearly visible on this overgrazed landscape. Photograph © Ian Teh.

PLATE 7
Desert, Bayin, Gansu. This desert landscape lies at the southwestern perimeter of the Loess Plateau. Photograph © Ian Teh.

PLATE 8
Yinchuan Plain (1). The Helan Mountains to the west of the Yinchuan Plain trap rainfall and create an arable basin in an otherwise arid region. Photograph by the author.

PLATE 9
Yinchuan Plain (2).
Photograph by the author.

PLATE 10
Yinchuan Plain (3).
Photograph by the author.

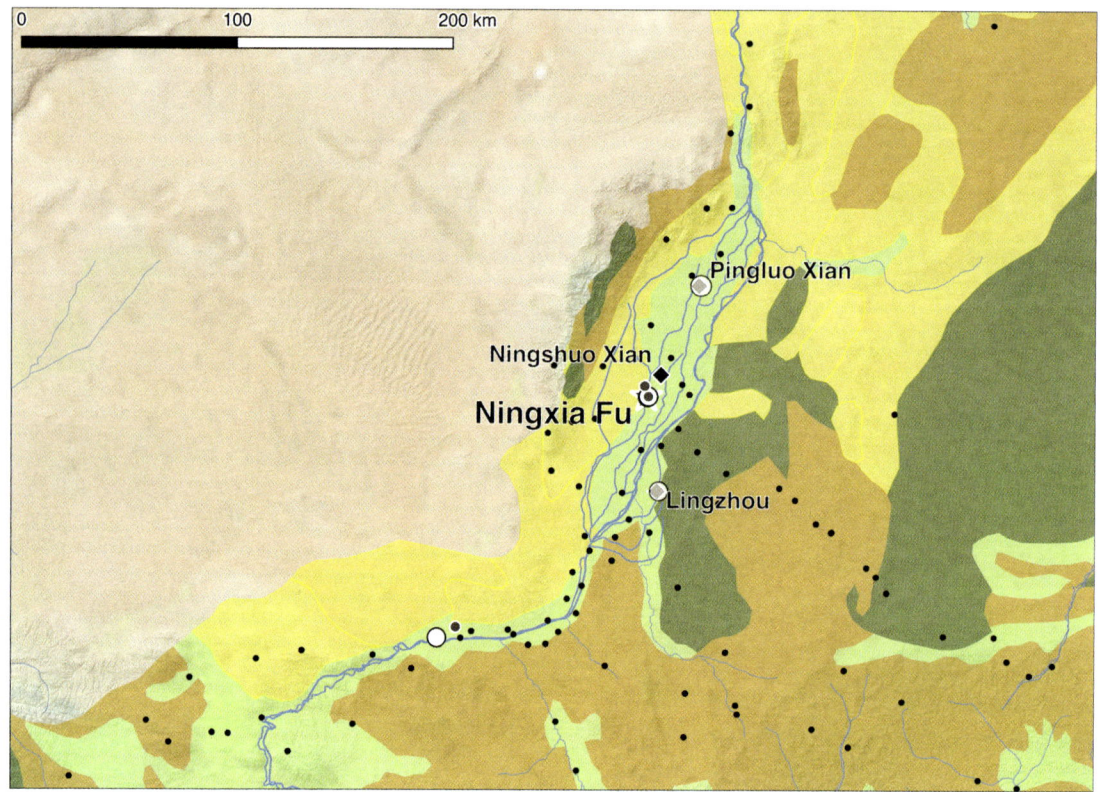

PLATE 11

Yinchuan Plain. The Yinchuan Plain, a small fertile region surrounded by deserts and mountains, was frequently contended in historical times. This map depicts its settlement geography in 1750.

PLATE 12
Han Dynasty Waterworks on the Guanzhong Plain. Irrigation works connected the Jing River to the Wei River from an early date. They permitted Chang'an to serve as an imperial capital but were frequently overwhelmed by silt accumulation. Today, the Han dynasty works lie far above the river level, which has incised its bed over the centuries. Photograph by the author.

PLATE 13

Terraces near Lüliang. The terraced landscape that farmers have created from the easily erodible loess soil lies near the eastern bank of the Yellow River where the Lüliang Mountains meet the Ordos. Agricultural intensification in this region has required generations of backbreaking labor. Photograph by the author.

PLATE 14
Gullies near Yulin. South of Yulin, the Wuding River, and the Mu Us Desert, the results of centuries of erosion are evident. Photograph by the author.

PLATE 15
Trees near Yinchuan. The moist Yinchuan plain includes forests and wetlands. Photograph by the author.

PLATE 16

Subtropical Forest and Loess Spires near Pingliang. This locale near the southwestern extremity of the Loess Plateau features the characteristic yellow soil of the plateau. Even protected by vegetation, it can easily erode as a result of high precipitation. Photograph by the author.

PLATE 17
A Pounded-Earth Wall at Yulin.
This site, a relic of the Ming Great Wall, is built from pounded earth, which is nearly as strong as concrete. Photograph by the author.

PLATE 18

The Disintegration of a Clod of Loess. These four images (a–d) show that loess soil easily forms firm clumps, but that they collapse into a fine powder even under the gentlest manipulation. Photographs by the author.

c

d

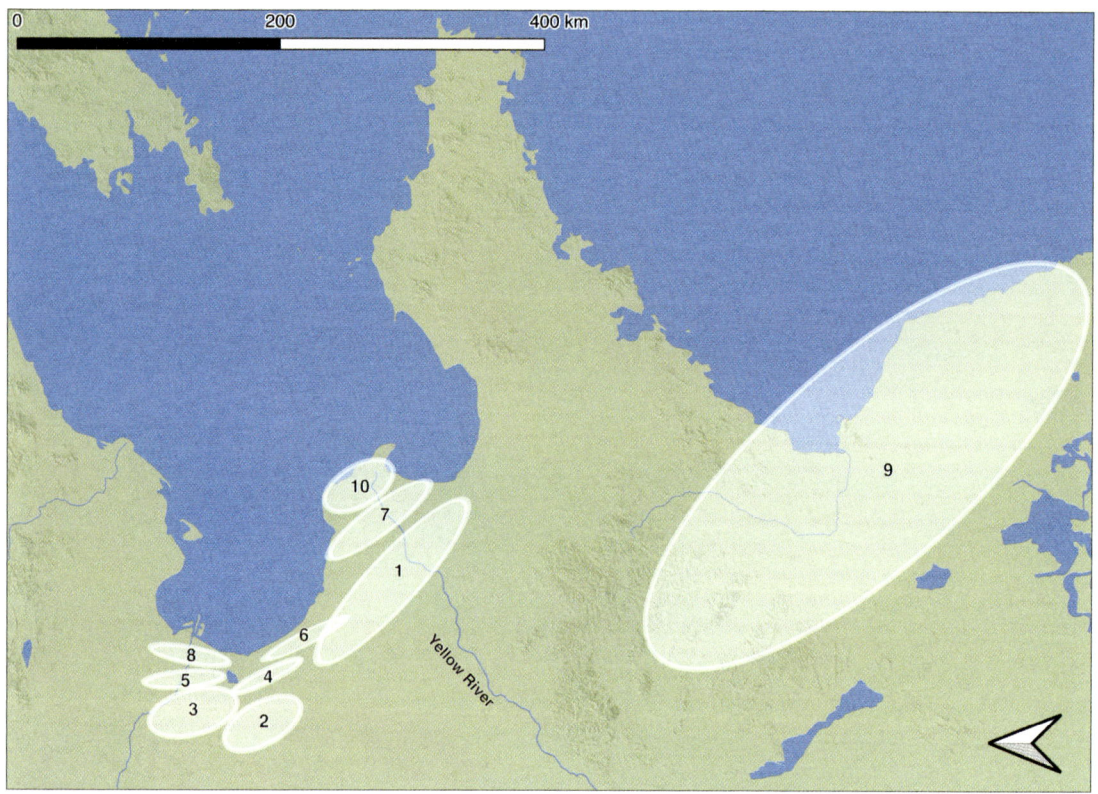

PLATE 19

Yellow River Paleodeltas. The paleodeltas and the Bohai Sea beyond them are one source of information about the changing rate of sediment accumulation over time. They also indicate the various routes that the river has taken to the sea during its history. The time periods represented by these deltas are (1) 6,000–5,000 BP; (2) 5,000–4,500 BP; (3) 4,500–3,400 BP; (4) 3,400–3,000 BP; (5) 3,000 BP–602 BCE; (6) 602 BCE–11 CE; (7) 11 CE–1048 CE; (8) 1048 CE–1128 CE; (9) 1128 CE–1855 CE; (10) 1855 CE–present. This map is oriented with east at the top. Based on Qiao et al., "Sedimentary Records."

PLATE 20
Henan Province. The modern province of Henan (dark gray boundary) straddles the eastern part of the Loess Plateau (white line) and the western part of the floodplain.

PLATE 21
Silt Watching at the Xiaolangdi Reservoir. This extraordinary image reveals the extent to which erosion from the Loess Plateau continues to challenge waterworks managers today. Photo by VCG/VCG via Getty Images.

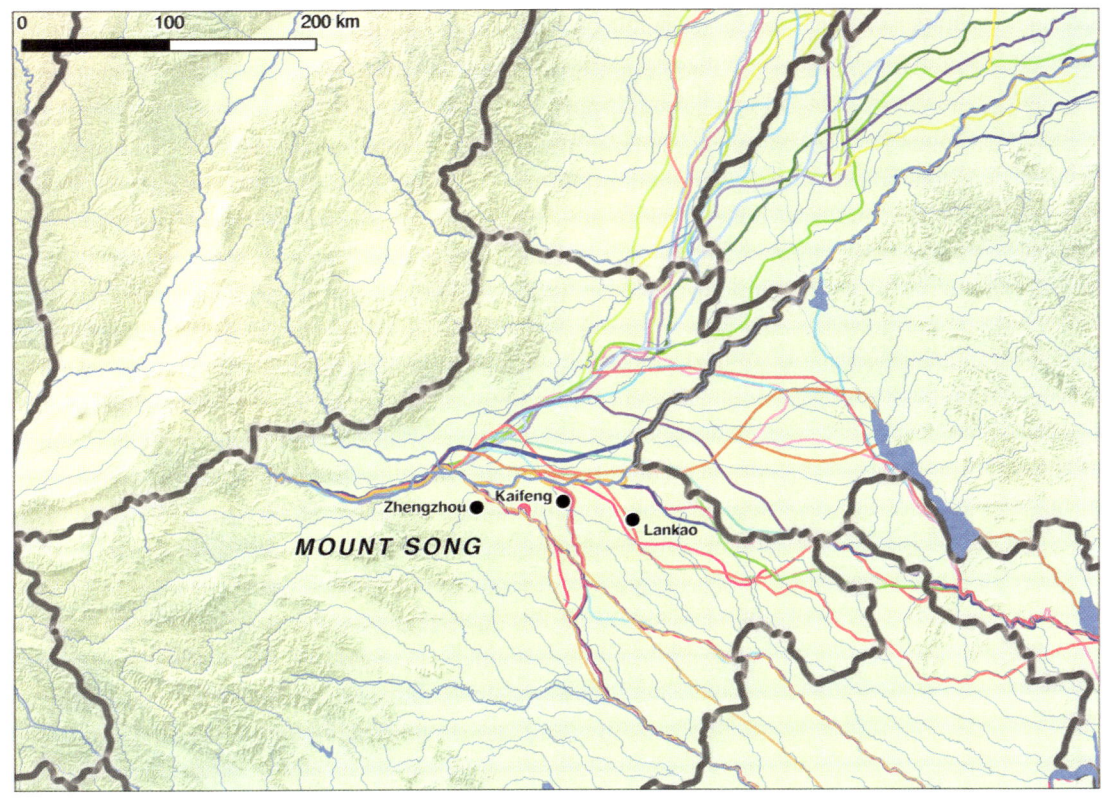

PLATE 22
The Course Change Region.
Almost all the historical courses of the Yellow River emanate from a low-lying region at the western perimeter of the floodplain in eastern Henan.

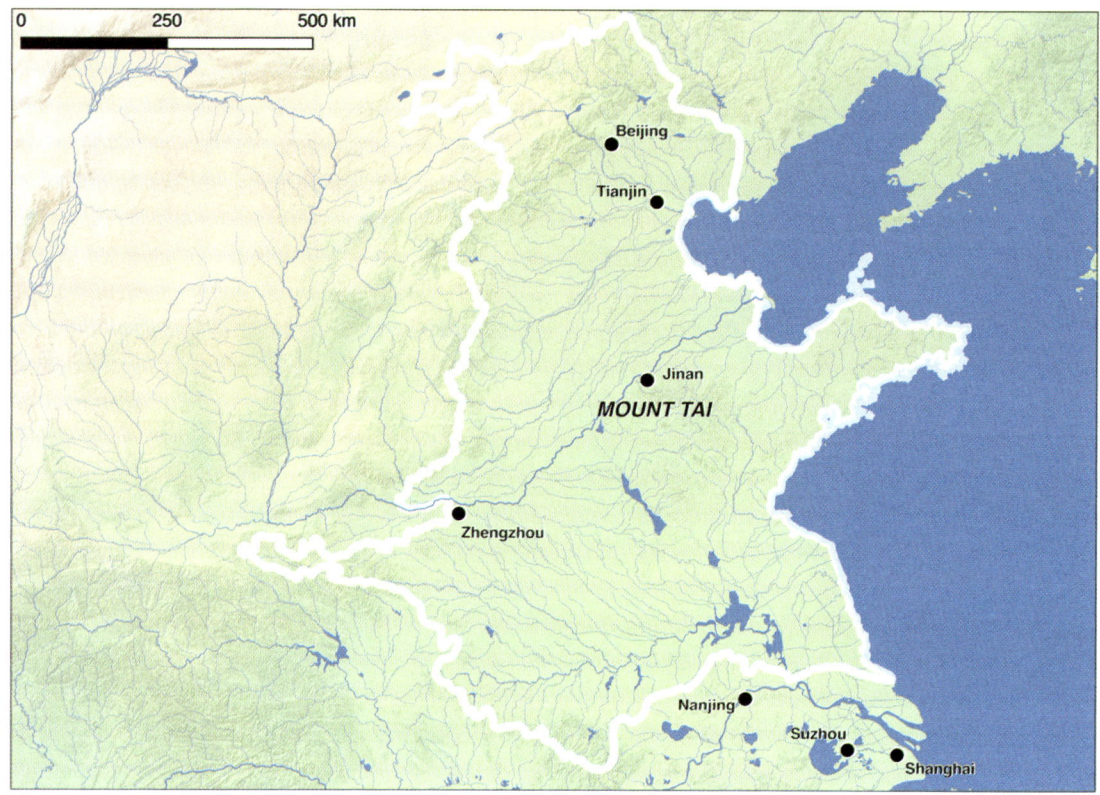

PLATE 23

The Floodplain. The Yellow River floodplain, crisscrossed with tributaries and historical lakes and wetlands, includes the Huai sub-basin to the south and the Hai sub-basin to the north.

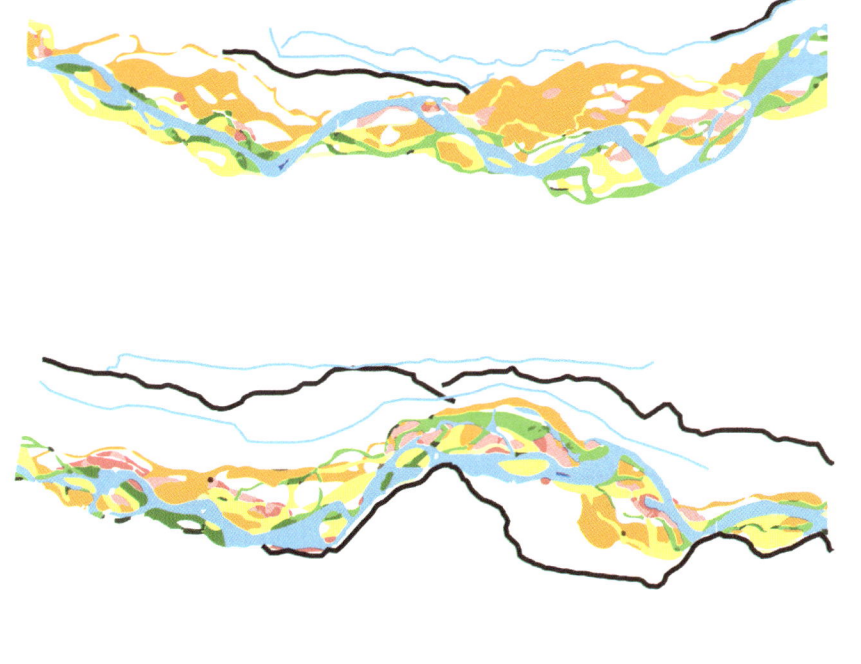

PLATE 24
A Diagram of Meanders within Far-Set Levees. The river was an actively alluviating landscape feature even when it was locked between levees. When its meanders ran into the earthworks that contained the river's flow, the dikes needed to be repaired or reinforced to prevent catastrophic breaches. Based on Shuilibu, *Huaihe liuyu ditu ji*.

PLATE 25
A Detail of a Nineteenth-Century Map of Embankments near Jinan. When the river meandered within its levee-confined course, it threatened the earthworks, which needed to be reinforced with the kinds of additional structures that are depicted here. From Arthur W. Hummel Sr., *A Map of the Yellow River in Shandong Province* (*Shandongsheng Huanghe tu*), after 1881. Courtesy of the Library of Congress Geography and Map Division, Control Number gm71005026.

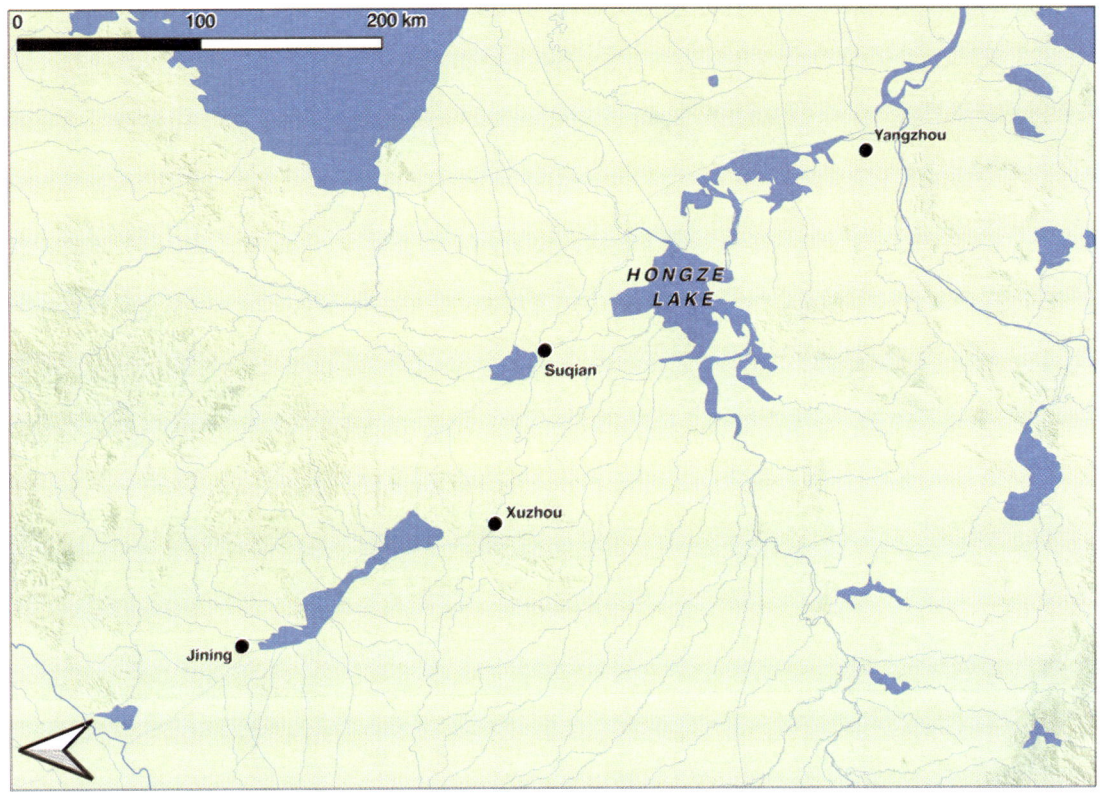

PLATE 26
The Lake District between the Yellow and Yangtze Rivers. A sequence of lakes punctuates the floodplain. Today they lie between the Huai River to the south and the Yellow River in its north-flowing course. This map is oriented with east at the top.

PLATE 27

The Tongwancheng Wall Today. Situated in a scrub and desert landscape, this is part of a wall that once surrounded a great medieval metropolis. Photo from Wikimedia Commons. By Cong— Own work, CC BY-SA 3.0.

PLATE 28

The Heng Mountain Landscape Today. The Heng Mountains, south of the Wuding River, once a centerpiece of the twelfth-century fortification strategy, are now deeply eroded and deforested. Photograph by the author.

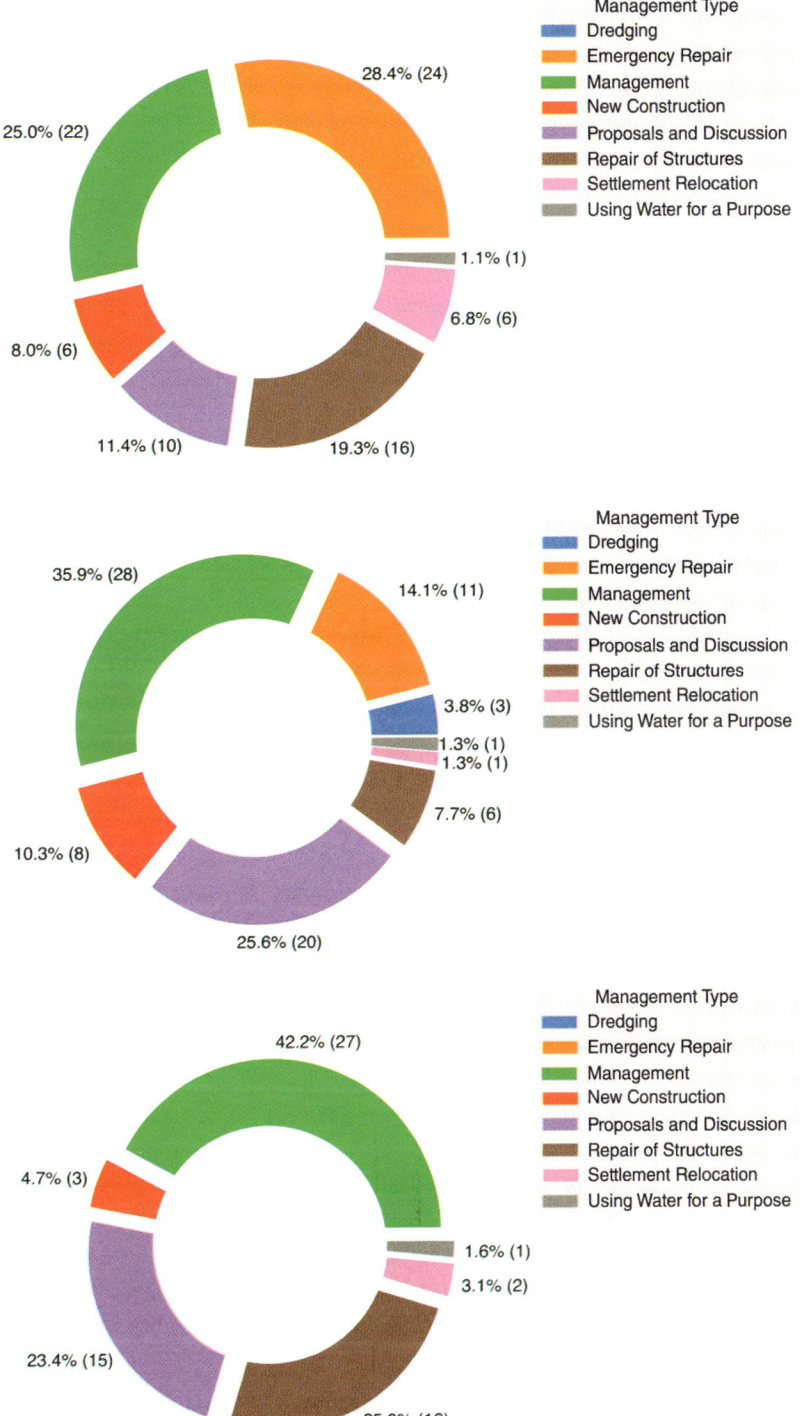

PLATE 29

Management Event Types, 920–1165. These three charts depict the change in the ratio of management event types from 920 to 1165, as the preponderance of infrastructure interventions shifted from emergency repairs, through proposals and discussion, and finally to routine repairs. The top graph covers the years 920–1029; the middle graph, 1029–1090; and the bottom graph, 1090–1165.

PLATE 30

The Jing River at the Site of the Zhengguo Canal. By the eleventh century, irrigating the Guanzhong Plain required engineers to cut a feeder canal through solid rock. Subsequent erosion is visible in the rocky landscape and clear from the fact that the river lies so far below the surrounding farmland today. Photograph by the author.

PLATE 31
The Yulin Garrison Today.
This Ming garrison, built out of brick and pounded earth, is well preserved today. Building this structure at the desert edge would have required an immense quantity of labor and an extraordinary number of trees to support foundations and to power kilns. Photo from Wikimedia Commons courtesy of Caitriana Nicholson, CC-By-SA 2.0.

PLATE 32

Li Mountain Village. This reconstructed nineteenth-century hamlet, located near the market town of Qikou in east-central Shaanxi, features cave dwellings and abandoned hillside terraces. The cave dwellings are comfortable homes with excellent temperature control. They are a form of vernacular architecture that is well suited to the environment. Terraces are well adapted to the loess plateau as well, since they assist with erosion control and expand arable land. Photograph by the author.

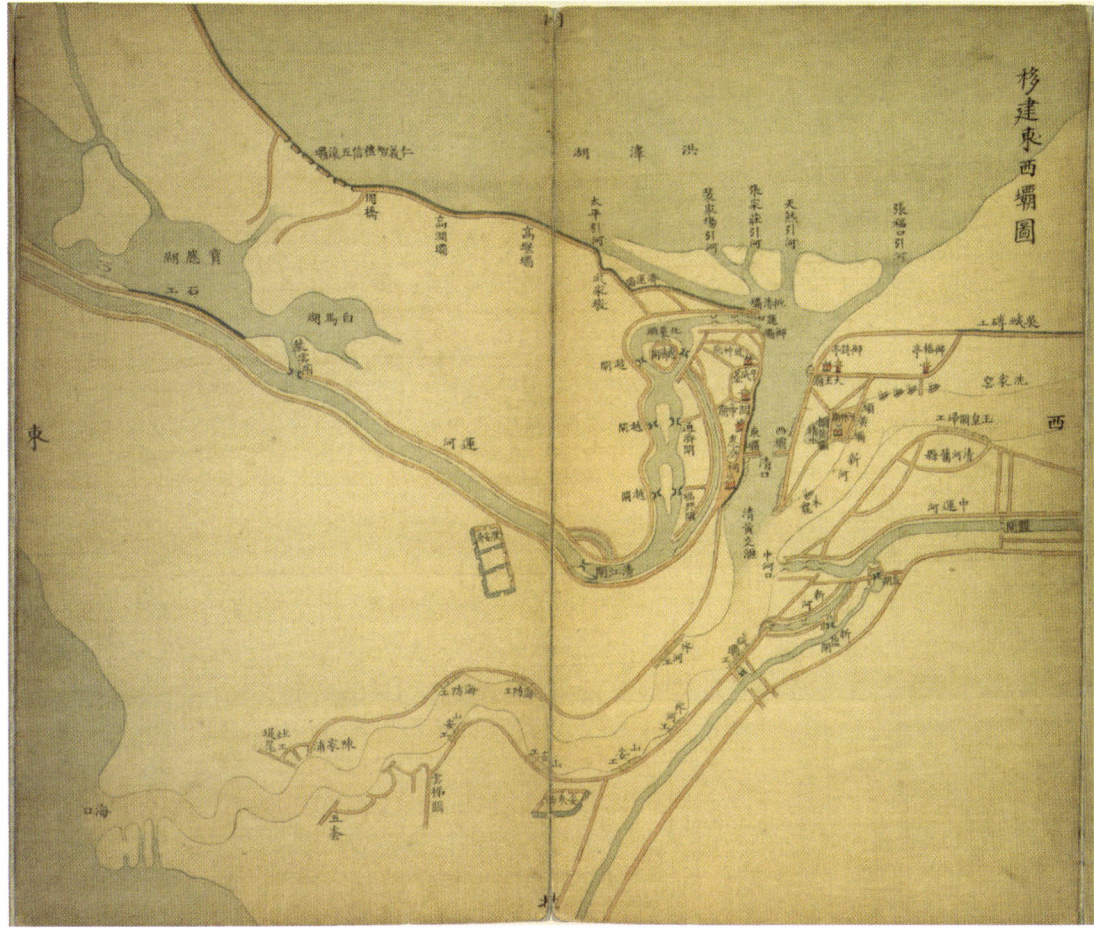

PLATE 33

Hongze Lake and Qingkou: An Eighteenth-Century Diagram. The full buildout of the Qingkou system did not occur until the end of the eighteenth century. This map, entitled *Relocating the East and West Dams* (*Yijian dongxiba tu*), depicts one instance of infrastructure improvement around the Qingkou structure in 1779. Boats traveling north on the Grand Canal passed north through three locks, looped south through the Qingkou passage propelled by strong currents from Hongze Lake water, and then crossed the Yellow River to enter the canal once again. This part of the journey took approximately one week. Plate 34 shows how the spillway depicted on the upper left (southeast) corner of this map appears in the Hongze landscape today. Courtesy of the Library of Congress Geography and Maps Division, Control Number gm71005017.

PLATE 34

A Qing Spillway at the Southern Part of Hongze Lake. This eighteenth-century spillway directed high waters from Hongze Lake south into surrounding farmland, inundating disfavored regions in the Hongze Lake vicinity to protect Qingkou at the opposite end of the lake. Although some parts of the Hongze waterworks network were advanced technology, others were simply graded earth and stone walls. This spillway appears in the upper left (southeast) corner of the map in Plate 33. Photograph by the author.

PLATE 35

Archaeological Excavations at the Site of the Kaifeng City Wall. There were meters of sediment accumulation during the final imperial centuries of Yellow River history. This excavation has also exposed the remains of a slack-water lake, where water pooled during a Qing-era flood. Refugees fled the city whenever there was a major flood. This image includes strata from Yuan, Ming, and Qing times. Photograph by the author.

PLATE 36
A Beached Ship on the Desiccated Yellow River near Kaifeng. Today, the Yellow River is a tame, shallow body of water that can no longer support much transportation or irrigation. Photograph by the author.

PLATE 37
A Reforested Terrace near Pingliang. Contemporary China is pursuing rapid reforestation on the Loess Plateau, but many sites feature single species planted in equidistant grids. Photograph by the author.

PLATE 38
A Check Dam near Pingliang.
Small dams to prevent flooding and soil loss in the heart of the erosion zone are a contemporary part of Loess Plateau landscapes. They have not been extensively deployed until recent decades. Photograph by the author.

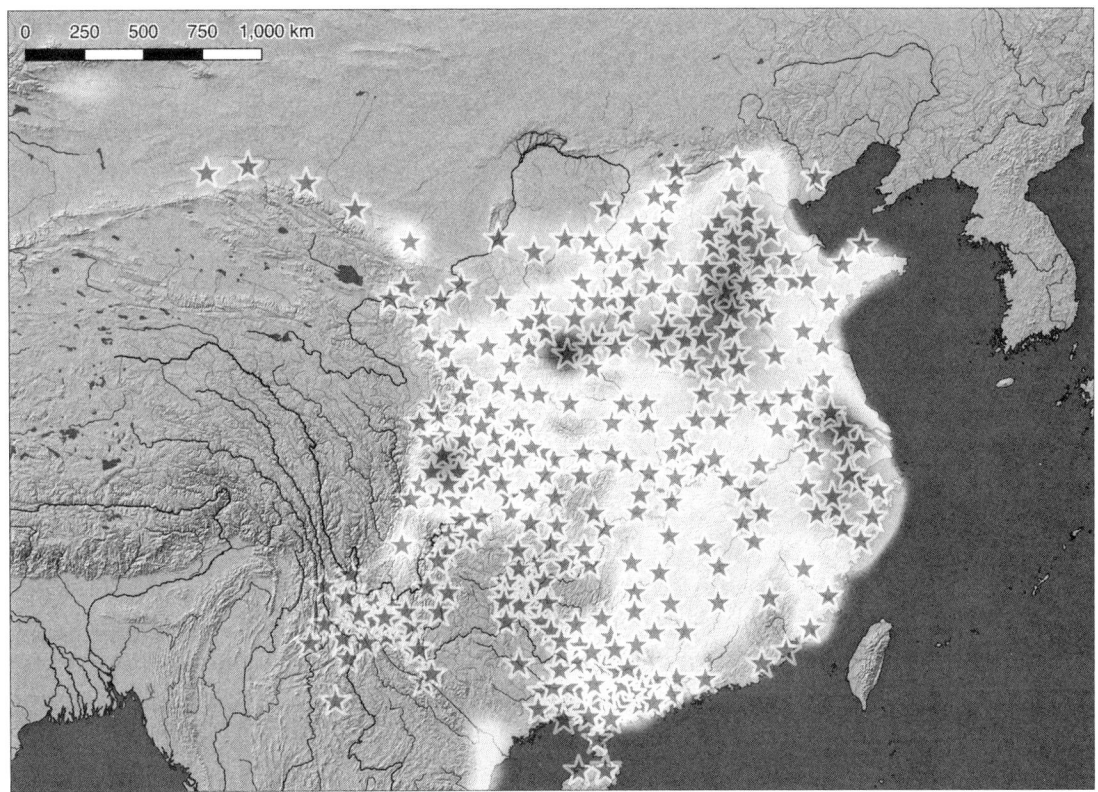

FIGURE 3.5
Tang Dynasty Population and Prefectures in 742. Population remained dense on the floodplain and in Guanzhong in 742 CE. By then, there were also significant settlements around the Ordos and throughout the floodplain.

of business. The pressure to provision the city with food, fuel, and timber, almost all of which came from the southern Ordos, affected ground cover there. The Qinling Mountains, whose northern slopes drained to the Wei River, were one target. The Liupan Mountains at the southwestern edge of the Loess Plateau were another. There is anecdotal information that by the tenth century, these sites were largely deforested.[9] In the northern reaches of the Lüliang Mountains, east of the Ordos, there are reports that forests were seriously compromised by the middle of the eighth century and gone by the middle of the tenth century.[10] However, there are no such reports from earlier times.

At first, Tang rulers felt no need to fortify the central Ordos, and climate conditions would have made it difficult for them to do so in any event. Only in the early eighth century, after a century of near-annual raids by Turks, Mongols, and other nomadic residents of the Loess Plateau and the grasslands and deserts to its north, did Tang commanders vociferously demand that the regime establish an arc of garrisons along the loess frontier. Several dozen new settlements were founded as a result.[11] Around the same time, the regime established salt-mining colonies on the loess in Shaanxi. These were organized like military colonies, with forced migrants compelled to farm the land to support themselves while also engaging in extractive mining.[12]

Nevertheless, Tang sources still offer numerous reports of large flat tablelands (*yuan*) on the Loess Plateau. The great poet Bai Juyi (772–846) referred to the Wei as a clear-flowing river.[13] As figure 3.5 reflects, the eighth-century population density on the Ordos was lower than it had been at the time of the 2 CE census. State policy in the early Tang favored large, intact grasslands for vast horse pastures on the Loess Plateau, not agriculture.

At that time, the Ordos was an ethnically, culturally, and linguistically diverse region that supported big-game hunting, pastoralism, and agriculture. Turco-Mongol peoples were fully incorporated into the northern tier of the empire. North China supported large populations of animals that relied on forests and grasslands. Deer were common, and Tang prefectures on the Ordos offered tribute goods in the forms of forest and grassland commodities like animal horns, pine resin, musk-deer glands, birch bark, and eagle feathers, as well as horses. Elites and commoners alike hunted deer and ate venison. The sheep was the most important domesticated animal, with large grazing herds stretching as far south and east as Shandong until the decline of grasslands quality set in during the eighth century.[14]

At the edge of the Mu Us Desert, near the modern city of Yulin, on the banks of the Wuding River lay a city called Xiazhou. A moist and hospitable site, it supported ten thousand permanent residents who relied on irrigated agriculture and a large population of animals. With the expansion of agriculture, it grew to fifty-three thousand residents in 742. A government horse pasturage lay nearby, with at least 185,000 horses around the year 680 and an affiliated pastoralist population of 14,320 people.[15]

In short, prior to the military conflicts and the intensive agricultural and forestry initiatives that began in the eighth century, most of the Loess Plateau remained as sparsely populated as it had been for hundreds of years. The environmental impacts of the vast metropolis in Chang'an seem to have been limited to the Guanzhong region and without basinwide consequences.

THE DAWN OF THE EROSION REGIME

By the eighth century, population was rising on the plateau, agricultural techniques became more intensive in core regions, and farming extended to new locales.[16] Both historical records and pollen evidence from this period reflect intensive land use and land clearance for agriculture on previously forested and sloping land as the average amount of cultivated land per person diminished from 2.81 hectares per person to 1.8, or a decline from almost seven acres per person to less than four and a half.[17] Cultivation moved into upland areas, a permanent Loess Plateau farming population emerged for the first time, and horse ranches began to be converted to farms.[18] For the first time since the turn of the Common Era, settlements extended out of the major tributary valleys and onto more erodible upland regions. Although the northern part of the Loess Plateau remained sparsely populated until the tenth century, significant areas of natural forest in the central and western Loess Plateau began to be logged, with timber transported to the capital at Chang'an.[19] This is the starting point of the new life span of the Yellow River.

Figure 3.6 is a series of three maps depicting the expansion of settlement geography on the Loess Plateau between the years 612, 742, and 1111. In 612, there were 35 prefectures, 182 counties, and 47 documented settlements without administrative status. Most of them were located to the east of the Ordos or in Guanzhong. In 741, there were 49 prefectures, 212 counties, and 164 other settlements, including dense outposts on the Yinchuan Plain and along Wei River tributaries. By 1111, the numbers had expanded to 62 prefectures, 186 counties, and 674 other settlements, including a dense arc of forts along the frontier with the Xi Xia.

Peace and prosperity extended through the moist decades of the 600s and into the droughts of the 700s. Beginning in the mid-700s and continuing for three centuries thereafter, the military geography of the Loess Plateau was a profound driver of ecological degradation. The exigencies of colonialism and war pushed forts and garrisons onto fragile grasslands, first into the tributary valleys north of the Wei, and then onto the hills south of the Mu Us.

FIGURE 3.6

Loess Plateau Settlement Geography. This series of maps depicts the dramatic expansion of settlement in the Ordos between (a) the middle of the eighth century, (b) the middle of the eleventh century, and (c) the early twelfth century at the conclusion of the fortification campaign on the Xi Xia frontier.

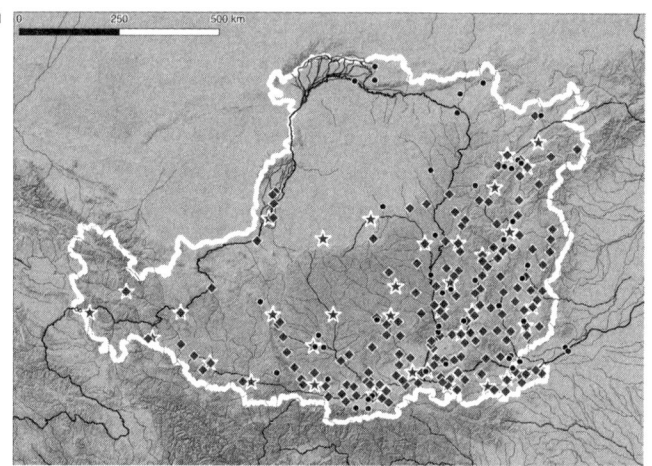
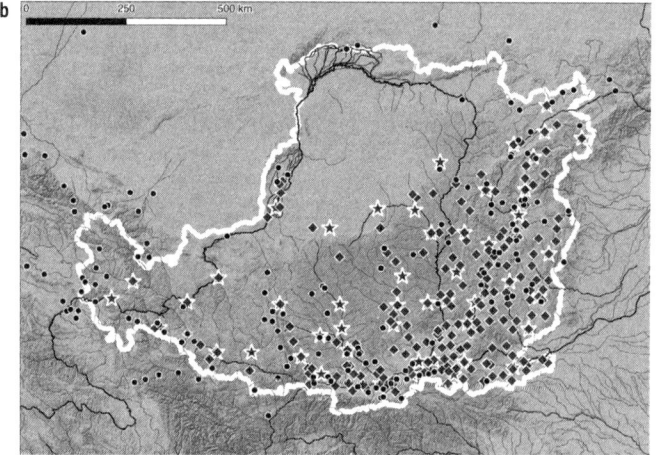
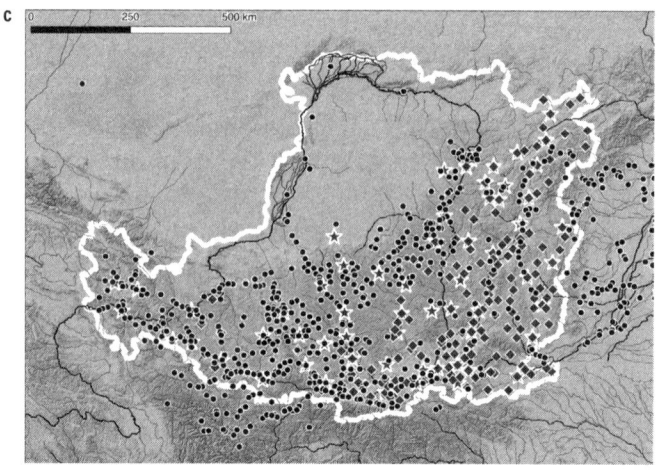

In 755, the Sogdian-Turkish Tang General An Lushan (703–757), who had amassed loyal troops in northeastern China and political influence in the capital, declared himself monarch and rebelled against the throne. The conflict spanned the reigns of three emperors. It was one of the most violent conflicts in history. War lasted seven years. Battles raged across northern China, and troops besieged, captured, and pillaged both of the two great Loess Plateau cities, Luoyang and Chang'an. Both imperial and rebel troops drew in allies and proxies from as far away as the Abbasid Caliphate. The Tang regime owed its survival largely to the intervention of the Uighur Khanate, which sent cavalry across the Ordos to Chang'an and Luoyang. By the end of the war, thirty-six million people had vanished from the census rolls, lost to a combination of death, dislocation, and the breakdown of record-keeping protocols.[20]

At this point, the Ordos, the most significant part of the potential loess erosion zone, remained a multicultural and unfortified region. Tang allies among the Uighurs, along with Tibetans, Tanguts, and other steppe peoples, had survived the seventy-year drought of the previous century. Now, with climate conditions improving, a new era of grassland political expansion was under way.[21] After the An Lushan war, in the eighth and ninth century rival claimants of the grasslands contended with one another for supremacy on the Ordos with local tribal peoples and Chinese interests, including rebels who were transformed into sanctioned hereditary warlords (*jiedushi*) after having been pardoned. The regime lost political control over the northeast. On the Loess Plateau, the later eighth and ninth centuries were the first time in eight hundred years that the fragile Ordos grasslands had been seriously contended among rival claimants, all of whom occupied and fortified the region.

During the An Lushan conflict, Ordos-based troops moved into the core battlefields far to the south and east. The power vacuum permitted Tibetan forces to occupy the northern Ordos as well as the entire Hexi Corridor, the narrow arable plain that lay to its west between the Gobi Desert and the Tibetan Plateau. Tibetan troops even captured the capital at Chang'an for fifteen days in 763 before they were repulsed by Tang troops. In 783, the Tang court recognized a fortified border that ran longitudinally through the mountains and grasslands, although periodic hostilities persisted until the Sino-Tibetan treaty of 821. This was the origin of a long wave of

Ordos fortification that would not end until the twelfth century. The Tang recognized a tripartite balance of power in the Ordos, occupied by soldiers and colonists, and divided between Tang, Tibetan, and Uighur territory. The situation lasted until the Tibetan and Uighur empires collapsed almost simultaneously during a period of drought in the 840s.[22]

Soon thereafter, various Tang allies brought troops to the Ordos to suppress another conflict. They included the Tanguts, who were Tibeto-Burman-speaking Tang allies with a homeland west of the Ordos, and Shatuo Turks, based north of the Ordos. The Huang Chao (835–884) Rebellion, after the name of its leader, was an uprising of salt smugglers, impoverished farmers, and tax-burdened landowners and merchants. It consumed the weakened Tang regime from 874 to 884. Rebel troops captured Chang'an in 881. After the war, the Tanguts took administrative control of the northern and western Ordos and established outposts on the Yinchuan Plain and across the grasslands south of the Mu Us Desert. The Shatuo Turks occupied the Loess Plateau east of the Ordos.

The grasslands began to become degraded.[23] The Dongzhi Tableland in Gansu, which today occupies only 412 square kilometers, just over 150 square miles, and is fragmented by gullies, is estimated to have covered about 1,050 square kilometers in the Tang, an area almost the size of Delaware. The Luochuan Tableland in Shaanxi, also known as the High Tableland of Northern Shaanxi, one large plateau in Tang times, has today been dissected into six small pieces, which are still retreating, at a rate of three to six square meters per year. The eighth- and ninth-century destruction of native ground cover is what caused the primordial gentle and flat ground of the Loess Plateau to begin its transformation into the deep-gullied and steep-highland landscape characteristic of the contemporary Loess Plateau.[24] Prior to that time, the grasslands and riparian forests had been healthy and capable of absorbing moisture.

In Tongwancheng (fig. 3.7), situated on the Wuding River, garrison troops and peasants had long cleared natural land cover to plant grain and other crops. The river lies at the grassland-desert ecotone at the edge of the Mu Us Desert. Tongwancheng probably had a permanent population of around ten thousand, supplemented by periodic troop surges and nomad encampments. In the midst of warfare between the Tang, the Uighurs, and the Tibetans, its agricultural

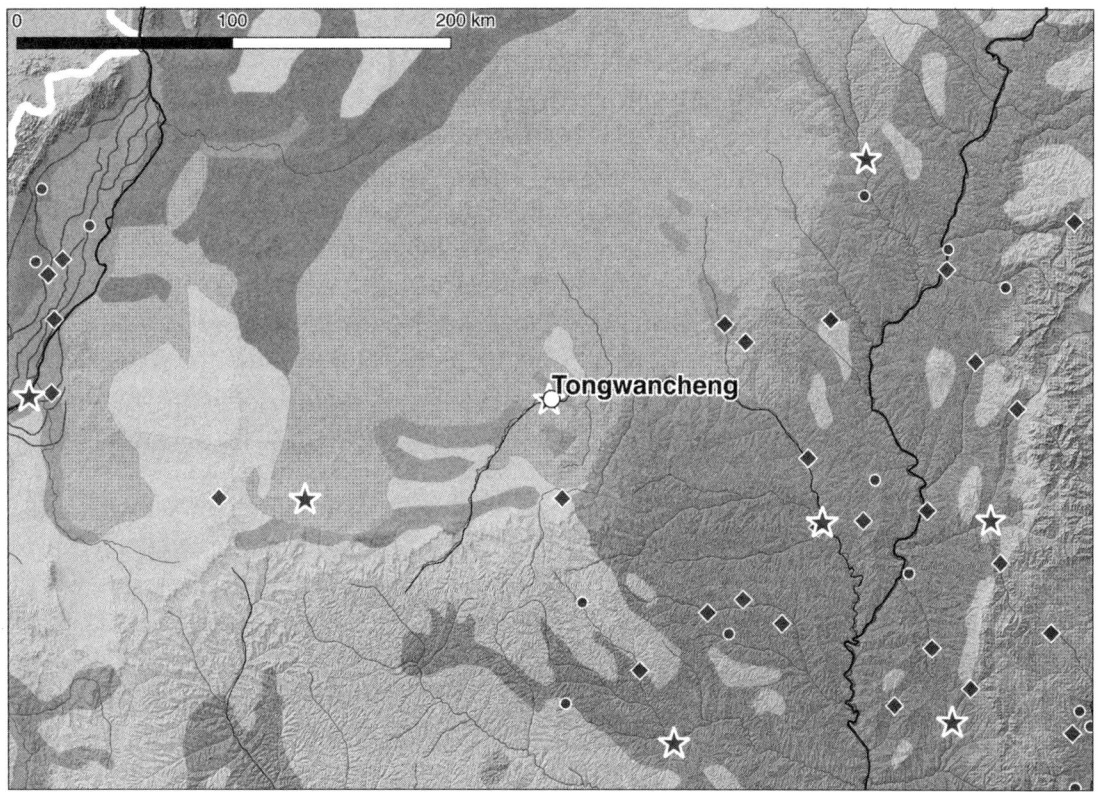

FIGURE 3.7
Tongwancheng. The medieval city of Tongwancheng lay on the banks of the Wuding River, at the southern edge of the Mu Us Desert and the intersection of multiple ecological regions.

hinterland, stripped of native ground cover, was abandoned, armies burned forests, and warhorses trampled the remaining grasslands. On degraded land, in the absence of active management in a marginal climate, the farmland and abandoned irrigation ditches became covered with sand transported by the winter monsoons. Drifts came to reach as high as the Tongwancheng wall (see plate 27 for a photograph of what remains of the Tongwancheng wall today).[25] Sand and loess sediment that washed into the Wuding River made its way downstream into the Yellow River.

By the late ninth century, when an uneasy peace had returned to the Ordos, Chinese and successor regimes sponsored major colonization campaigns to foster settler colonialism and agricultural expansion.[26] Around Tongwancheng, the steppe-desert transition zone was once again exploited for pastoralism and grain farming. More land came under cultivation, and more timber was hewed for fuel. Postwar struggles for suzerainty "played havoc with the natural landscape processes, resulting in the complete desertification of the Mu Us area."[27] There is no record of Tongwancheng in Chinese sources after the early fifteenth century, although its vicinity, the Wuding basin and the Yulin region, remained important sites in later times.[28]

Far from the Wuding River, at the southwestern margin of the Ordos, a high-resolution pollen record from a sediment core at Tianchi Lake confirms a similar chronology of sudden ecological collapse around the same time. Situated at a gateway between northwestern China and Sichuan, Tianchi Lake was a strategic site contended by Chinese, Tibetan, and Tangut forces in the ninth century. Around the turn of the tenth century, tree pollen began to disappear, and vegetation commenced a "rapid shift" to the steppelike ecology that still dominates today. The event was, as paleoecologist Ke Zhang and his coauthors put it, "unquestionably" a result of human activity.[29] The shift in vegetation type was accompanied by an increase in anthropogenic indicators such as pollen from cereal and fiber crops as well as pond algae. The pollen tells the story of new populations clearing forests for grain cultivation and growing grain on mountain slopes fertilized with manure. An increase in microcharcoal particles is evidence of deforestation and "intense human-induced fire activities."[30] In 841, a report from that region reported a "flow of sand" (*liusha*) that washed away forts, residences, and livestock corrals.[31] Although the boundary between pastoralism and agriculture fluctuated north and south during the course of subsequent historical and climate vicissitudes, ground-cover disturbance rarely contracted after the late Tang.[32]

In short, in the eighth and ninth centuries, the dynamics that initiated the era of large-scale vegetation destruction on the Loess Plateau included the geopolitics of an ethnically diverse region on the borderlands between agricultural and pastoralist lifeways and at a time of dramatic swings in rainfall and temperature. A few decades later, the resulting erosion came to be reflected in the floodplain event record.

CANAL ECOLOGY AT THE BEGINNING OF THE FLOOD ERA

Until the early eighth century, the North China Plain was the most prosperous and densely populated part of the Chinese empire. Figure 3.5 depicts the population at the time of the 742 census, a distribution similar to what it had been at the time of two previous extant censuses from 2 CE and 609 CE. Less than half of China's eighth-century population lived south of the Yellow River basin, but by 1550, over two-thirds of its population did. By the time of the 1080 census, the southern shift in population density was well under way. In fact, the eleventh-century census maps more closely to those in 1200, 1391, 1542, and 1948 than to the ones that preceded it.[33]

The demographic transformation is another proxy for the new Yellow River life span that began just over one thousand years ago. The historian Robert Hartwell believes that the persistent catastrophes that began to plague the floodplain beginning in the tenth century are a major cause of demographic changes in north China. Some people who abandoned the floodplain moved to relatively sparsely populated hinterland highlands nearby, exacerbating local contributions to erosion. Others moved northwest into the Ordos, or to the south. It was not until the 1550s that population density on the floodplain returned to its previous high point, met in 1080.[34]

Prior to the An Lushan Rebellion, a time of overall unity and prosperity, precipitation fluctuated far above and below its long-term mean. Figure 3.8 depicts the relative scarcity of attested floods and engineering initiatives during the period from 500 to 900, the centuries immediately before the dawn of the eventful era on the floodplain. Prosperity, hydrological stability, and unpredictable rainfall combined to favor significant investment in canal infrastructure, which supported transportation and agriculture. This was an era of limited erosion, few floods, and low rainfall. Defensive earthworks on the floodplain were few and far between.[35]

Extensive construction refined and ramified the canal system from the seventh century onward, interlocking the river's own hydrology—which connected the floodplain to the Loess Plateau and the Tibetan highlands—and a new anthropogenic water system that connected the North China Plain to the waterways of the south.

One canal linked Luoyang to the Yangtze delta metropolis of Suzhou via the flat alluvial plain of the Huai River. Completed with the recorded labor of five million men and women, the canal largely followed older routes and replaced a Han-era canal that had filled with silt and become impassable.[36] Even in this period of low erosion and low alluviation, the shallow canals required consistent maintenance in order to remain usable. Another canal linked Luoyang with the Korean frontier near modern Beijing, transporting grain

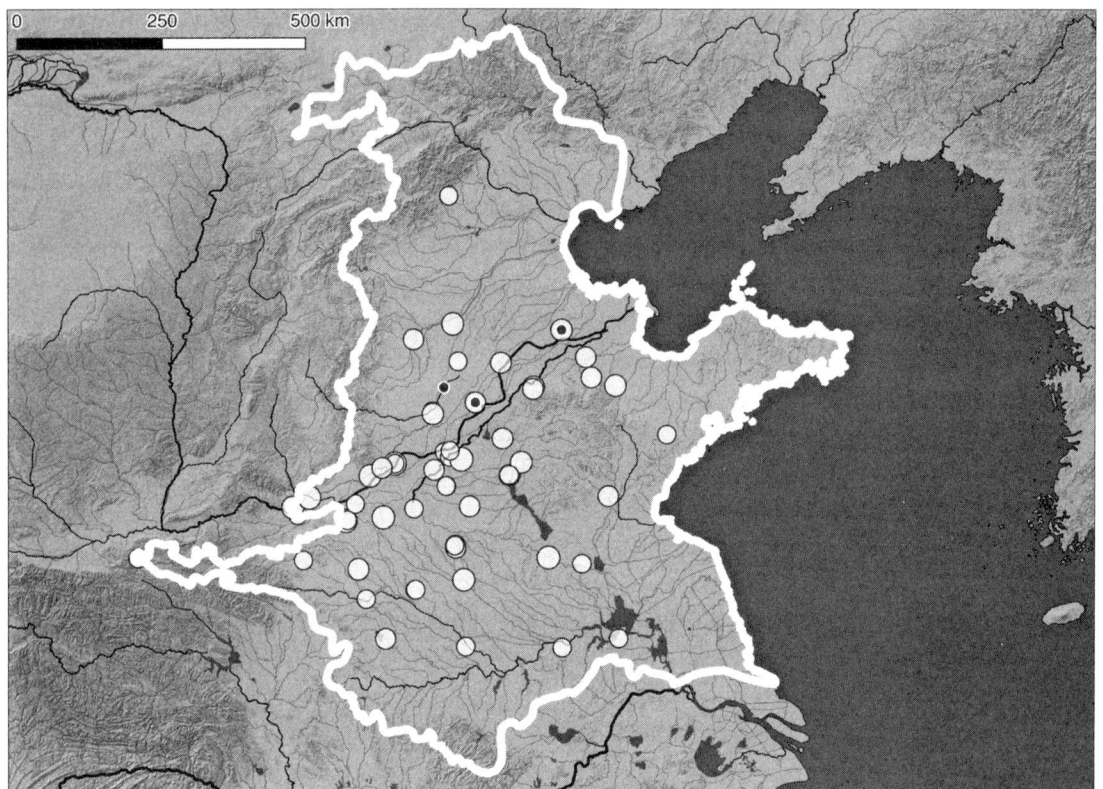

FIGURE 3.8
The Geography of Floodplain Events, 550–750. Relatively few events transpired on the floodplain between 550 and 750 CE. Historical sources reflect an almost complete inattention to waterworks infrastructure development during that period.

and personnel to a series of military campaigns there. Diverted Yellow River tributaries brought water to the canal as it traversed the northern floodplain. A third canal connected the Huai River with the Yangtze. Hongze Lake, previously a small reservoir in a depression on the floodplain adjacent to the Huai River, expanded significantly as a result of the new canal geography. Named for the first time in the early seventh century, it would gradually be surrounded with embankments as it became an increasingly important part of the hydrosocial system of the floodplain, eventually becoming the fourth-largest freshwater lake in China.

Collectively, the canals formed a triangle that roughly limned the floodplain. The Tongji Canal followed a route south of the main channel of the Yellow River, passing through the wetlands and shallow lakes around the Huai River. The Yongji Canal flowed to the north of the Yellow River. The two canals met in Luoyang. The Shanyang Canal approximately paralleled the coastline. Each canal paralleled existing courses of short tributary rivers of the floodplain, sometimes taking their routes, and passing through marshes and shallow lakes along the way. The canals created straight lines and relatively silt-free channels through the existing landscape.

Elsewhere, a seventh-century waterworks system in Shanxi irrigated more than thirty-two thousand hectares of land, and a new transportation canal was built the following year to carry crops to the capital.[37] The following century, the commissioner in charge of grain transport from the Yangtze region to the capital built a series of transshipment granaries along the route to Chang'an. He introduced a relay system whereby grain would be offloaded and reloaded onto the boats that were most appropriate to each segment and onto carts for the final leg of the journey from Sanmenxia to Chang'an.[38] Despite repeated efforts to build canals to bypass the Sanmen rapids, none of these initiatives was successful. Chang'an's foodshed became more precarious as time went on.

The canal infrastructure rerouted tributaries and created new channels for Yellow River floodwaters and sediments to traverse the alluvial plain. Canals transformed floodplain hydrology. On the Loess Plateau, after the medieval drought concluded, the canals exacerbated the region's vulnerability to flooding and sedimentation. The canal system long preceded the system of dams and embankments. By the middle of the eighth century, the well-being of the empire depended on a vast hydrosocial architecture of transportation and irrigation waterworks. The canals ensured prosperity on the engineered floodplain. The eleventh-century scholar Su Shi (1037–1101) wrote about the abundance of oranges, clams, and rice found on the Huai plain and proclaimed it the empire's preeminent region.[39] However, even in its earliest iterations, the canals required regular maintenance to avoid being overwhelmed with silt. Over time, efforts to sustain them would become increasingly ingenious and urgent.

After the An Lushan Rebellion in the mid-700s, the floodplain, like the Loess Plateau, fell to semi-independent competing warlords. They periodically damaged one another's waterworks in acts of war. There are few records of Tang floodplain management after the mid-eighth century, and attested flood disasters remained rare as well. However, there were still periodic attempts to maintain the canal system, which had become essential to the well-being of the empire. With the floodplain in turmoil, grain from the south could no longer be shipped via canal. Instead, it was transported through a long and arduous journey upriver on the Yangtze River and its tributaries and then north over the Qinling Mountains. Food shortages plagued Chang'an. Within a few years of neglect, the canals became impassable. When the floodplain briefly came under control of the regime after the war, its residents were impressed into corvée labor, dredging the silt-clogged waterways between the Huai and Yellow Rivers to restore the transportation routes.[40]

These gains were short lived. By the early ninth century, the canals had unequivocally filled with sediment and were no longer usable. As flooding accelerated, water management efforts pivoted away from the canal system in favor of the first concerted efforts to create drainage systems and levees.[41] An enormous flood in 858 killed tens of thousands of people and inundated thousands of acres of farmland. Thus began a new floodplain regime.

The rate of flooding continued to accelerate against the backdrop of an era of political upheaval in the tenth century, known as the Five Dynasties. Rulers of the de facto independent states of the late Tang, Chinese and Turkic alike, declared themselves emperors after the formal collapse of the Tang regime in 907 CE. The seventy years that followed were a time of near-constant warfare throughout northern China as various claimants struggled to reunify the territory of the former empire. On the Ordos frontier, the Tanguts and Shatuo Turks expressed nominal loyalty to all five successor courts, but the borders that ran across the loess remained fortified and there were sporadic campaigns there. At the eastern edge of the Loess Plateau, a Shatuo regime known as the Northern Han founded a capital at the Fen River city of Taiyuan and controlled Shanxi from 951 to 975.

On the floodplain, the Five Dynasties claimants destroyed levees for the purpose of displacing and blocking one another's troops, rendering canals useless by flooding them with silty water.[42] One general breached levees in Shandong in 918 to block the advance of rival troops. Another commander used the same tactic in 923.[43] By engineering these intentionally catastrophic breaches, regimes began to see the river as simultaneously deadly and tractable. The earliest references to the Yellow River as problem and peril date to the ninth century.[44]

Between accelerating erosion upstream and infrastructure neglect and destruction downstream, the Five Dynasties era was one of significant flooding, with an average of one breach every three years and one flood every other year. In 893, the course shifted across the delta by 100 *li*, about 50 kilometers, or 30 miles, near the river mouth. An unmanaged bank stretched for over 60 kilometers, almost 40 miles. Refugees migrated across the plain. Unable to farm for fear of inundated crops, they caught fish and gathered plants for sustenance. In the 950s, the chancellor of the Later Zhou regime, which controlled the floodplain at the time, oversaw a process of shoring up breaches and extending the levee system.[45] As that regime increasingly stabilized its control over the population, the frontiers, and the domestic warlords of North China, an important objective was to fix in place both the water and the population. In 960, the Later Zhou ruler declared himself ruler of the Great Song dynasty, which would soon unify all of China.

A PRECARIOUS CAPITAL AND ITS TIMBERSHED

The census of 742 recorded a population of around fifty million people, a number that would double to one hundred million by the 1080 census. War, commerce, and environmental transformations drove technological innovation. New varieties of fast-ripening rice from Southeast Asia made it possible to harvest two or even three crops each year in the south. Farmers hybridized these with existing strains to develop new crosses that could grow in saline, swampy, alkaline, or dry conditions. Supported by government incentives, farmers brought marginal lands under cultivation for the first time, and in the most productive regions, especially in the south, grain surpluses supported an urban, commercial, and industrial economy.[46] When trees became scarce, coal mining in northern China fueled an iron and steel industry that grew rapidly around the turn of the millennium.[47] An unprecedented shift to a market economy meant that land, labor, natural resources, and harvested and manufactured goods were all commodified. New arrangements in political economy reshaped hydrosocial dynamics across the Yellow River basin.

Transformations in the political economy of the Chinese empire encouraged farmers to engage in new kinds of erosion-causing practices. The technology of advanced steel-blade, curve-beam plows permitted more intensive hillside cultivation.[48] The first reference in the Chinese literature to terraces (*titian*), the characteristic mountain-slope fields of East Asia (plate 13), dates to the late twelfth century, although the existence of this technique can also be inferred from Tang poems.[49] The private market in land combined with government policies that favored colonization and agrarianism to bring new lands under cultivation.

The Northern Song regime sited its capital in the metropolis of Kaifeng, which was located in the low-lying basin at the floodplain's volatile western periphery, 200 kilometers (about 125 miles) east of Luoyang. The regime excavated a new canal that terminated in Kaifeng. Known as the Bian, it took a shorter route across the floodplain than the canals to Luoyang had, allowing people and goods to arrive more quickly at the new capital city. Although the canal took a new route, Song planners adopted its name from a series of canals that had connected the Yellow River and the Huai since before the Common Era. Indeed, the regime chose the city because the revived canal network through the Huai basin connected it so effectively with the affluent south, and Kaifeng could be provisioned without traversing any difficult stretches of the Yellow River. In this marshy region of flat terrain and poor drainage, channels were easy to create but difficult to maintain, even without the added challenges of Yellow River sedimentation and flooding. With the addition of tree-lined banks and locks that raised and lowered the water level, canal channels readily linked existing streams and lakes. These ventures required massive labor but presented few engineering challenges.

The Bian Canal and the Yellow River offered Kaifeng water for transportation, irrigated agriculture, and household needs. However, Kaifeng was in an exceedingly precarious location, highly susceptible to flooding. During the 167 years of the Northern Song, Kaifeng experienced 279 river-related disasters of varying intensity.[50] Nonetheless, in 960, when the regime decided to situate the capital there, the river had not changed course for eight hundred years. Although the era of frequent flooding had begun, planners did not realize how profoundly it would disrupt the stability and well-being of the region. An essay entitled "Establishing the Capital" (*An du*) explicitly contrasted Kaifeng with Chang'an, which lacked a reliable network of waterways and suffered from persistent water shortages.[51]

By the eighth century, Chang'an's timbershed had disappeared. In contrast, the new imperial capital was located near hills that were still forest covered in the tenth century. These were soon stripped as well. The Lüliang Mountains, east of the Ordos on the Loess Plateau, were a major Northern Song timber site. Merchant interests dragged large trees from there by the tens of thousands. They sold some to the court to furnish the capital while others found their way to a private timber market. Woodcutters clear-cut sites and then moved on.[52] The twelfth-century writer Hong Mai was implicitly criticizing the capital construction projects of his own day when he lamented that during Sui times, in the eastern capital at Luoyang, "there were no tall trees to be found in the near mountains, [and all needed timber] had to be transported from afar." In his own time, timber for palaces and temples in the capital had to come from all over the empire. Imperial decrees frequently banned firewood timbering in hills and mountains near the capital, and there was a persistent shortage of charcoal and firewood in the capital until coal replaced wood for heating and cooking in the 1070s.[53]

Far beyond the capital, prairies and riparian and mountain forests in the Yellow River basin faced serious peacetime pressures during the era of rapid commercialization and population growth. Timber, supplied on a commercial market, went to build houses, palaces, and cities; to produce pine soot for ink; and to serve industry. As levee construction on the Yellow River burgeoned in the era of frequent floods, water-control structures themselves became a significant source of deforestation. For example, the repair of a single breach near Kaifeng in 1019 required sixteen million units of timber poles, brushwood bundles, and bamboo gabions. Tens of thousands of laborers were tasked with scouring mountain forests for brush to make into fascines. The eminent eleventh-century diplomat and polymath Shen Kua (1031–1095) reported that all the pine forests in Shandong had disappeared and that those in the Taihang Mountains and to the west of Luoyang were being stripped bare.[54] By the time of the great course changes in 1048 and 1128, timber to repair waterworks structures on the lower course of the river had to be floated all the way downstream to the alluvial plain from Shaanxi, Shanxi, and the uplands of Hebei, which further exacerbated deforestation pressures upstream.[55]

The commercial market in timber was a new feature of the Northern Song, a consequence of China's medieval economic revolution. This was an era of severe ground-cover collapse on loess slopes for many reasons.[56] Commentators were well aware that they were seeing old-growth forests being razed for the first time.[57] *The Song History* (*Song shi*) refers to the commercial timber market with the information that "it is possible to profit from the annual harvest of gigantic trees" (*sui huo jumu zhi li*).[58] Ian Miller dates the widespread commodification of timber to the twelfth century, citing texts from that time that encourage the use of timber as an investment. He identifies the eighth to the eleventh centuries as a "wood crisis" in China.[59] The days of easy timber resource extraction in northern China quickly came to an end. Shen Kua writes, "It has gradually reached the stage" that in the Taihang mountains north of the capital "at least half the slopes have been denuded of pines."[60]

The environmental historian Mark Elvin has commented that "military power, economic development, and pressure on resources" were interlinked beginning as long ago as the Bronze Age. In earlier times, the exploitation of land and resources for military advantage transpired under the direct command of the state. However, in the new millennium, commercial and imperial incentives intertwined with one another in new ways, which often amplified the pace and intensity of environmental degradation.[61]

THE MILITARY CONTRIBUTION TO ENVIRONMENTAL COLLAPSE

The Tangut court declared formal resistance to the new Northern Song state in 982 and proclaimed the sovereign nation of Xi Xia in 1038. The boundary between Song and Xi Xia was never well defined by treaty, territory changed hands often, and both sides frequently transgressed the border. Almost all Chinese references to surveying and mapmaking from any place in the empire during this era refer to the remote and unstable frontier through the central Ordos and the Liupan Mountains, which were coming under thorough state scrutiny for the first time. Survey documents detail the arable land, potable water, and timber resources available on the northern Loess Plateau.[62] During the moist tenth century, the boundary between pastoralism and agriculture had moved toward the north. Moisture levels then fell precipitously during the first half of the 1000s and remained well below the long-term mean for most of the century. Herding peoples, pushed out of their ancestral lands, expanded toward the north and west, into hills, and intensified their land use there as Song policy aggressively favored colonization of the region.[63]

After eighty years of skirmishes, colonization, farming, salt mining, horse pasturage, logging, and earthworks, full-scale war ultimately broke out in the 1040s amid a decade of drought and a spate of earthquakes in Shaanxi and around the realm. The war ended inconclusively in 1044. Tangut and allied armies won all four large battles between the two adversaries but at great cost. They did not succeed in dislodging Song armies or garrisons from the Ordos or the Liupan Mountains. The war effort depleted Xi Xia treasuries and personnel. Under the terms of the 1044 truce, the Xi Xia rulers recognized the Song sovereign as an emperor, and the Song began to send annual gifts to Xi Xia in recognition of its power and legitimacy. This pact did not significantly reduce hostilities or inhibit the Song quest to colonize the Ordos or to fortify the border there. After the war, the Song regime closed the border to trade and initiated costly annual incursions into Xi Xia territory. Early Song military planners used old Qin walls to mark the border with the Tanguts.

The Song colonial frontier in the Ordos was precariously maintained, often transgressed, and periodically refined amid skirmishes, battles, and diplomacy. The actions of a multicultural population of farmers, merchants, bandits, herders, and soldiers in the region, occupying the frontier in pursuit of their own interests, contributed to border instability. Around the turn of the millennium, the Ordos was a salt-producing region, a gateway to long-distance trade routes, a home for large herds of sheep and goats, the breeding grounds for warhorses, and the portal to Guanzhong. The sparsely settled central Ordos and the Liupan Mountains at the western Ordos periphery were of great interest to the Han Chinese, the Tanguts, and Tibetans, and the Uighurs as well. All had historical claims to the region.

The Liupan Mountains were a natural barrier between the Northern Song and the Xi Xia. Contended between the two regimes, both Chinese and Tangut armies deforested the mountain slopes when they gathered in the region, opened new lands to cultivation, and farmed crops in time of peace.[64] When the prefecture of Qinzhou (fig. 3.9) became a site of intensive frontier fortification, its tree cover was rapidly destroyed. Its timber was earmarked for military purposes, and civilians, with no way to acquire firewood, had to sneak across the border into Xi Xia territory in search of fuel.[65]

Once the capital moved to Kaifeng, its easternmost locale thus far in imperial history, the court was physically remote from the Loess Plateau. In contrast to times when the court had been situated at Chang'an, there was limited scrutiny of the Loess Plateau and the deteriorating environmental conditions there. The Ordos became a

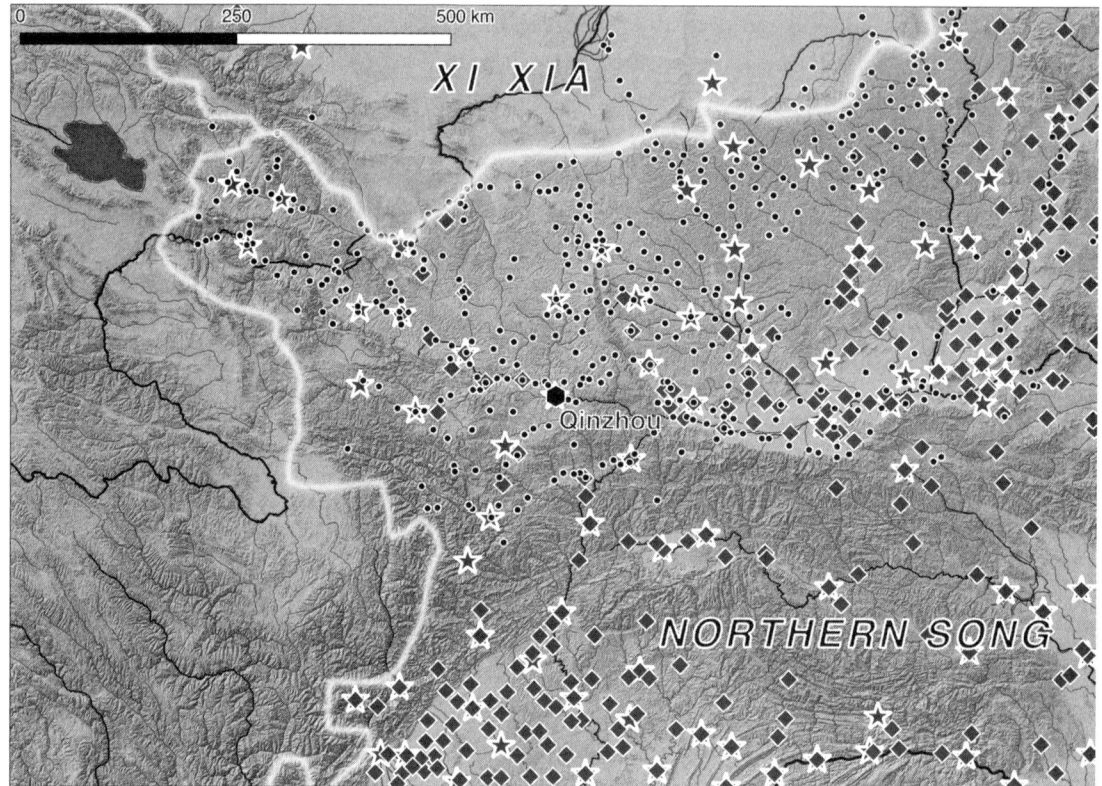

FIGURE 3.9
The Qinzhou Region in 1111.
Qinzhou, near the headwaters of the Wei River, was a site of significant attention during the conflict between the Song and Xi Xia regimes. It lay at the heart of the southwestern region of the fortification zone.

relatively autonomous region trapped in a cycle of low state attention, high population, limited commerce, declining agriculture, and food scarcity. The regime required peasant settler-colonists and billeted soldier-farmers to claim marginal loess land for agriculture to support self-sufficiency and to feed campaigning armies.

The war transpired on the Ordos Plateau and in the Liupan Mountains. Figure 3.10 depicts Xi Xia territory around the year 1111, including the sites of major battles and the massive fortification on the Song side of the border. By the time of the war, the Xi Xia domain had grown from a handful of towns near the Mu Us Desert to an extensive empire that spanned the entire northern Ordos and the Hexi corridor. Battles, migrations, and troop maneuvers had direct and acute environmental impacts, and mass fortification along the Xi Xia border caused sustained land degradation. Until the eleventh century, Chinese settlement on the Ordos concentrated largely in the Wei floodplain of Guanzhong. Water sources north of the Wei were

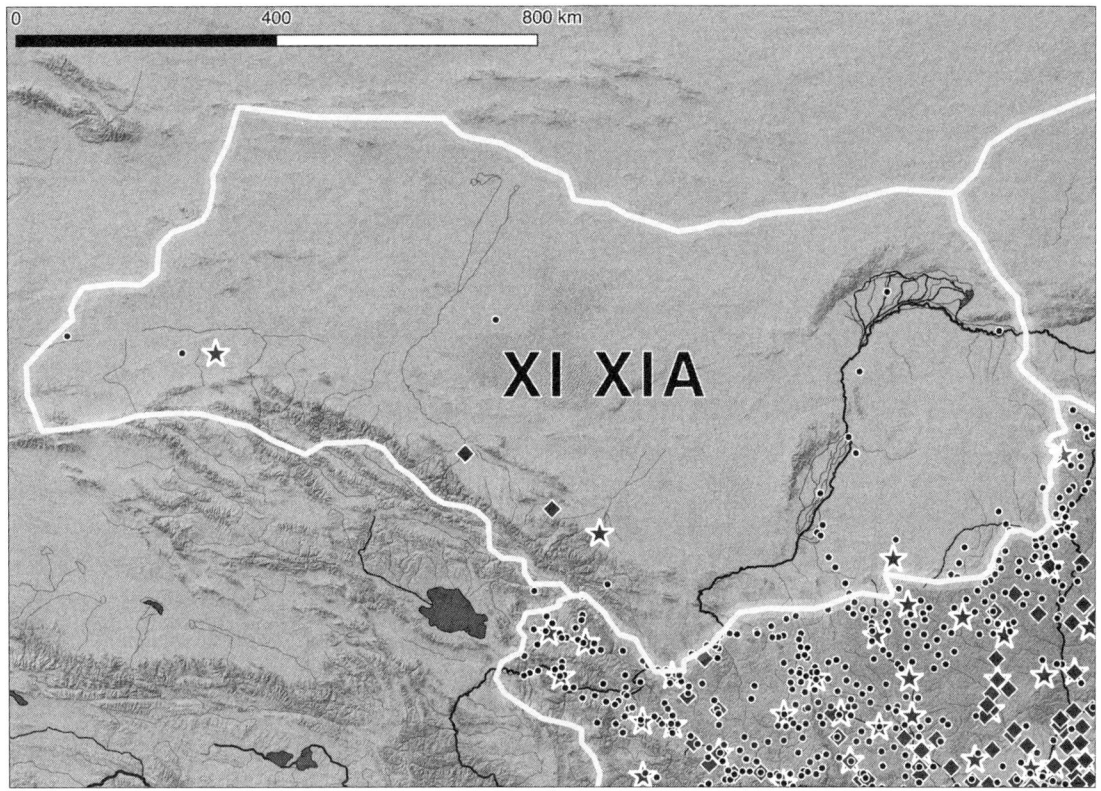

FIGURE 3.10
The Xi Xia Regime and the Song Frontier. Administrative geography data for Xi Xia territory are limited, but the highly fortified Song frontier across the Ordos and in the Liupan Mountains is well attested in historical sources.

too scarce for large-scale irrigation and rainfall too unpredictable for rain-fed agriculture. The central Ordos had no major cities and few sedentary elites. It supported a small and multiethnic population pursuing diverse and multiple subsistence strategies.

The Song regime conducted a massive fortification campaign on the Ordos over the course of the eleventh and early twelfth centuries (fig. 3.11), founding more than three hundred new garrisons in the course of a few decades of the eleventh century to resist Tangut incursions and to colonize this multiethnic region. They arrayed the new forts every few kilometers across the grassland-desert ecotone, ideally within signaling distance of one another. Settlements were densest in the loess hills of Shaanxi and Gansu. Each new fort had a substantial effect on previously undisturbed native ground cover. Documents from the mid-1040s onward also call for the construction of trenches (*hao*) or markers (*hou*) every few kilometers across the entire Ordos.[66] Earthworks cut into the easily eroded soil. Continued high levels of military

LOESS IS MORE

FIGURE 3.11
The Song Fortification Campaign. These snapshot dates depict the years (a) 741, (b) 1111, and (c) 1189. After the Song retreat from north China in 1127, the successor Jin regime defortified the Loess Plateau.

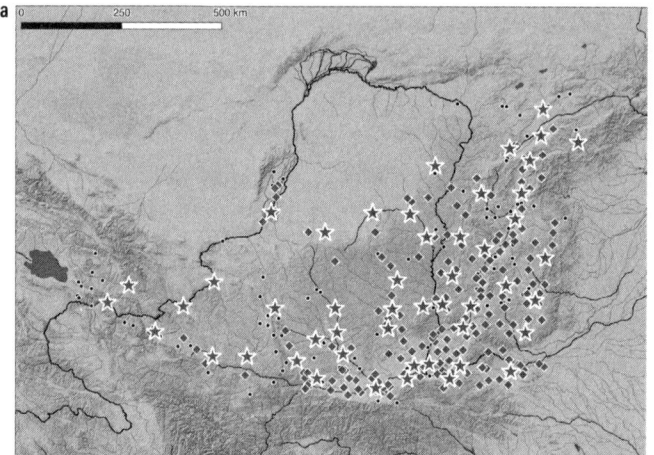
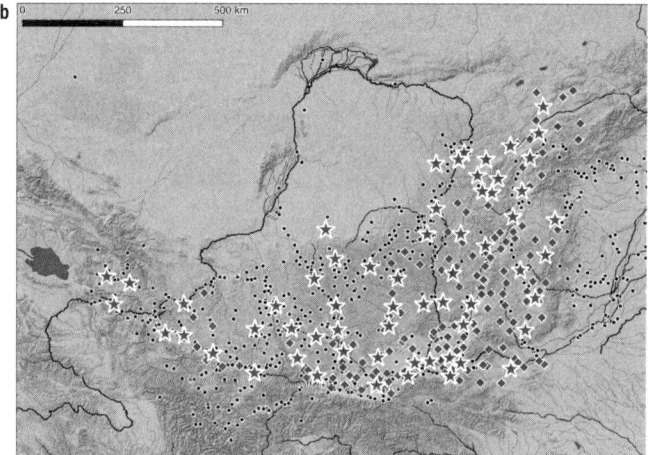
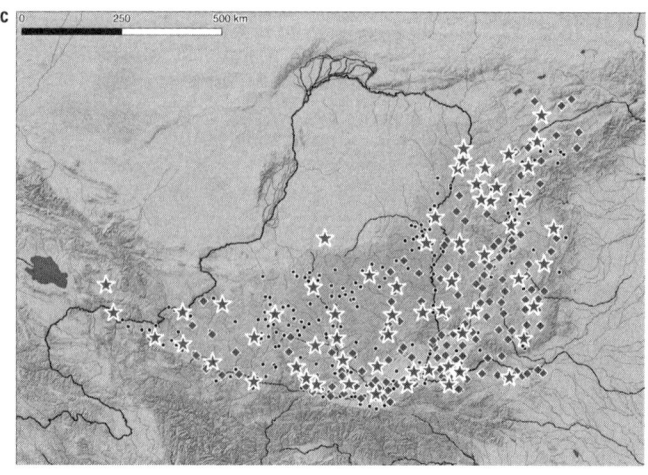

spending after the end of the war in the mid-1040s went into projects of surveying, mapping, and construction on the Xi Xia frontier.[67]

The centerpiece of the Song postwar military strategy in the Ordos and in the Liupan Mountains was to conduct a series of fortification campaigns. The objective was to establish forts and earthworks year after year that would gradually assert a more northerly border by populating the region with settlements. The Xi Xia did so as well, although the geography of the Tangut settlements is not as clear. Both sides contested the other's fortification initiatives and border markets in what the historian Ruth Dunnell has described as "a kind of guerrilla warfare."[68]

The aggressive fortification strategy (*chengshou zhanlue*) was official policy. Intended to defend against the mobile cavalry of Xia, new forts housing ideally self-supporting soldier-farmers (*tuntian*) punctuated frontier prefectures, often in lieu of constituent counties. In 1041, the Ordos frontier supported 34,000 horses, 155,600 people from 670 indigenous communities, and 32,580 imperial soldiers in 20 battalions, along with 900 additional battalions of provincial troops and militias. By 1044, there were 500 Song imperial battalions, 500,000 troops, and 300,000 Xia cavalry.[69] There was one massive fort- and wall-building campaign in 1040, followed by another in 1042 to "fill in the empty spaces."[70] Some of the forts were as little as 15 *li* apart (about 8 kilometers, or 5 miles), and none more than 80 *li* apart (about 40 kilometers, or 25 miles).[71] Although the Song never built a continuous barrier like the Ming Great Wall, the garrisons functioned in the same way. At least in the short term, no single outpost taxed the carrying capacity of the region, but the distribution of hundreds of small outposts meant that deforestation and ground-cover disturbance was spread widely across the Loess Plateau. Resources were allocated toward building additional forts and garrisons rather than consolidating troops or developing policies that allowed armies to become more mobile.[72] Civilian populations faced massive provisioning burdens, which themselves accelerated environmental degradation.[73] Erosion rose steeply, and springs, ponds, and tributaries collapsed.[74]

Along with hundreds of new forts and garrisons, the regime also founded new counties and prefectures. These were political and environmentally impactful walled outposts of regional government staffed with civilian and military commands and populations of hun-

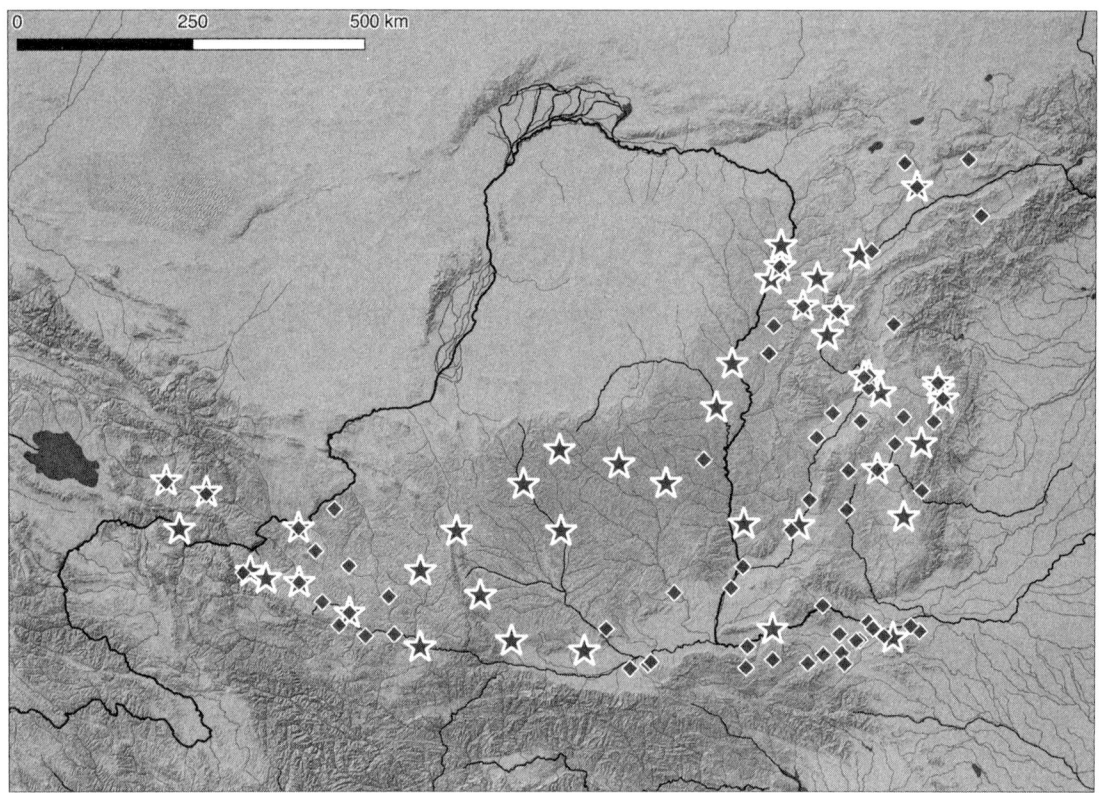

FIGURE 3.12
Newly Founded Counties and Prefectures on the Loess Plateau, 960-1127. In addition to establishing garrisons and forts, the Song regime established new prefectures, particularly to the north and west, which managed military affairs; and new counties, particularly to the south and east, which collected tax revenue in populous and prosperous locations. The Loess Plateau was a site of both such phenomena around the turn of the new millennium.

dreds or even thousands of people. Figure 3.12 shows that the new counties and prefectures stretched across the entire Loess Plateau, not just the Xi Xia front. On the Xi Xia frontier, the new jurisdictions were primarily prefectures, which were command centers for military enterprises. To the south and east, the new administrative seats also included numerous counties. Counties were primarily sites for tax collection and civic business, so the new districts there reflect the expansion of population and commercial activity on the hills and valleys of the eastern and southern Loess Plateau and the civilian production of iron, timber, and food for the war effort.[75]

The fortification campaign accelerated over the course of the eleventh century. During the 1090s there was a ceaseless campaign to make sorties into Xi Xia territory and establish Song forts. More than fifty forts were established during that decade, with twenty, many in putatively Xia territory, established in 1097 and 1098 alone.[76] The more stable of them were transformed into counties and prefectures, and thus the border crept northward. The fields, pastures, and lumber operations of these remote, largely self-supporting outposts destroyed fragile ground cover, exposed erosion-prone sand and soil that made its way into the Yellow River through wind and water deposition, and ultimately drove disastrous flooding downstream.

The war and its forts also imposed an onerous burden on the individuals conscripted to fight in it. Soldiers were often accompanied by wives and children, who were forced to assist them with work such as woodcutting. The war was an affair of forced colonization, with concomitant environmental impacts.[77] Desertion and flight to mountains and forests by both Chinese troops and tribal allies, and their attempts to self-transfer from certain "newly built forts and outposts" to other more desirable ones, reveal how ecologically marginal some of these garrisons were and the pressures they imposed on the land.[78] Because the central Ordos and the Liupan Mountains were relatively isolated from China's more productive agricultural zones, forced colonization also caused erosion by driving agricultural intensification in marginal regions where food crops needed to be grown for army rations. Trees fell to refugees and deserters who fled war zones and searched for firewood, putting ecological pressure on previously unpopulated locales.

Relatively few fortifications lay on open grassland. Such sites would have been difficult to provision and defend, and water would have been in short supply. Instead, the garrisons clustered around loess hillslopes, which provided good vantage points and were close to the strategic passes through which Tangut cavalries were most likely to travel on their way to Guanzhong.[79] Water and terrain were ongoing concerns when establishing the fortifications. The celebrated essayist and statesman Ouyang Xiu (1007–1072) was a sometime critic of the fortification campaign who advocated for a more aggressive stance against the Tanguts.[80] A close observer of policy on the Ordos, he explained that garrison sites needed to be rugged, surrounded by defensible ravines, situated near major rivers, and replete with springs, trees, and streams.[81] Steeply sloped locales, with dense forest cover and ample waterways, in the semi-arid central Ordos, were precisely the kinds of places at the highest risk of erosion when deforested.

Commentators recognized that they were making their decisions under dynamic environmental conditions. Shen Kua discussed one site that had previously been a garrison but had to be abandoned because of a shift in the course of the Wuding River. Now, he reported, the

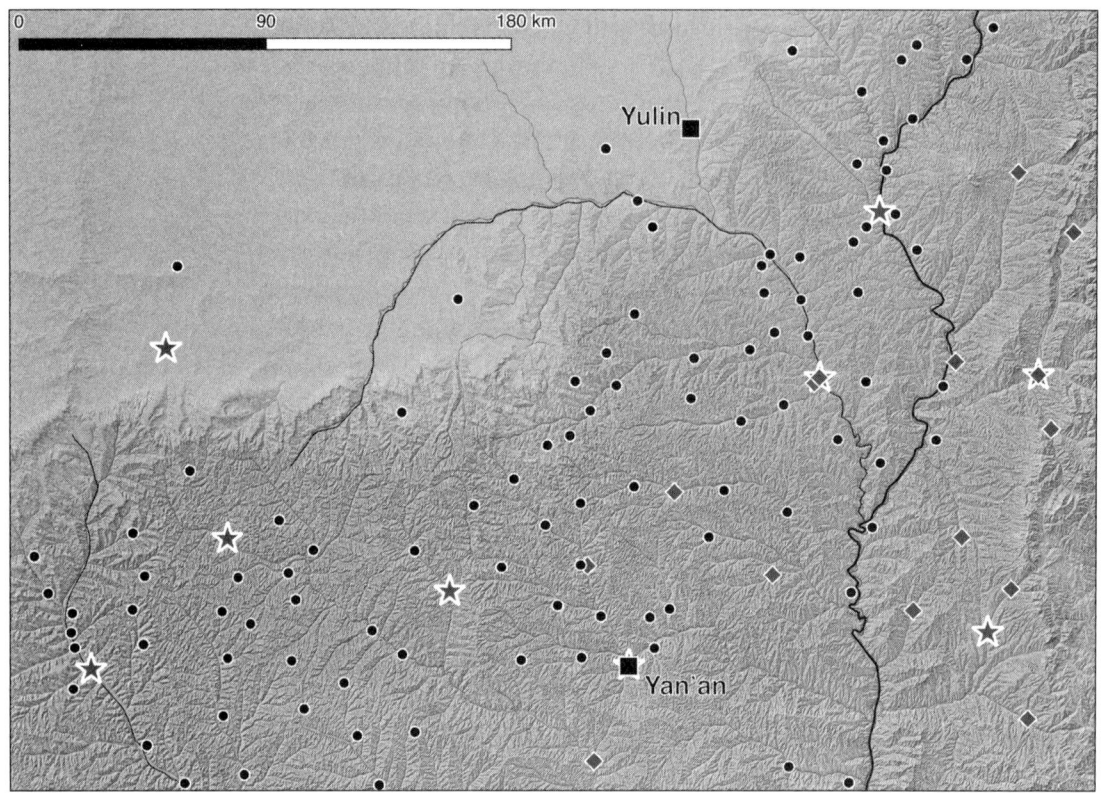

FIGURE 3.13
Fortifications in the Heng Mountains in 1111. The Heng Mountains south of the Wuding River were a major site of fortification during the eleventh and early twelfth centuries.

site had become "nothing but rocky cliffs, with precipices protruding more than 10 *zhang*, and no water at the foot of the cliff face."[82]

As figure 3.13 depicts, fortifications were dense in the Heng Mountains, a range of loess hills situated between Yan'an city and its loess grassland plateaus to the south and Yulin city to the north. The Heng Mountains were the watershed for the Wuding River. North of the Heng Mountains lies the Mu Us Desert, then the Mongolian steppe. These mountains were a critical focus of the intensive Song fortification strategy. They offered excellent vantage points, defensible passes, and ample water. The mountains themselves included suitable spots for farming and horse pasturage, and surrounding areas had salt and iron ore. Before their capture by Chinese troops during the 1040s war, the Heng Mountains were an important site for the Xi Xia. They were also home to an indigenous population (their ethnicity unclear, Song sources referred to them as foreigners, or *fanren*) who provided allied troops to the Tanguts until they were expropriated by Song armies.[83] The fortification campaign in the Ordos was as much about military support for settler colonialism as it was about the war against the Tanguts. After defeating the Xi Xia in the Heng Mountains in 1098, Song troops occupied these steep-sloped loess hills, built roads through them, and constructed numerous forts in rugged and defensible terrain with running streams and springs.[84]

Prior to their conquest, the Heng Mountain slopes had been covered by cypress forests and crisscrossed with valley creeks.[85] Recognizing reasons for concern about ecological stability, the Song regime forbade tree felling in the Heng Mountains.[86] Nevertheless, new garrisons continued to be founded in suitable spots throughout the region, occupied by a rising population of troops accompanied by their dependent families, who had no alternative but to travel up the small mountain streams to gather wood and brush for sustenance, denuding the slopes as they went.[87] Thereafter, "whenever there was a violent rain, the rapids carried large rocks."[88] Plate 28 depicts the gullied and deforested Heng Mountain landscape today.

Ouyang Xiu, Shen Kua, and other prominent contemporary figures were aware of the relationship among warfare, timber exploitation, and erosion. Indeed, they even legislated against it, but wider geopolitical pressures made anti-erosion policies impossible to enforce.[89] The population of Shaanxi in 1102 was 2,847,009, or 10 percent of the population of the whole empire. Although this was only half the population recorded at the time of the Tang census of 742, the Song residents had spread across the entire Ordos and were no longer clustered around Chang'an as their predecessors had been.[90]

Agriculture and pastoralism persisted on the Loess Plateau when the region came under the control of steppe-facing regimes after 1127. A government founded by the Jurchen people, whose homeland was north of the Korean Peninsula, governed the Yellow River basin from 1127 to 1234, when it passed into control by the Mongols, who held the territory until 1368. During those centuries, the Ordos frontier was no longer contended between multiple polities. Jurchen and Mongol rulers did not practice agricultural settler colonialism or privilege agriculture over pastoralism. They presided over multicultural alliances. For these reasons, they did not maintain a fortified or densely populated frontier. Coerced settlers departed from marginal terrain.

As figure 3.14 depicts, these steppe-facing regimes abandoned a great many Song forts and administrative seats on the Loess Plateau.

FIGURE 3.14
The Jin and Yuan Loess Plateau.
Between the (a) Jin and (b) Yuan eras, settlement on the Loess Plateau rolled back even further.

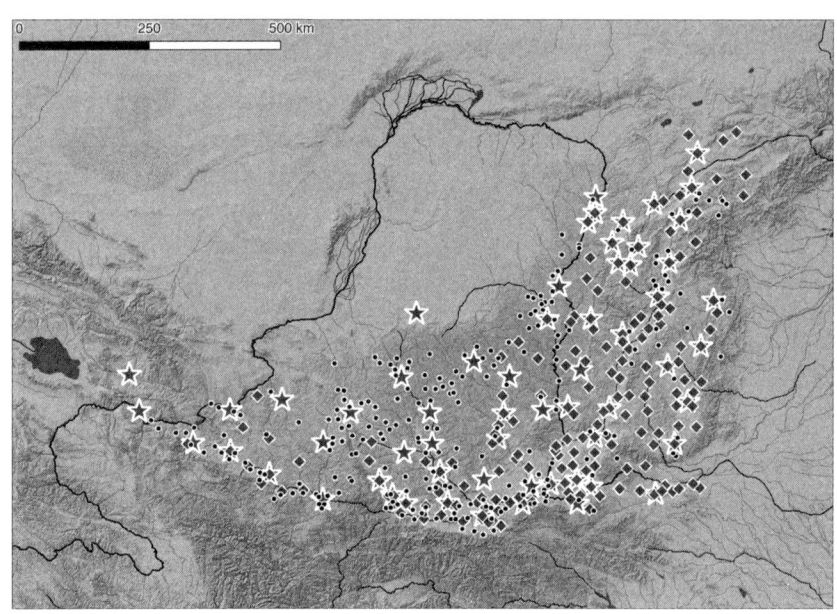

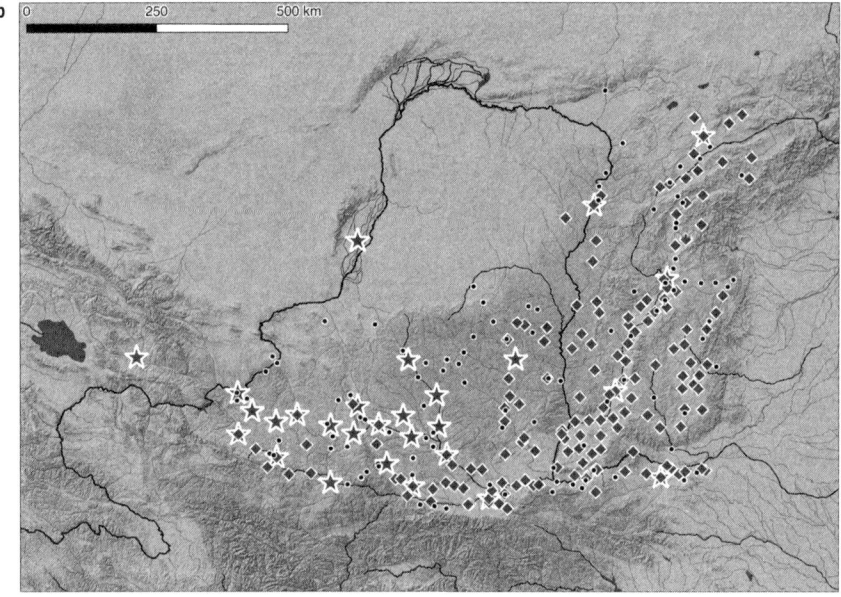

Population contracted significantly.[91] At the time of the 1312 census, the entire population of Shaanxi was only 449,045, or 0.8 percent of the empire's population.[92] For two and a half centuries, the forests and grasslands of the Loess Plateau were able to recover from the intense burdens placed on them during the long eleventh century—long enough that vegetation could have rebounded significantly, but not enough time to regenerate lost soil or remodel gullied terrain.

DOWNSTREAM: NORTH, THEN SOUTH

During the long eleventh century, colonial ambitions on the central Ordos and in the Liupan Mountains, along with farming, commercial mining, and forestry across the Loess Plateau, gave rise to an extraordinary amount of erosion. By the middle of the tenth century, people who wrote about the river on the floodplain recognized that they were seeing unprecedented amounts of sediment accumulation there. They realized that the historical sources they consulted for guidance about waterworks maintenance no longer offered accurate advice.[93] The riverbed perched above the level of surrounding houses in Shandong as levees were built higher and higher above the sediment-filled river.

As the collection of timelines in figure 3.15 indicates (see also figs. 3.3 and 3.4), sources reported river-related disasters in a way that reveals two large peaks: one in the late tenth century and another in the second half of the eleventh century. Events of infrastructure construction and management became far more common throughout these centuries as well. Plate 29 depicts the changing ratio of the management event types. Not counting the large proportion generically designated "management," the event types chart a vivid story of people's changing relationship to the engineered river. Between 920 and 1029, more than a quarter of management events were emergency fixes. From 1029 to 1090, the largest portion of reported management events were proposals and discussions. Emergency repairs declined as a proportion of events. Finally, from 1090 to 1165, repairs of existing structures were the most common type of events. A simplistic but broadly correct version of the story of these centuries is this: first river managers reacted to unexpected catastrophes; then they debated possible strategies for living with the newly silt-filled river; finally they prioritized renovations and maintenance to the system that had come into being.

FIGURE 3.15

The Eventful Era on the Floodplain, 920–1165. This series of timelines depicts an era of generally rising disaster on the floodplain and generally increasing attention to waterworks management, although with some variation from one decade to the next. Figure (a) depicts floodplain events between 920 and 1029, and figure (b) depicts the ratio between management events and disaster events attested in historical sources during those years. Figures (c) and (d) cover the years 1029 to 1090, and figures (e) and (f) refer to 1090 to 1165.

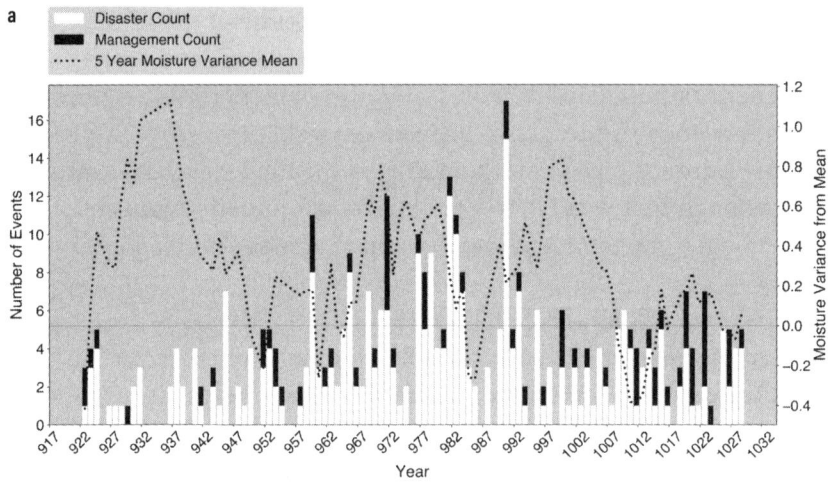
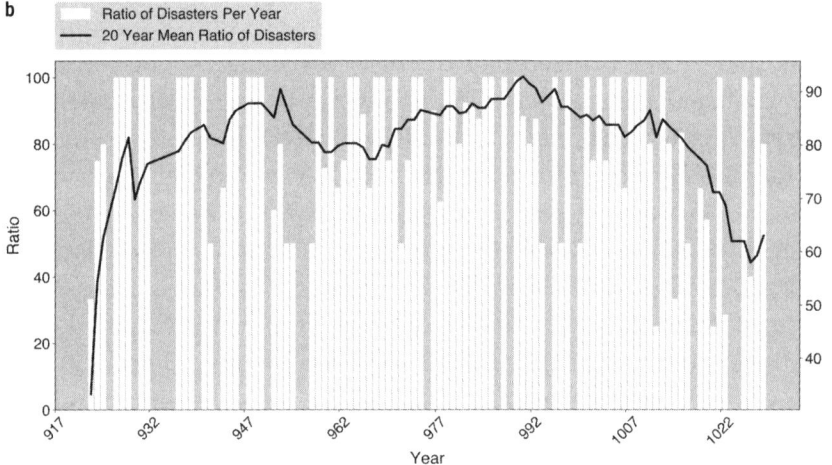
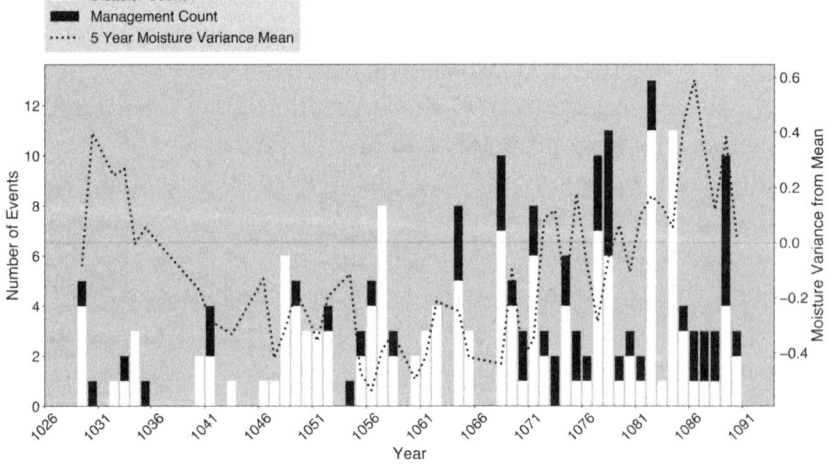

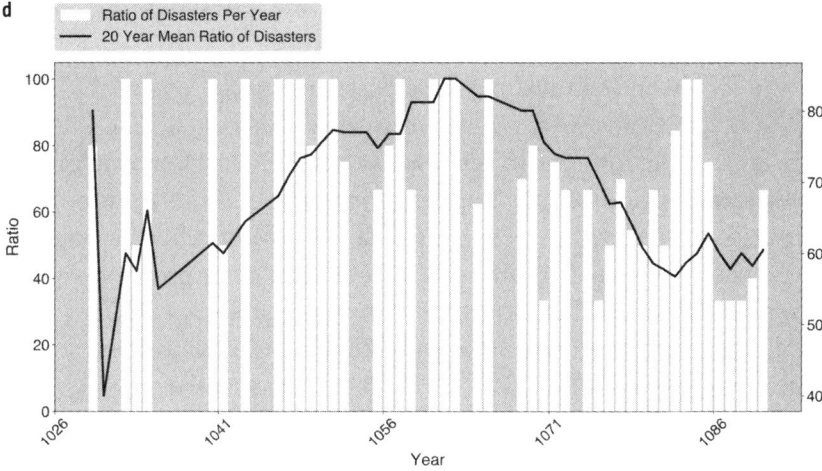

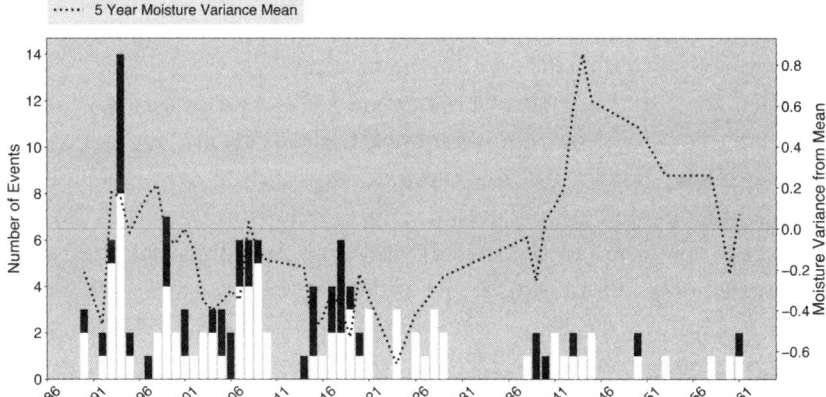

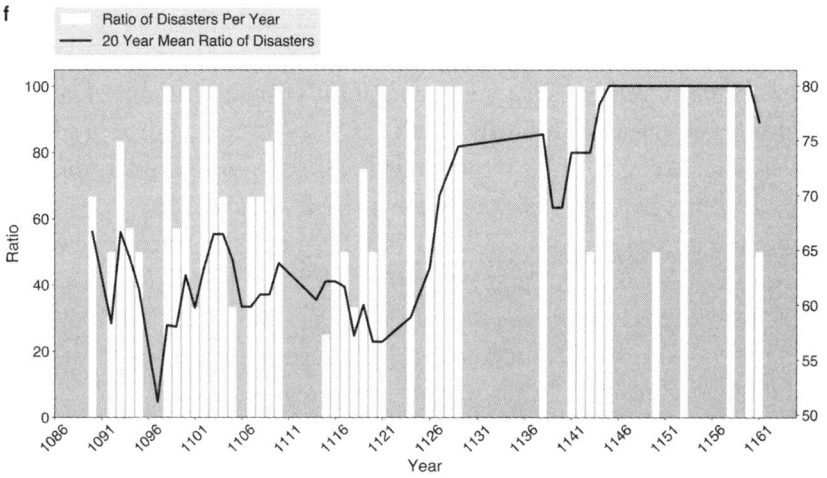

Severe erosion reached as far upstream as Guanzhong, where hydrocrats had revived the Zhengguo Canal to irrigate land there, in the fertile agricultural region closest to the Xi Xia front. To do so, they cut a new feeder canal through solid rock several kilometers up the Jing River from the original gate. The ancient earthworks downstream had become clogged with sediment and were no longer viable.[94] Maintaining the canal also required timber for support structures. The canal, "cramped . . . between a river prone to violent floods and steep, gullied slopes with unpredictable torrents," required earth and wood structures that would be replaced almost every year. Infrastructure maintenance was another driver of deforestation.[95] Despite these efforts, the canal system so quickly filled with silt that it only irrigated two thousand *qing* of land, a small fraction of the forty thousand *qing* that followed its original completion. Plate 30 shows the site of the eleventh-century feeder canal. Today, it is shallow and clogged with boulders, and the riverbed has eroded to a level far below that of the surrounding farmland.

The Bian Canal and the rest of the water transportation system that crossed the Huai basin required constant dredging and levee maintenance to keep the capital provisioned. For hundreds of years, few regimes had devoted attention to the floodplain, and emperors did not routinely visit waterworks sites. In contrast, the Song—founded during a time of frequent rainstorms and rising erosion, and with a precarious capital at Kaifeng—hearkened back to the majestic *feng* and *shan* sacrifices around the turn of the Common Era and the legendary work of Yu the Great before that. Both of the first two Song emperors visited flood control sites near the capital and commissioned paintings of themselves inspecting the sites in the fashion of Yu the Great. In 983, the Taizong emperor wrote a poem commemorating the successful repair of a levee breach not far from the capital, a task that had required ten months of effort by one hundred thousand impressed laborers and fifty thousand soldiers. He compared himself to the Han emperor who famously performed rituals at the side of the river, and he resurrected the notion that imperial flood control was a fundamental task of good rulership.

Three years later, in 986, twenty-six years into a regime that was becoming increasingly attuned to unprecedented flood risk, the Taizong emperor's successor, the Taizu emperor, issued a document that called for anyone in the empire with water conservancy expertise to share ideas about how to solve a flood problem that had reached crisis levels throughout the interconnected floodplain network of rivers and canals: "All court and government officials, scholars, recluses and commoners who have knowledge of historical texts on the Yellow River and its canals and who have ideas and strategies of water conservancy, especially long-term strategies that will put an end to the recurrent floods, you are permitted to come up to court to submit your proposals, or to let the postal messenger bring your details suggestions [to court]. I will read them all personally, adopt those innovative and useful, encourage all the others, and publicize and reward those selected."[96] In 1012, the third Song emperor, Zhenzong, eulogized the river as an animate entity and made sacrifices to it, holding celebrations whenever there was a positive turn in river management.[97]

As flooding became more frequent, efforts to engineer the floodplain by digging drainage canals and creating levees became more frequent and extensive. By the middle of the tenth century, there were already laws requiring floodplain residents to plant trees on the tops of embankments to stabilize the soil and to gather flood control materials and repair levees every spring. It became routine to maintain parallel levee structures—one set of embankments close to the river to straighten its course and speed its current, and another farther away to protect farmland.

Ling Zhang's extraordinary and prize-winning 2016 book, *The River, the Plain, and the State*, discusses the decades that led up to the river's catastrophic 1048 course change, the first one in eight hundred years. The river, increasingly engorged with sediment, gradually began breaching to the north in Hebei. Even then, the regime was intentionally inducing flooding, digging ditches, and creating ponds of standing water north of the river. Their objective was to create artificial ponds, wetlands, and canals to slow the progress of a potential invasion by the Khitan (Qidan) Liao, China's adversaries to the north.[98] This was an era of immense and ambitious activity in political ventures, environmental engineering initiatives, and ritual practice: massive undertakings to turn nature toward political needs on the floodplain and the Loess Plateau alike.

As flooding and waterlogging increased in Hebei while soil quality declined, waterborne disease became a regular feature of life there, and social structure and land use became unsettled.[99] The 1004 treaty between the Song and the Liao explicitly abjured the two signatory regimes from moving watercourses. By the turn of the millennium, it was thinkable (as a matter of ideology) and feasible (via engineering expertise and labor deployment) to do just that.[100] Box 3.1 details some of the expertise, equipment, and techniques they utilized.

Flood management in the new era required monumental feats of labor exploitation to move earth, dredge waterways, tend animals, provision rations, and scour hills for building material. The corvée labor system allowed the government to impress civilian subjects into public works employment. Large waterworks projects assigned soldiers and civilians to labor by the tens of thousands for months on end, and the court sometimes pardoned prisoners in order for them to aid in waterworks labor.[101]

The engineering projects came at a price of hardship for the people who worked on them. As the river became both more unstable and stringently managed during this era of rapid erosion and high rainfall, riverbank residents were forced to labor at waterworks maintenance along with enlisted soldiers. Hard labor on the waterways was the most dangerous and terrible duty for civilians and prefectural soldiers alike. In 1016, numerous soldiers working in the water in the Yangtze delta prefecture of Hangzhou contracted disease, and many of them died. In 1023, at the height of summer, many troops assigned to waterworks labor died of thirst. In 1076, troops digging a canal in Huainan became so ill and weak that they could no longer handle the labor. Many had to be carried to the work site for required headcounts, and some died along the way. They were flogged, forced to pay extortion to their commanders, and often had their wages garnished. The documentary record includes accounts of waterways workers mutilating themselves to avoid service as soldiers or laborers.[102]

While laborers suffered in the trenches, magistrates and prefects in jurisdictions along the river, empaneled as ad hoc river levee commissioners, discovered that their career advancement depended in part on what transpired along the river within their jurisdictions. This encouraged local rather than systemic responses, and it fostered inconsistency between one district and the next. A new bureaucratic infrastructure emerged to gather, store, and protect material and mobilize workers. There were forty-eight storehouses in two prefectures near the capital in 1027.[103] Sometimes waterworks efforts were managed by the governors of quasi-provinces. Periodic imperial edicts demanded annual dike repairs and maintenance efforts on the part of regional officials and founded ad hoc regional offices to manage the efforts. As floods became more frequent, and waterworks more jerry-rigged and convoluted, the existing bureaucratic system proved poorly equipped to handle the situation. In 1058, the fourth Song emperor founded a directorate of waterways, the Dushuijian.[104]

Even though the river was a large-scale hydrological system, Song water engineering and statecraft were largely the purview of local officials, and their career incentives focused on their ability to collect taxes and maintain public order in the jurisdictions where they were billeted for three-year terms. The water, with its increasing sediment load, had to go somewhere. A given prefect's excellent levee maintenance simply accelerated the river's current through his jurisdiction and passed the problem along, exacerbating flooding further downstream. For instance, in 993 one local official created a diversionary canal near the capital in Huazhou. His efforts were effective but only caused more breaches downstream.[105] In 1004, a central court official was put in charge of inspecting dikes and writing reports about their conditions. The information in these documents reflected on personnel records of local officials.[106] Officials were promoted in part on the basis of the success of their waterworks efforts. The benchmark accomplishment was to efficiently maintain waterworks while limiting their cost in money and requisitioned labor. All county magistrates had to submit a waterworks audit each spring. Certain officials achieved renown based on good hydraulic works. However, at this time indirect incentives for local officials took the place of coordinated river management policy, which only sporadically came into focus.[107]

Chanzhou, located about 120 kilometers northeast of the capital, was the site of some of the most persistent breaches in the early eleventh century. In 1034, the river breached its embankments there and broke into several streams that captured the course of two Yangtze River tributaries (fig. 3.16). Divided in this fashion, the current slowed and the water settled, depositing sand and sediment on former farmland as it receded.[108] As the possibility of a major course change became more likely on the silty and perched river, the regime made the explicit decision that when breaches occurred, they should be directed north into the waterlogged Hebei on the Khitan frontier

Box 3.1
Managing the Floodplain in the Long Eleventh Century: Tools, Texts, and Techniques

Even as the Yellow River was becoming increasingly choked by silt and progressively more prone to flooding, engineers and other people who interacted with the river devised new means of living alongside it. The eleventh century opened with debates at court about whether to breach more levees and open more drainage canals in order to reduce the force of the river's current—in the language of the time, whether to "divide the water's power" (*fen shui shi*) to reduce flood risk—or whether to maintain and improve the levees in order to speed up the current and scour the sediment from the river bed even though this risked more catastrophic breaches.[1]

Levee technology improved dramatically throughout the eleventh century as hydrocrats learned new ways to protect cities, towns, farms, and the canal network from encroaching silt and sand. The court gradually created a river management bureaucracy at the capital to gather, store, and protect materials, mobilize workers, and develop regional-scale solutions. Engineers used sediment deposited by earlier floods to build setback dikes (*yaodi*) far from the stream. They built "sawtooths" (*juya*) and "horseheads" (*matou*), too—stone jetties placed across the current to break its flow. They installed fascines (*sao*), bundles of sorghum stalks used in lieu of increasingly scarce timber to protect banks and fill breaches, and "wooden dragons" (*mulong*), floating structures to protect dikes. In the 1070s, the famous political reformer Wang Anshi (1021–1086) empaneled the Yellow River Dredging Commission, which deployed an "iron dragon-claw silt dispersing wheel" (*tie longzhua yangni che*) to trawl up and down the river, keeping silt entrained in the current and flowing downstream instead of settling on the riverbed.[2]

By the end of the thirteenth century, the full preindustrial tool kit of techniques for harnessing rivers was in place. To manage seasonal differentials in flow rates, engineers devised dams with spillways. They captured water in reservoirs when water levels were low and disgorged excess water when levels were high. The reservoirs were connected to irrigation canals and drainage canals. Retention basins, fields surrounded with low dikes, could be seasonally inundated to water crops and to take the pressure off the river and the main dikes

Box 3.1 (continued)

when water was high. To facilitate transportation, the main channel could be canalized, with systems that included weirs (submerged embankments) aslant the main axis of the channel to level it, often paired with slipways and various types of locks that raised and lowered the water level for the purpose of transferring boats from one elevation to another. Where the main stream was not suitable for transportation, lateral transport canals with carefully controlled gradients ran parallel to it, bringing in water from tributaries or from the main stream itself. In the upper reaches of river valleys, lateral irrigation canals used gravity to draw water downhill through channels other than the main stream. Contour canals were lateral canals that terminated in a river other than the one in which they originated, and they were known as summit canals when they crossed a watershed range.[3]

The first hydraulic engineering treatises also date from this time. The first one, by one Li Chui, presented to the court in 1012 after a season of multiple levee breaches, was a three-chapter illustrated treatise called *Daohe xingsheng shu*. The title as a whole could be translated as "Documents about the Circumstances and Advantages of River Guidance."[4] As the title of this lost text implies, Li argued that it was possible to guide the river in accordance with its circumstances, but not to transform it fundamentally. Li Chui proposed dividing the river into multiple channels that would follow ancient courses, but his plan was rejected as prohibitively expensive and unfeasible.[5] Northern Song polymaths like Su Shi and Shen Kua discussed how to manage sluice gates for the purpose of moving silt-laden water out of the main channel and onto agricultural fields, and farmers demanded silt distribution at ideal times during the agricultural year.[6] Shen Kua also wrote about how to use fascines to repair levee breaches. He described one case in which engineers tried and failed to install a giant three-hundred-foot fascine rather than a collection of standard twenty-foot ones.[7] Song hydrologists developed a set of classification systems for communicating about types of floods, levels of water quality, and varieties of sediment deposition.[8]

Middle-period farmers and hydrocrats sought to put in place a stable system that would distribute both water and sediment in ways beneficial to society and to waterways. At this time, the guiding

> Box 3.1 (continued)
>
> principle of politics, river management, and frontier management was *anjing*, or stability. As the rate of levee breaches continued to accelerate, the state was forced into active management mode at a precarious and fractionated time.
>
> Statesmen and engineers also referenced the mythic postdiluvian river system as it existed during the reign of the legendary culture hero Yu the Great. The ideological and moral dimensions of the legend provided justification for an activist approach to river management, and the spatial and technical parts of the legend were a road map for approaches to river engineering.[9]
>
> ---
>
> 1. For the philosophical and technological background to this debate, see Mostern, "Mapping the Tracks of Yu." Some of the material in this section was previously published in that article.
> 2. Needham, "Hydraulic Engineering," 335.
> 3. Needham, "Hydraulic Engineering," 214–15.
> 4. I am intentionally, perhaps tendentiously, breaking up the common binome *xingsheng* (an advantageous or strategic position) and creating a less felicitous translation in order to make a point about the term *xing*.
> 5. Huanghe shuili weiyuanhui, *Huanghe dashi ji,* 89.
> 6. Needham, "Hydraulic Engineering," 230–31.
> 7. Needham, "Hydraulic Engineering," 342–43.
> 8. Huanghe shuili weiyuanhui, *Lidai zhi Huang wenxuan*.
> 9. Zhang, *The River, the Plain, and the State*, 107–40.

rather than south to the canal lands of the Huai basin, the most important transportation artery for the regime.

Tenth-century Song officials, courtiers, and hydrocrats, well aware that flooding was becoming more frequent, engaged in "forceful intervention" to produce a desired outcome.[109] Faced with an increasingly unruly and silty river, hydrocrats developed plans about what they wanted to accomplish and how to achieve the desired results. In the face of rapidly changing environmental conditions, the regime had not yet developed a consistent fiscal or bureaucratic response or a full tool kit of engineering practices. However, they were beginning to respond, attempting to mitigate floods and breaches as they occurred, to debate policy, and to try new technologies. Rethinking the colonization of the Loess Plateau was not considered an option. As Ling Zhang explains: "[The state] neglected the serious environmental

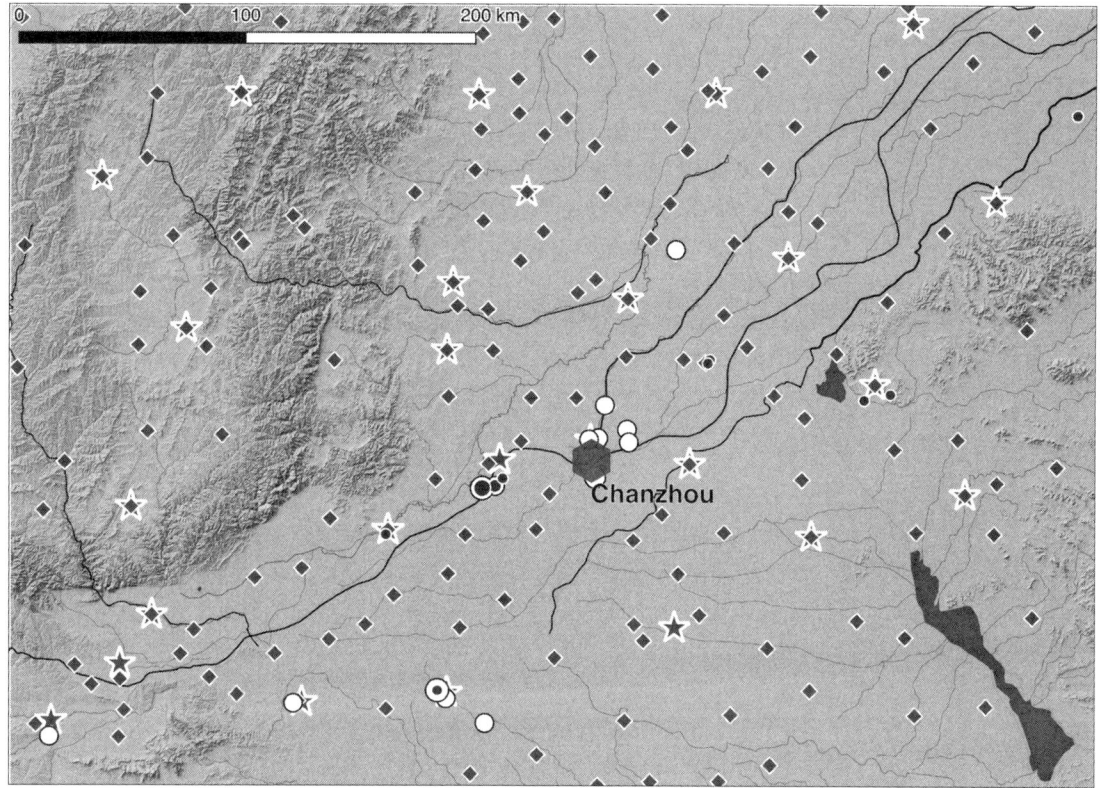

FIGURE 3.16
The 1034 Chanzhou Breach.
The Yellow River avulsed from Chanzhou, north of Kaifeng, in 1034.

problems that were then arising in the middle reaches of the Yellow River, such as the exacerbation of soil erosion and loss of control over water currents. Downstream, however, the consequences of these problems were less readily ignored."[110] Although there was a growing awareness of the problem of silt in the lower course of the river, there was not yet any coherent strategy for a response, other than a decision to disgorge water onto the already-waterlogged Hebei plain.[111]

This turned out to be a decision with long-term consequences. For the next three hundred years, the river would disperse across the floodplain and change course frequently. The canal and levee network would fill with sediment and become largely unusable, and the economic and social rise of the Yangtze delta would accelerate relative to the Yellow River basin. The situation of the floodplain at the end of the first millennium of the Common Era mirrored the one that had existed a thousand years before, but this time, the river carried an order of magnitude more sediment than it had then. Determining

LOESS IS MORE

how to excavate the canals and reconstruct a hydrosocial system on the floodplain would be the task of the next millennium.

THE FLOODPLAIN AWASH, 1048–1351

After four dry years, the year 1048 saw slightly above-average rainfall and the Yellow River breach its northern bank near Chanzhou. The river, engorged with sand and silt, made an unprecedented turn forty-five degrees to the north into Hebei, inundating dozens of counties, hundreds of cities and towns, and thousands of farms. From Chanzhou, the river took a route running north of the Shandong Peninsula.[112] Although nobody at the time could have predicted the precise timing or impact of the event, its occurrence and direction were the result of decades during which engineers and bureaucrats had been fortifying levees on the south bank of the river to protect the canals through the Huai basin and to plan for just this eventuality. Refugees fled the flood zone, and government officials, at odds with one another and unaccustomed to flood control, argued about policy, requisitioned funds, and embarked on a range of construction projects. Figure 3.17 shows the new course as an array of north-facing disasters and infrastructure initiatives that lasted until the Jurchen took control of the northern part of the floodplain in 1127 and established the Jin regime there. Figure 3.17 also depicts the middle of the eleventh century as a period of almost entirely unmanaged flooding.

Following sudden and significant disruption to the grasslands around the middle course of the Yellow River and a rapid increase in the amount of sediment flowing into the river, the river had changed course for the first time in eight hundred years. The regime took little action for eight years: historical sources report only one emergency repair effort in the vicinity of the breach in 1049. Some one million people died or fled Hebei, silt blocked lakes and tributaries throughout the province, and land became waterlogged and sandy. The whole hydrology of the province changed.[113] Mosquito-borne illness rose, and agriculture contracted. Hebei's share of the empire's population collapsed after the flood.[114] Refugees roamed the countryside and surged into cities, prompting anxiety about social instability and inspiring new and futile policies of social control to count and constrain the desperate and mobile population.

The human and environmental effects of the 1048 course change were immense and long-lived in a way that the catastrophes a

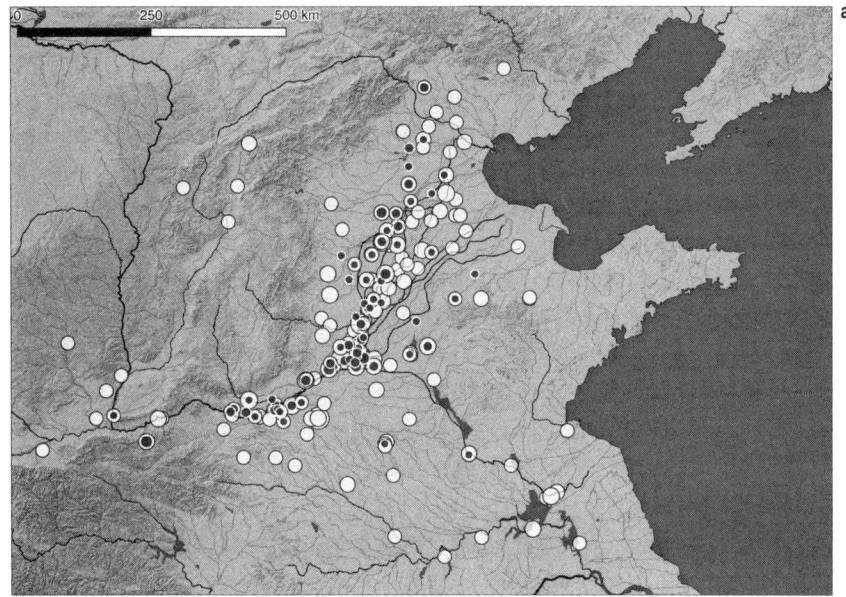
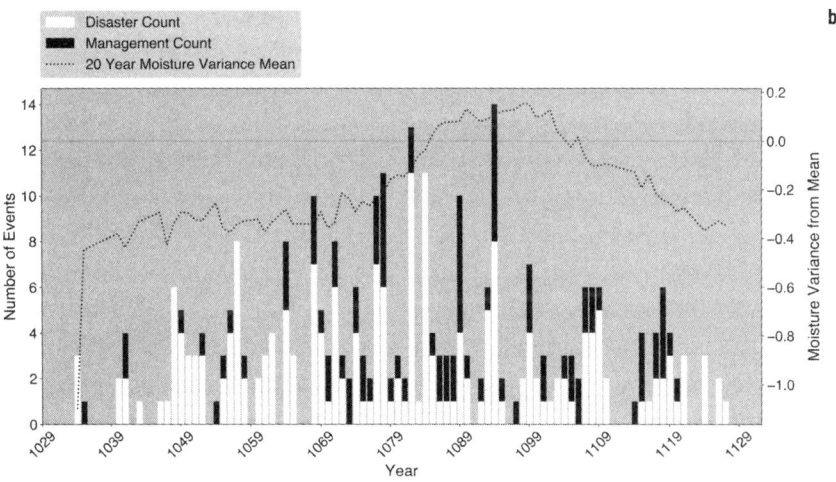

FIGURE 3.17

Map (a) and Timeline (b): The Floodplain, 1034–1127. Between 1034 and the end of the Northern Song regime in 1127, there was a high frequency of both disaster and management, associated primarily with the river's northern course.

millennium earlier during the low-sediment era had not been. Contemporaneous sources reveal that when the Yellow River flooded, its entrained gravel, sand, and loess sediment blocked the streams and creeks of Hebei, causing them to change course or dry up. They could not be dredged: the amount of material was too overwhelming. Canals, clogged with sediment, had to be abandoned, and the cities along their shores collapsed economically. Some streams carried more water than they had before, and new swamps formed in the land around the flood zone, areas that were also filled with silt and sand. Silt filled in old lakes and ponds while new ones emerged in low-lying, poorly drained places. Sandstorms became a feature of the region from that time onward.[115]

As a result of these short- and long-term effects, the population declined by as much as 90 percent through death and departure. Where residents had once practiced agriculture, those who remained turned to handicrafts, fishing, hunting, and gathering wild herbs. Tasked with gathering timber and other vegetation to build embankments, laborers deforested the flood zone. Residents were compelled to deliver immense quantities of timber and fascines. This further destabilized the very banks they were trying to build and exacerbated future erosion.[116]

The abandoned river course became a site of sand and dust, which also blew into cities and onto farmland. The course change rearranged sites of prosperity and sites of misery. The region around the former southern course saw population growth, prosperity, and the expansion of silk production in the eight decades after the course change, with capital and energy infused by refugees from the flood zone who discovered unclaimed fertile sites that had been newly liberated from the risk of inundation.[117] The course change also inspired new grassroots efforts to reengineer landscapes, and new kinds of local landscapes came into existence.

The 1048 flood changed the fluvial geography of the floodplain and launched a conversation about how the regime wished to coexist with the high-sediment waterway that flowed through its heartland. The flood also marked the dawn of what Zhang refers to as the environmental mode of consumption. From this time on, imperial regimes could opt to spend vast amounts of money to maintain subsistence and stability in the engineered and silt-clogged landscape they had created. However, no regime chose to do so for centuries to come. In fact, hydrocrats argued about whether to return the river to its former course, reinforce its new course through Henan, or divide it into many channels.[118] As we will see below and in the next chapter, it was not until the seventeenth century that the high-interventionist approach to floodplain management completely prevailed.

From 1059 onward, the Yellow River could no longer be used for grain transport because it was too prone to flooding and too clogged with sediment. The same was true of the canal network that linked the Northern Song capital at Kaifeng to the fertile rice farms of the south. At this moment the dominant elite discourse about the Yellow River shifted toward the late-imperial doctrine that the river was a scourge: unpredictable and hard to domesticate, useless for transportation, a massive sponge of labor and goods, and a carrier of sediment always at risk of clogging more useful waterways—natural streambeds and engineered canals alike.[119]

The statesman Ouyang Xiu, a prominent commentator on the Loess Plateau during the Tangut wars of the previous decade, wrote a series of essays in 1055 that drew a direct line between sediment accumulation and flooding. In the first essay, he criticized people who advocated engineering the river into its former course. He accused them of making ambitious plans without understanding geography or comprehending how silt deposit raised the channel bed. He explained that terrain determined drainage and that deposited silt had blocked the former course. He advocated strongly against spending vast sums from the state treasury and exploiting coerced corvée laborers to perform the futile task of returning the river to its former course, especially during a time of disaster and drought (*tianzai suihan zhi shi*). In a best-case scenario, workers would create a shallow, narrow channel that would be inadequate to the river's power (*shi*). "This is [even] beyond the abilities of Yu the Great," he exclaimed.[120] He was emphatic that the "fundamental nature" (*benxing*) of the problem was silt, which prevented the current from moving rapidly at times of high water and ultimately was the reason the river had changed course. Ouyang Xiu did not name the Loess Plateau as the origin of the sediment, but he wrote eloquently about the ways that rapid silt accumulation constrained future options for river management.[121]

Likewise, after a 1056 levee-building debacle that resulted in an earthwork that held for only one day, then emperor Shenzong demanded that the river improvers "go with the flow" (*shunshui*) when it came to river management. Sediment was never effectively handled at its source. On the floodplain, the Bureau of Rivers and Canals began to manage river statecraft and engineering under the aegis of the Finance Commission. Local circumstances came to be sidelined in favor of large-scale solutions, but these never included serious discussion of circumstances on the Loess Plateau.[122]

The river did not stabilize after the 1048 course change. Even as its path through Hebei remained unmanaged and the consequences of the flood were not ameliorated, its flow broke into two branches in 1077. One flowed south of the Shandong Peninsula and disgorged into the Huai River. The other continued to follow a northerly route through the Hai River sub-basin. The full northern course, restored a year later, held for only a few decades.[123]

By the early twelfth century, several ways of thinking about the river were fully ingrained. One was that it flooded frequently and sometimes catastrophically. Another was that people could direct it into particular channels as needed. A third was that military considerations were one objective for directing waterways in particular fashions. In the winter of 1128, after the defeat of Hebei, the Kaifeng governor and general Du Chong (d. 1141) sought to halt the southern advance of the Jurchen. Former vassals of both the Khitan Liao and the Goryeo regime in Korea, the Jurchens declared the regime name of Jin in 1115, overthrew the Khitan Liao in 1120, and captured Kaifeng in 1127. As Du Chong and his troops retreated south to a redoubt on the southern banks of the Yangtze in what is now the modern city of Nanjing, they intentionally breached the south bank dikes and dams near the capital, diverting the Yellow River into the course of a Huai tributary to slow the invading Jin cavalry. All the waters flowed to the south of the Shandong Peninsula, taking a new course into the Huai, and devastating farmland, cities, and communities of the vast plain in between.[124] The Yellow River resumed its southern course. The Huai River, overwhelmed by the silty water of the Yellow River and no longer able to reach the sea, backwashed to flood the swamps and lakes of its former course, expanding Hongze Lake and unleashing environmental catastrophes with short- and long-term consequences.

The most unstable era in all of recorded floodplain history began with the 1128 breach. From then until 1276, the Huai River marked an international boundary. The Huai basin became both a frequent theater of war and a site of persistent flooding, and the canal transportation routes that transected it ceased to be maintained. As figure 3.18 shows, following the 1128 breach, the river took a series of multiple and unimpeded southern courses through the low-lying, marshy Huai basin. The Huai River, by then incorporating the channel of the Yellow River as well, formed the border between the Song and Jin

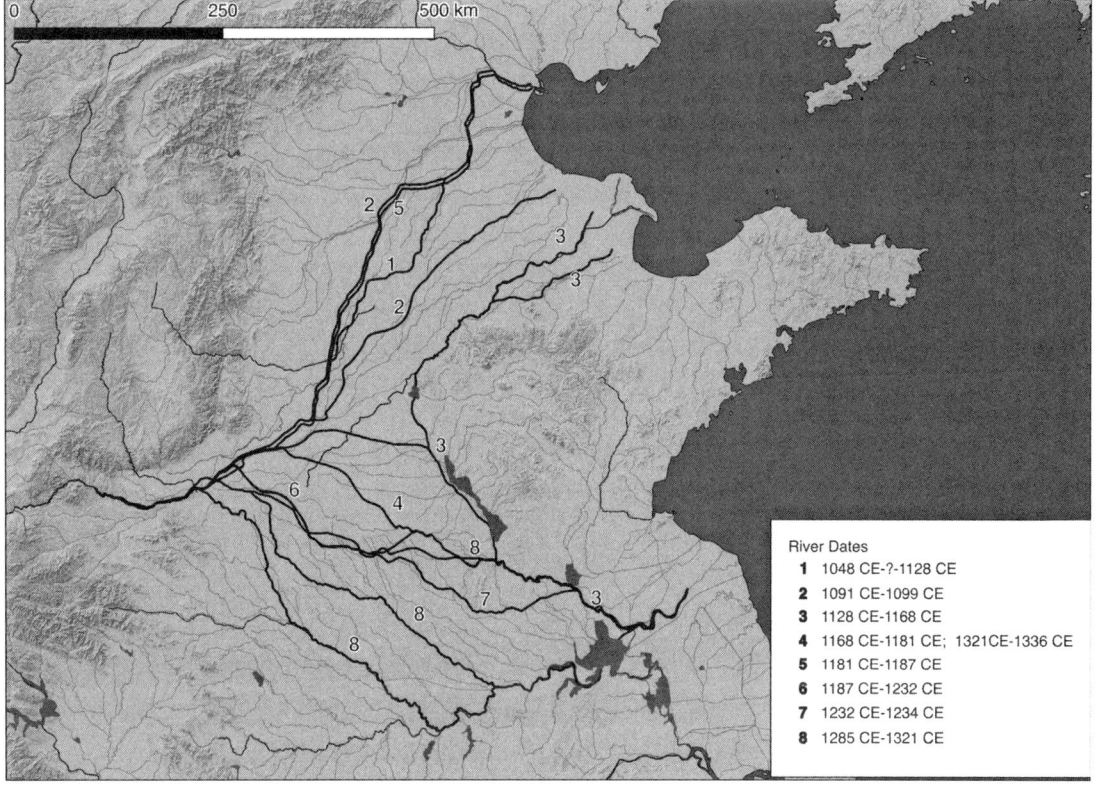

FIGURE 3.18
The Yellow River Floodplain during an Era of Division. The river changed course frequently between 1048 and 1321, and it occupied multiple courses during much of that time, including simultaneous courses north and south of the Shandong Peninsula in the middle of the twelfth century.

regimes. The marshlands to the south of the Huai were the battlefield between the two for decades. The river was untrammeled and unmanaged. It changed course frequently, flooded annually, and frequently occupied multiple different paths to the sea at the same time, depositing sand, silt, and gravel along the way. In some places, sandy sediment covered fertile farmland and clogged springs and tributaries. In others, stagnant pools bred disease and caused salinity. Water courses routinely overtopped their banks and spread across the plain with little effective intervention. According to the "Rivers and Canals Monograph" of the *History of the Jin*, "For several decades . . . after Jin overcame Song, sometimes the river breached [its embankments] and sometimes [the breach was] filled in, and it migrated without taking an established [course]."[125]

With political control split between multiple regimes governing northern and southern China, the canal network—overcome with silt, and an outdated infrastructure with transportation between north

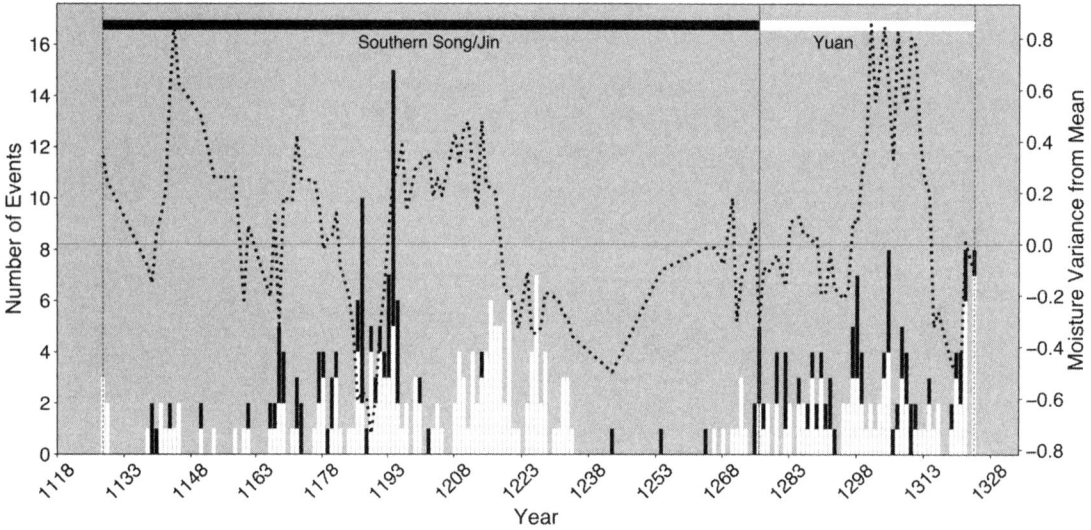

FIGURE 3.19
Timeline of Floodplain Events, 1128–1234. There are few discrete management and disaster events attested during some of the river's most tumultuous eras, during which the floodplain was a battlefield in the mid-twelfth century and the mid-thirteenth century.

and south curtailed—fell into disuse. Along with the breakdown of engineered waterworks, careful documentation of flood events collapsed as well. As figure 3.19 reflects, this is the most poorly documented era of floodplain history during the imperial era. Despite the visible misery and chaos in long-term population trends, for example, the historical record records only a small number of disaster and management events. This is the period in which the methodology of this book to count and analyze the attestations of river-related events is least reflective of the reality of river history. The river had reverted to an ancient pattern, meandering freely across sparsely populated wetlands; periods of high water were not remarkable and were not attested as distinct disastrous events, since urban and agrarian life had become precarious and unpredictable in general. These decades, among the most disruptive in imperial times in terms of people's livelihoods and political economy, are not reflected that way in the annals of disaster history. Figure 3.20 depicts the floods, avulsions, and breaches on the floodplain between 1128 and 1300, including the lack of documentation during earlier decades and the later expansion of management events and documentation.

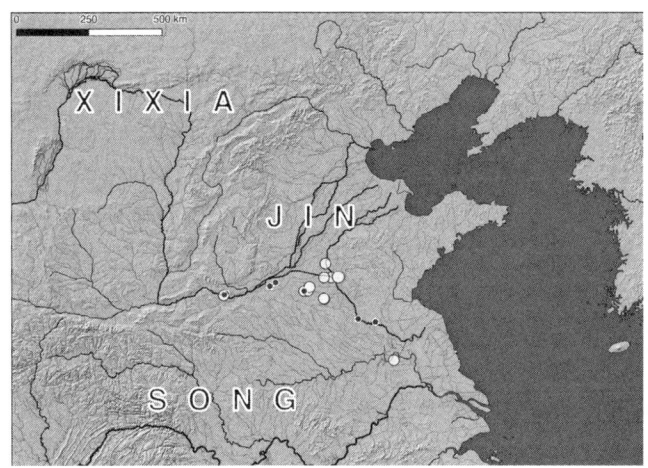

FIGURE 3.20

The Floodplain, 1128 to 1368. The period from 1128 to 1276, when the floodplain was divided between regimes and canals fell into disuse, are poorly documented in the historical records. Management and scrutiny resumed after the unification of the realm in the late thirteenth century. Figure (a) depicts the years 1128 to 1142. Figure (b) depicts 1142 to 1176. Figure (c) depicts 1176 to 1368.

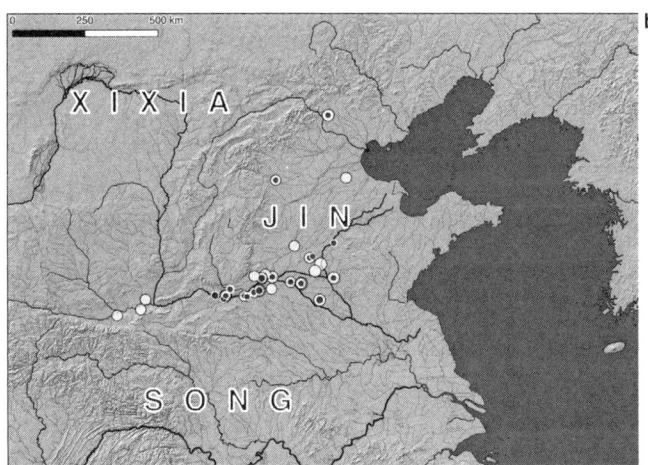

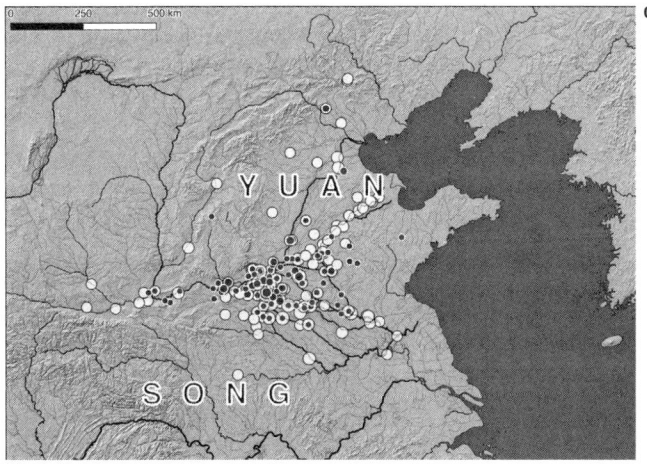

Flooding and fighting proved a deadly combination for the residents of the Huai basin. The two Song provincial circuits south of the Huai, with a combined population of more than 1 million households in 1102, had only 150,000 households in 1162. Even a century after the Jin invasion, the population had rebounded to only one-third of its prewar level.[126] The area of cultivated land, 120,000 *qing* at the end of the Northern Song, was only 3,000 *qing* in the 1170s. Famine and depopulation remained widespread.[127] Dozens of county and prefecture seats were temporarily abandoned.[128] By Yuan times, the Huai basin, once prosperous and densely populated, was best known for poverty and banditry. Large regions of northern China, both around the abandoned northerly courses in Hebei and around the new course in Huaibei, were set on a path of sandification, tributary extinction, waterlogging, and peripheralization. Millions of people died or lost their homes and land. Waterworks were reduced to ruin, the floodplain became barren, and Hongze Lake continued to expand and to flood frequently. From then onward, the floodplain "suffered a long-term economic decline."[129]

In spite of all this, there were efforts to govern the floodplain. The Jin regime innovated a military-style river management system that included twenty-five locations set for installing stalk fascines, each one overseen by a patrol officer (*xunheguan*), with a superior officer overseeing every four or five sites. Twelve thousand soldiers were assigned to river management work on the Yellow River floodplain. Their efforts required more than one million bundles of wood every year and close to two million bundles of stalks.[130]

Jin river management debates focused on subdividing the river in order to minimize heavy expenses on hydraulic works and determine locations for breaching the river's dikes to create the new courses in places that would minimize damage to cities and farmlands that lay in the way of planned new routes. The goal was to institutionalize and stabilize a multiple-course river network and not to maximize population and agricultural production or to support a canal network. In other words, the regime did not favor levee construction. As the floodplain population had dispersed and canal transport had become obsolete, policy pointed away from flood prevention and toward harm reduction. The activities that constituted good river management shifted from era to era.

The principle that governed these discussions was known as *fen sha qi shi*: "to divide [the river] in order to extinguish its power." After one occasion when the river breached in 1168, one Jin interlocutor proposed to make it policy for the river to be divided between its two courses to the sea, with 60 percent of the water directed to the new course and 40 percent to the new course. The emperor agreed to the plan and requisitioned labor to strengthen the southern levees.[131] In 1193, another regional official observed that in the past, the Yellow River had branches and discharge ponds off its southern bank, and he proposed channeling it into these historical courses, noting that "if it were possible to lead it off in such a way that it spread out, this would be enough to drain away the power [*shi*] of the river." In 1194, the southern levees were intentionally breached once again, this time by Jin forces seeking to contain Song forces to the south of the Huai.[132]

In 1234, the Mongol Yuan regime took control of north China. The Mongol conquest included yet another intentional breach, this one in eastern Hebei in 1232, to inundate Jin troops, a tactic that inflicted high casualties.[133] The river course shifted again in 1233, persistent breaches occurred around the avulsion site, and the invaders breached the levees again in 1234, this time near Kaifeng.[134] The Yuan rulers now had authority over a Yellow River that spread out, with its entrained sediment, over multiple slow channels across the alluvial plain. Defensive dikes and diversion canals only protected particular fields and population centers and not the whole river course.[135]

Governing a continent-spanning, multicultural empire, the Mongols commissioned the first scientific expedition to the headwaters of the Yellow River. In 1279, a year after the conquest of China, Khubilai Khan commissioned an officer to lead an experienced group of Jurchens and Uzbeks west from the city of Lanzhou at the foot of the Liupan Mountains to Amdo in Tibet. They were charged with finding the river's source and reporting back their discoveries. Although earlier expeditions had traveled to the western regions, this was the first that was tasked specifically with seeking the river's source. After the travelers returned, an official and author named Pan Angxiao (fl. 1289–1302) wrote a short book entitled *A Monograph of the River's Origin* (*Heyuan zhi*).[136] This text marks a break with prior traditions on how to depict the northwest. An earlier trope, which extended throughout the first millennium of the Common Era, designates the origin of the river according to the waterways at its source. From the fourteenth century onward, it is structured instead around a desert zone.[137] The Yuan expedition was a notable exception to the general rule of the imperial era that the Yellow River watershed was often appertained as a series of disconnected parts.

The new regime established its capital at Beijing, at the northern perimeter of the floodplain. At that time, the canal network had become unusable, and goods from the south were being transported by sea. Once northern and southern China were unified under a single regime after 1276, the equation changed completely. The new Mongol regime took great efforts to restore the Grand Canal and to reengineer the neglected floodplain to ensure its stable and predictable functioning. As figure 3.21a reflects, the Yuan regime revived the tradition of carefully documenting floodplain events. Figure 3.21b shows that this was a period of ambitious floodplain management. Khubilai Khan commissioned Guo Shoujing (1231–1316), an inventor, astronomer, mathematician, engineer, and leading polymath in an era of Eurasian scientific interchange, to design reservoirs and canals around the new capital at Beijing to provision water there and to survey the floodplain between the capital and the Yellow River in order to plan for a transportation canal. The long-term results of his proposal are a subject of the next chapter.[138]

CONCLUSION

This chapter has described the most significant turning point in Yellow River history. It is a story that pivots, again and again, on warfare. The An Lushan Rebellion set the Yellow River on a course of frequent flooding. Later conflicts only accelerated the process. The Song–Xi Xia conflict wreaked unprecedented havoc on the forests and grasslands of the central Loess Plateau and the Liupan Mountains, and the Song-Jin and Jin-Yuan conflicts destroyed floodplain waterworks.

Upstream, under conditions of deforestation, these centuries mark the beginning of an era of rapid erosion and high sediment deposition. Downstream, on a canal-crossed floodplain, historical annals and soil cores alike bring the inception of the disaster era into precise, vivid focus. Political responses to the new reality took hold only slowly, amid a profound degree of human suffering and labor exploitation. Hydrocrats launched numerous experiments and wrote

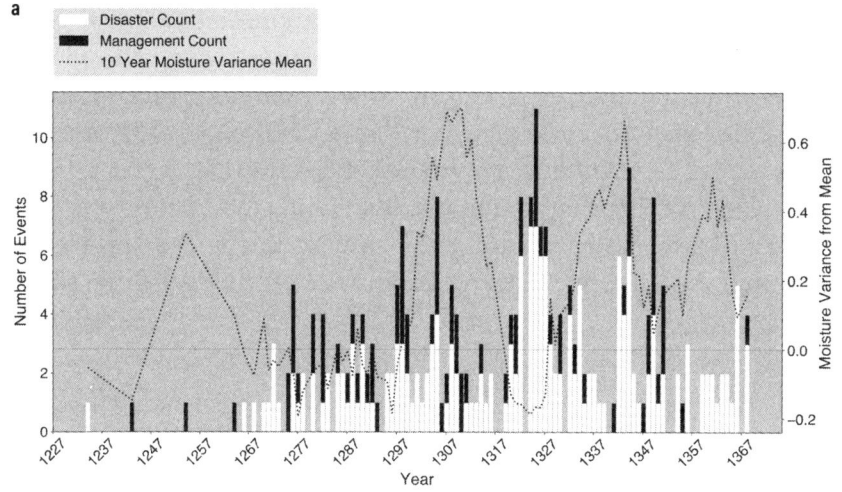
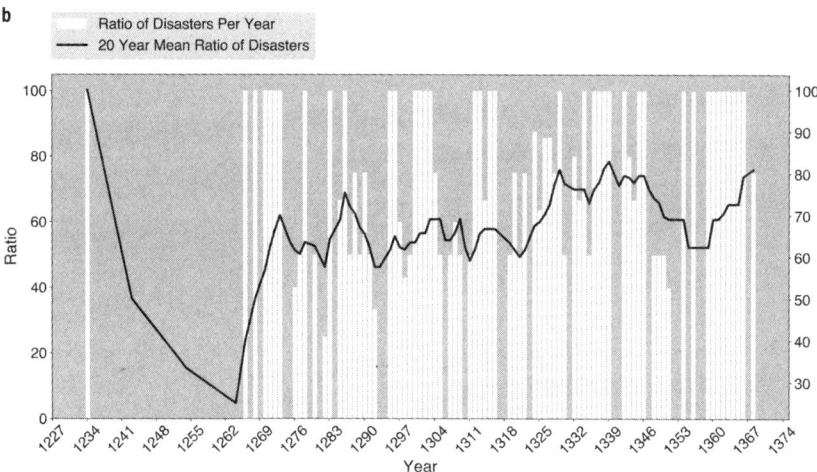

FIGURE 3.21

Floodplain Events, 1234–1368. With the reunification of the Yellow River basin under the control of the Mongol Yuan regime in 1276, attention to floodplain events and to waterworks management resumed. Figure (a) depicts the number and type of events that were recorded each year, and Figure (b) depicts the ratio between management events and disaster events.

plenty of texts, but the regime was caught by surprise, fragmented and localized. Systematic policy emerged only after the division of the empire among multiple regimes and the abandonment of the transportation and irrigation canals, when the Jin regime that had founded a military bureau for river control settled on an approach that favored a low floodplain population and a dispersed wetlands environment.

Just as the Tang-Song transition has not been studied from an environmental angle, so too historical writing about the emergence of the south as China's center of wealth, population, and urban life generally points to its natural advantages. These are significant indeed. South of the Yangtze, farmers can harvest multiple rice crops each year and transport it easily along a dense canal network. However, this was also an era of multiple rounds of collapse in the north, from the decline of Chang'an during the Tang to the inundation and sandification of the alluvial plain. Ecological decline in the north goes hand in hand with the transformation of the south during the same centuries.

CHAPTER 4

Levies and Levees: The Engineered River, 1351–1855

The final life span of the imperial river was its shortest one, but as the most recent one, it looms largest in the historical imagination. In this era erosion from the Loess Plateau rose again by an order of magnitude, and the pressures of massive sediment deposition on the floodplain gave rise to both catastrophic floods and extraordinary accomplishments in hydrological engineering. This life span of the Yellow River began in the middle of the fourteenth century and ended in the middle of the nineteenth century. It is indelibly linked with the rise and fall of the Grand Canal (Dayunhe), the final iteration of the transportation networks that traversed the floodplain.

Throughout these centuries, the imperial capital lay for the first time at the northeastern periphery of the Yellow River basin, in the city of Beijing. To its west and northwest, Beijing is sheltered by mountain ranges that form the boundary of the Yellow River watershed. Short streams north and south of the city join the Hai River and drain to the Bohai Gulf north of the Shandong Peninsula. The coastline is about 150 kilometers from the capital. The North China Plain stretches south from Beijing. Figure 4.1 depicts this terrain.

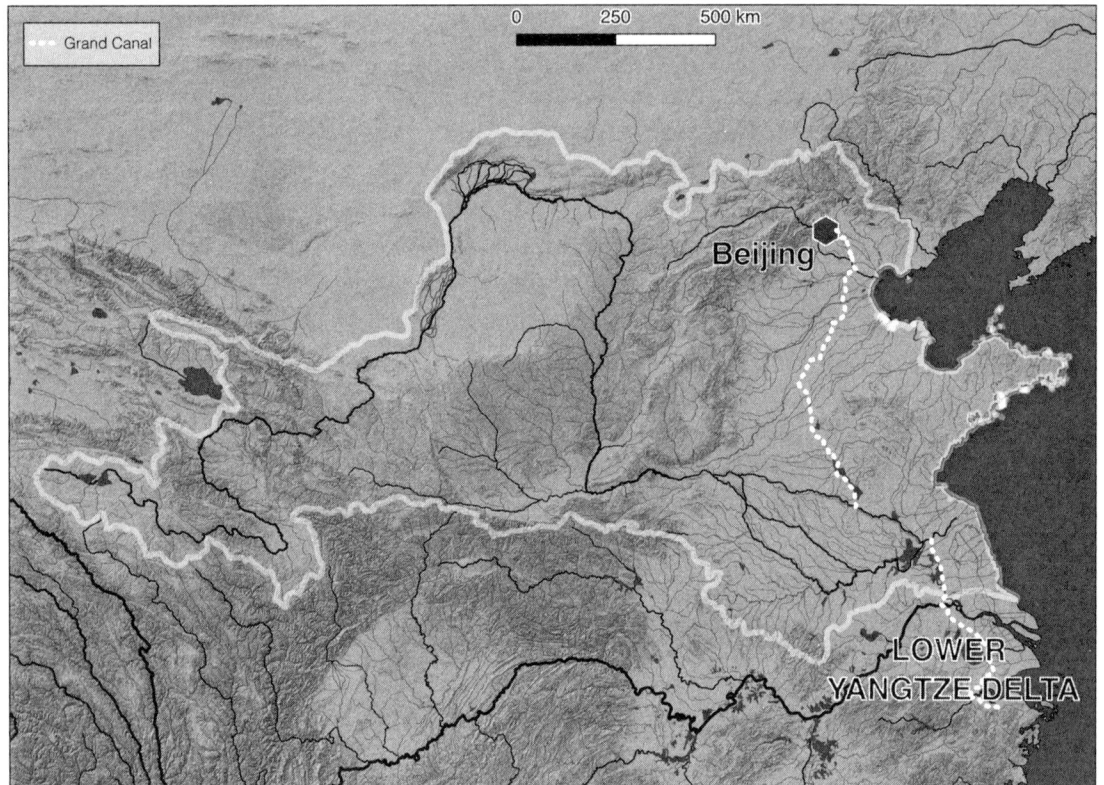

FIGURE 4.1
Beijing and the Grand Canal. In its early centuries, Grand Canal transportation combined travel on the engineered waterworks with transit on existing water courses. These shallow courses across flat terrain changed with the seasons and carried silt whenever the Yellow River overtopped or breached its levees.

With the capital located in the northeast, distant from any previous sites, the geography of political scrutiny and state investment shifted from the middle reaches of the river and the central expanse of the floodplain. The foodshed for Beijing lay hundreds of kilometers to the south in the Yangtze delta, where initiatives to build drainage, irrigation, and transportation networks fostered a commercial economy and a large population of craftspeople and farmers who could harvest up to three crops of rice each year.

Every late-imperial regime expended great effort to design and maintain the vast infrastructure known as the Grand Canal, literally "the Great Transport Watercourse," for the purpose of conveying resources north to the capital from the empire's center of wealth, culture, and population in the lower Yangtze. Despite its name, the Grand Canal was not really a single waterway; rather, the term was shorthand for an exceedingly complex network of personnel and funding regulations, labor and expertise, levees and other earthworks, drainage and transportation canals, locks and sluices, and reservoirs and wetlands. The Grand Canal system preoccupied the court, crisscrossed the Yellow River floodplain, and transected the river itself.

From the perspective of Yellow River history, the turn of imperial geography away from the Loess Plateau and toward the floodplain and its peripheries meant that, to an even greater extent than in earlier periods, the river was never treated as a single fluvial system during late-imperial times. Meanwhile, the erosion rate from the plateau continued to rise. Sediment, increasingly dominated by sand and gravel rather than fertile soil as ecological conditions deteriorated, clogged even the best-engineered river courses and canals. The vast floodplain waterworks system was designed with the requirements of the Grand Canal at its core. However, the infrastructure of embankments, reservoirs, and drainage channels transformed the risks and opportunities for farming, commerce, urban life, and transportation throughout the floodplain. As infrastructure investment revenues flowed to the floodplain, canal-side merchants and construction contractors flourished, even as farmers lived on sandy and saline plots in increasingly precarious circumstances and southern taxpayers grew increasingly resentful over subsidizing a region far from home. Upstream, a fifteenth- and sixteenth-century era of military exigency and compulsory garrison agriculture gave way to a subsequent period in which large civilian populations—settler colonists and their descendants who received tax incentives to cultivate even the most ecologically fragile and marginal land—relied on new dryland crops. There, generations of farmers and merchants made their homes in eroded and arid terrain. As conditions further deteriorated, few could emigrate or alter their mode of subsistence.

Figure 4.2 depicts moisture levels plummeting from a moist fourteenth century to an exceptionally dry period that spanned decades around the turn of the sixteenth century. The political crises of the mid-seventeenth century included a devastating drought of epic proportions. Moisture levels during the final three centuries of the imperial era fluctuated around the long-term historical mean. Upstream, there were two peaks of erosion-causing activity during this same period. The first coincided with Chinese recolonization of Mongol territory in the late fourteenth century and culminated with the construction of the Great Wall across the Ordos in the 1470s. The Chinese military retreated from the Ordos in the early seventeenth

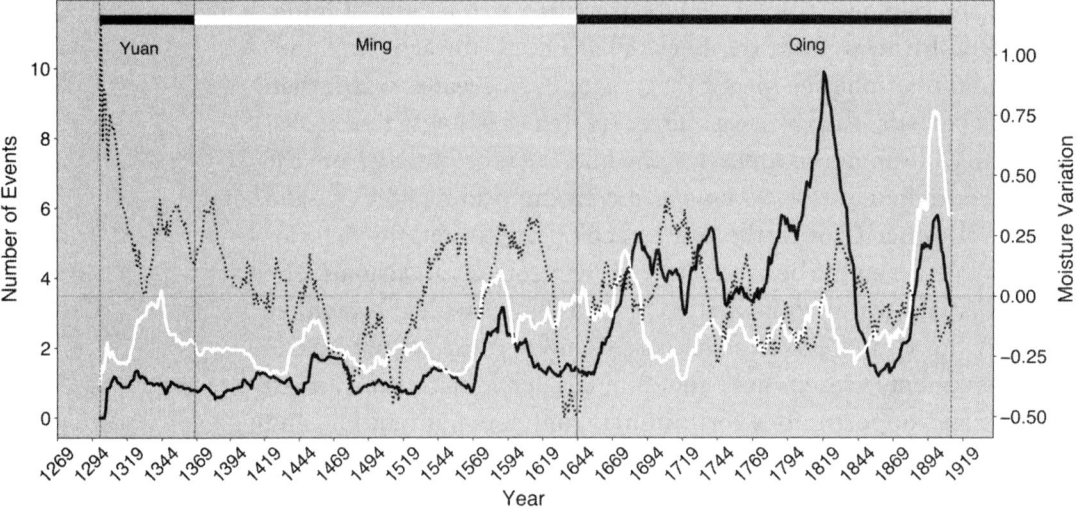

FIGURE 4.2
Events and Moisture Levels, 1300 to 1911. Until the seventeenth century, the rate of attested management events and the rate of reported disaster events tracked fairly closely to one another, with the number of management events consistently below the number of disaster events. That relationship changed dramatically in the late seventeenth century, the beginning of more than a century of vigorous infrastructure development that successfully suppressed flood rates. The moisture timeline and the event timeline are seldom related during these centuries.

century under escalating pressures of desertification, a major earthquake, and fiscal collapse. Late in the century, the new Manchu Qing regime launched a policy that offered tax breaks to agricultural colonists, who received incentives to plant new imported crops—maize and tubers—on dry and upland regions that had still maintained their original ground cover. Populations more than doubled.

Downstream there were three periods in which the proponents of aggressive engineering interventions prevailed in court debates over rivals who argued for less investment in the floodplain and more in ocean transportation (fig. 4.2). The first was in the mid-fourteenth century, the second was in the late sixteenth century, and the third was in the long eighteenth century. As figure 4.3 reveals, the eighteenth-century version of the hydrosocial floodplain involved a level of interventionist engineering that was entirely unprecedented in Yellow River history.

There was no single normative version of success on the floodplain. Each round of support for earthworks construction initiatives over wetlands maintenance represented a victory for one side in a debate that stretched back to the mythic era of Yu the Great. Each such incident produced an increasingly manufactured floodplain

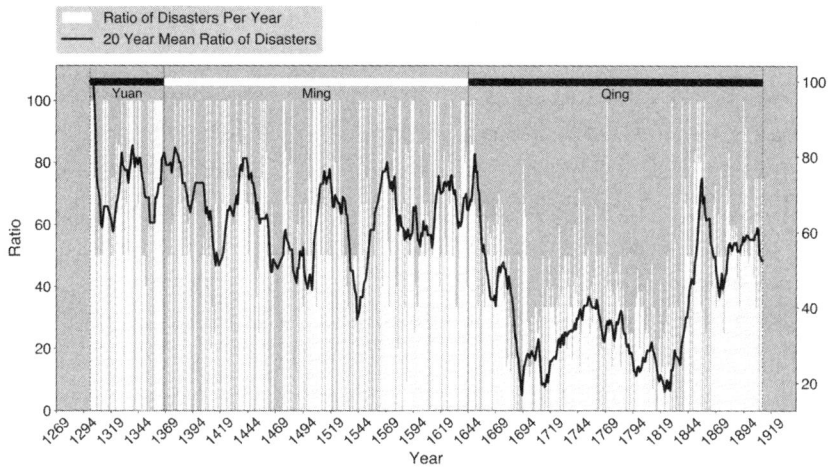

FIGURE 4.3
Disaster to Management Ratio, 1300–1911. During the long eighteenth century, the Qing regime paid an unprecedented amount of attention to floodplain management.

ecology, always entangled with the Grand Canal and transportation to the capital. Each time, at a profound cost of lives and money, the floodplain eventually devolved back to a version of the primeval wetlands system. The historical Yellow River of contemporary imagination, locked entirely behind levees and venting frequently in devastating floods, actually existed for only a few decades in the aging nineteenth-century waterworks system.

The final life span of Yellow River history exemplifies the high-revenue, high-investment model of early modernity that spanned Eurasia, transforming its ecologies and its societies. Nature was increasingly commodified both as private property and as a source of marketable resources in what Kenneth Pomeranz refers to as a "developmentalist project," a package of unprecedented state power, sedentarization of mobile peasants and nomadic herders, and intensive resource exploitation.[1] During the early modern centuries that followed the Mongol Empire and the Black Death, powerful governments used incentives and coercion to pressure people to replace the flora and fauna of forests and marshes with their own preferred species, to expand their populations, and to use land in increasingly intensive ways.[2] As territorial states grew larger and stronger and collected more surplus as tax revenue, they managed finally to vanquish other kinds of polities, including the pastoralist societies that had long occupied the Ordos and that until then had been able to successfully resist extermination and expropriation. Agrarian empires became paramount within their own borders. They fielded bigger armies, maintained more records, and collected more revenues. Growing and militarily secure populations pushed agriculture into peripheries such as hillslopes, wetlands, and forests—precisely the places that either were prone to erosion or had until that point served as drainage sinks in times of high water, and for those very reasons had not previously been exploited.[3]

The new farmlands were marginal places that only the state could love. As Pomeranz puts it, the late-imperial Chinese state also promoted "notions of statecraft in which the central government actively prop[ped] up regions in which the ability to live a normative family life [was] economically and ecologically vulnerable, while intervening less in the economies of richer areas (except to tax them)."[4] Throughout the Yellow River basin—on the floodplain, in Henan, and on the Ordos, "this economic vision created a particular idea of environmental management in which the central task within China proper was the reproduction of the material basis of modest agrarian prosperity in ever-more-precarious ecological conditions as Chinese settlers moved up hillsides, pressed closer to riverbanks, lowered water tables, and moved onto marginal lands."[5]

THE ORDOS ENCLOSED: GREAT WALLS AND NEW CROPS

The large-scale transitions of political history, including changing notions of political economy and evolving ideas about the imperial spatial imaginary, were highly consequential for the late-imperial Loess Plateau. The Mongol Yuan regime (1234–1368) supported a small, multicultural population that practiced diverse forms of subsistence. The Ming rulers (1368–1644) mounted a massive colonization campaign on the Ordos, successfully extirpated its indigenous pastoralist residents, and moved in a large settler population. The Loess Plateau was not a site of military conflict during the Manchu Qing regime (1644–1911); however, the court aggressively supported the colonization of new agricultural land and the commodification of timber, salt, cotton, and other commercial crops, and other goods.

Following the Jurchen Jin defeat of the Northern Song in 1127, the Ordos and the rest of the Loess Plateau experienced almost three centuries during which human populations and farms were sparse and regimes maintained few or no fortifications. As table 4.1 shows,

TABLE 4.1
Late Imperial Population in Shaanxi

Year	Regime	Population	Percentage of total imperial population
1102	Song	2,847,009	6.28
1312	Yuan	449,045	0.75
1393		1,805,661	2.71
1491	Ming	3,048,057	5.20
1578		3,505,791	5.25
1783	Qing	8,260,000	3.00
1854		12,059,000	2.92

Source: Cao, *Shaanxi shengzhi renkou zhi*, 331.
Note: Note the dramatic population reduction in the Yuan and the rise in the Qing. Shaanxi populations did not exceed their early imperial peak until the late eighteenth century. Consistent with these numbers for Shaanxi, another source estimates a total Loess Plateau population of 14,732,742 in the middle Ming dynasty and 40,315,284 in 1820.

the Mongol-era population in Shaanxi contracted to a number lower than it had been at any time in recorded history.[6]

During the Ming era, agricultural colonization and military ambition drove the Ordos population far above its previous historical peak. The population of Shaanxi nearly tripled between the census of 1393 and the census of 1491. This is despite the fact that Chang'an, by then known as Xi'an, was no longer an imperial capital. Indeed, it was no longer a major metropolis at all. During an era of robust commercial exchange centered far to the southeast, Xi'an was too isolated from the rest of the imperial transportation network, and its foodshed and timbershed were inadequate. In 1556, an earthquake, estimated at a magnitude of around 7.9 on the modern Richter scale, struck the Wei River plain. According to imperial records, approximately 830,000 Loess Plateau residents lost their lives.[7] Soon afterward, in the early seventeenth century, catastrophic drought coupled with high erosion drove desertification that shuttered many of the northerly Ordos settlements and garrisons. The Ming regime fell in the mid-seventeenth century. Population and state investment skyrocketed thereafter. The nineteenth-century population of Shaanxi was more than triple that of its Ming peak, and the ecological consequences of this were predictable. The early modern story of the Loess

FIGURE 4.4
Loess Plateau Political Geography Timeline, 1127–1644. Although the number of counties on the Loess Plateau declined somewhat between the twelfth and seventeenth centuries, the number of settlements outside the administrative system increased to an unprecedented number.

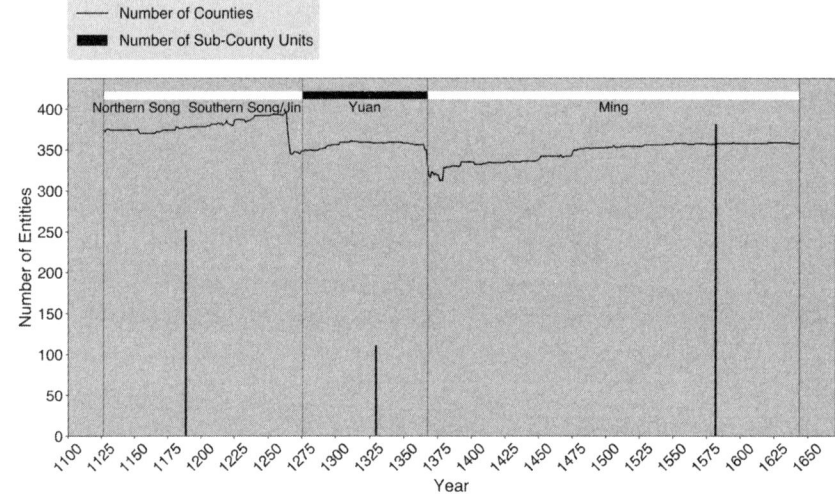

Plateau is one of complex currents in human activity entangled with contingencies of climate, geology, biology, and geopolitics.

The defeat of the Northern Song in the early twelfth century ushered in more than two centuries of Central Asian rule in northern China. Fortification ended and agriculture receded in the middle reaches of the Yellow River. Most of the Song roads, walls, forts, settlements, and farms were abandoned. Small, mobile, multiethnic populations of pastoralists and salt miners reclaimed the region with their herd animals. Intact forests persisted through this period and into the early Ming. Those that had been burned or logged rebounded, although the timeframe was too short to rehabilitate prehistoric soils or restore the deeply gullied terrain. Jin and Yuan sources describe bamboo cutting on the Guanzhong Plain, but there is otherwise little reference to timbering.[8] Figure 4.4, a timeline of settlements on the Loess Plateau between 1127 and 1644, shows that Yuan-era counties and nonadministrative units dropped far below historical levels, and it depicts the expansion of those sites by the late sixteenth century. Figure 4.5a depicts the distribution of counties, prefectures, and other settlements in 1330, and Figure 4.5b shows the distribution of those settlements in 1582, with the Yinchuan Plain densely colonized, an arc of garrisons and prefectures following the line of the Great Wall across the Ordos, strings of counties along all the tributary rivers of the Loess Plateau, but less settlement in Guanzhong than there had been in previous eras.

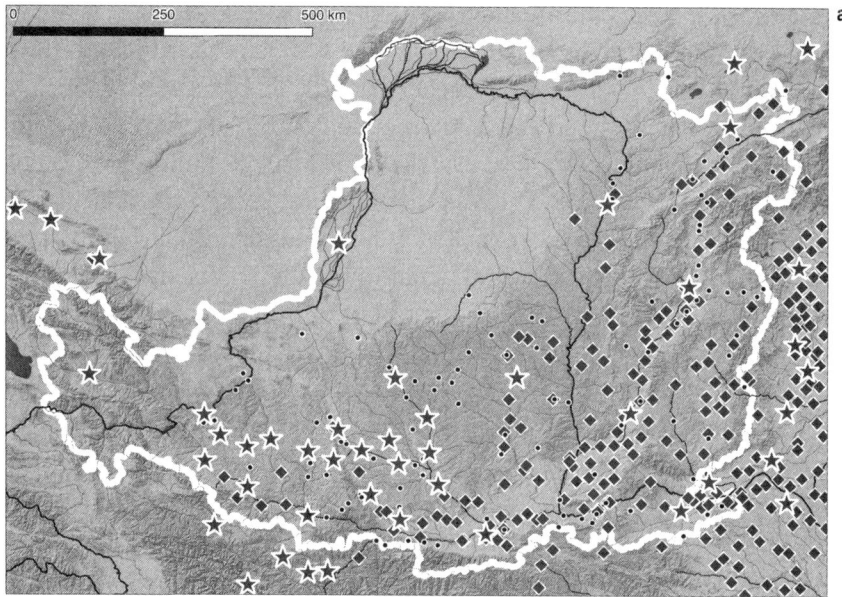

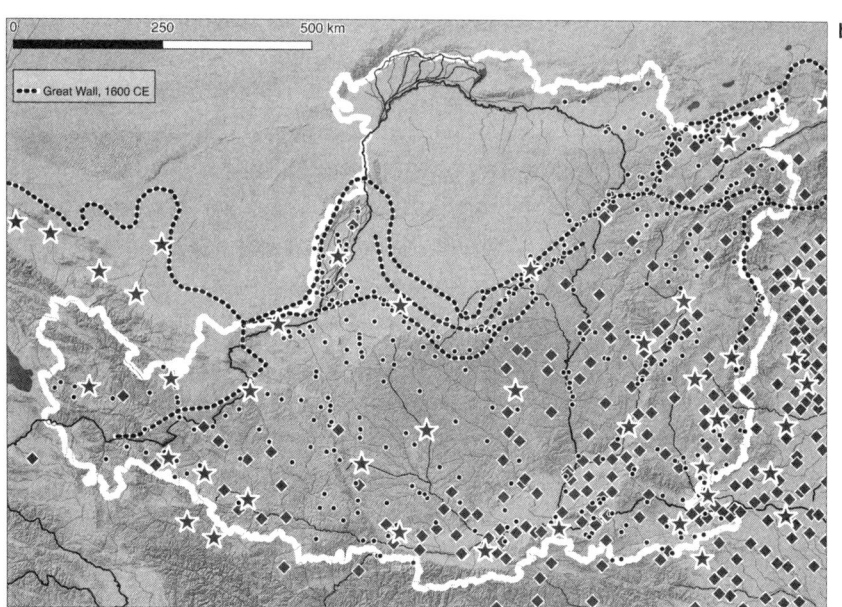

a FIGURE 4.5
Loess Plateau Political Geography during the Ming Dynasty. Along with a renewed campaign to establish forts and prefectures across the central Ordos, the three parallel lines of the Great Wall brought hundreds of thousands of soldiers and laborers to the fragile terrain at the edge of the Mu Us Desert. Figure (a) depicts political geography in 1330, near the end of the Yuan era. Figure (b) depicts the political geography from 1582 and the wall network from around 1600, reflecting a significant expansion of settlements on the western part of the Loess Plateau as well as along the Great Wall.

The Yuan government faltered badly in the mid-fourteenth century amid epidemics, failures in governance and fiscal policy, and cataclysmic social unrest. In 1368, the rebel leader Zhu Yuanzhang captured the capital and declared the Ming regime. On the Ordos, adjacent to the Mongolian homeland, internal power struggles among Mongol clans added to the regime's difficulties in maintaining suzerainty. Even before the Mongols were expelled from China in 1368, the descendants of Chinggis Khan had difficulty maintaining hegemony among their people. The frontier between agrarian Ming China and the nomadic Mongol lands was a zone of intense interaction. Populations from multiple groups—various clans of Mongols, Han Chinese, and their allies—resided in one another's territories, traded extensively, sporadically contended for hegemony, and lived in culturally hybrid ways.[9] Five separate federations of Mongols, internally divided further into often contentious tribal associations, occupied different regions of the steppe and desert of inner Asia. Among them, the Chahar Mongols occupied territory north of Shanxi, the Khalkas lived in today's Mongolia, and the Tümeds lived in the Ordos. These three groups were the dominant peoples along the northern perimeter of the Loess Plateau.

Until the 1420s, the Chinese Ming regime attempted to maintain a border to the north of the Ordos and to practice multiethnic steppe confederation diplomacy. The new regime did not fortify the Ordos grasslands or dislodge its pastoralist residents and their animals. Fourteenth-century Ming military policy favored mobile campaigns into Mongol territory over the defensive fortification of a border. The court also established a limited number of defensive forts inside the Great Bend of the Yellow River.[10]

The second half of the fifteenth century was a period of deep and lengthy drought that upended politics on the Ordos. While campaigning against the Mongols, Chinese troops suffered a dramatic defeat in the mountains west of Beijing in 1449. The emperor himself was taken captive. The regime considered retreating to the south, as the Song had done in the twelfth century. By the middle of the century, Mongol detachments had moved into the Ordos: they included mounted warriors and also groups of herders seeking water and pasture during an era of scarcity. Advocates for a fortification strategy spoke of the fertile fields, rich soil, and abundant salt lakes that persisted in the Ordos. They recommended not only building garrisons but also moving in settler-colonist farmers. Opponents of

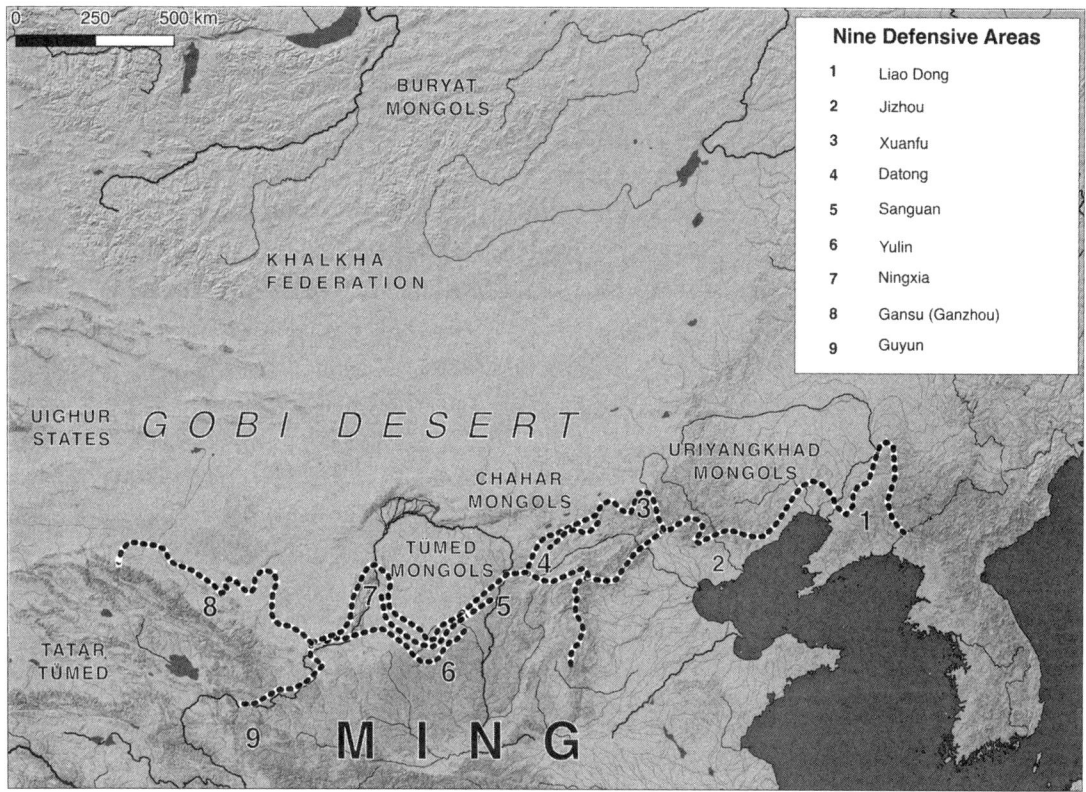

FIGURE 4.6
The Nine Defense Areas. In the context of geopolitics during the fifteenth and sixteenth centuries, thousands of kilometers of wall came to crisscross the Loess Plateau. Based on Mote, *Imperial China*.

the fortification position countered that during an era of drought, the farmers who already lived in the region were being overtaxed and excessively impressed into corvée labor. Through the early 1470s, the court rejected the fortification approach in favor of support for mobile troops deployed against Mongol detachments.[11]

Ultimately, the wall-building advocates prevailed even as drought conditions persisted. Beginning in 1474, a construction crew of forty thousand laborers built long walls. The term *Great Wall* only came into use later. The walls were often manufactured from pounded loess soil piled on top of stone foundations. Where no stone was available, laborers substituted large bricks, fired in kilns set up nearby and fueled with local wood.[12] By the middle of the sixteenth century, the wall network had become extensive, with two defense lines, numerous signal towers, and detachments of soldiers deployed across the Ordos.[13] The long drought came to an end in the early 1500s, and moist conditions prevailed during most of the next century.

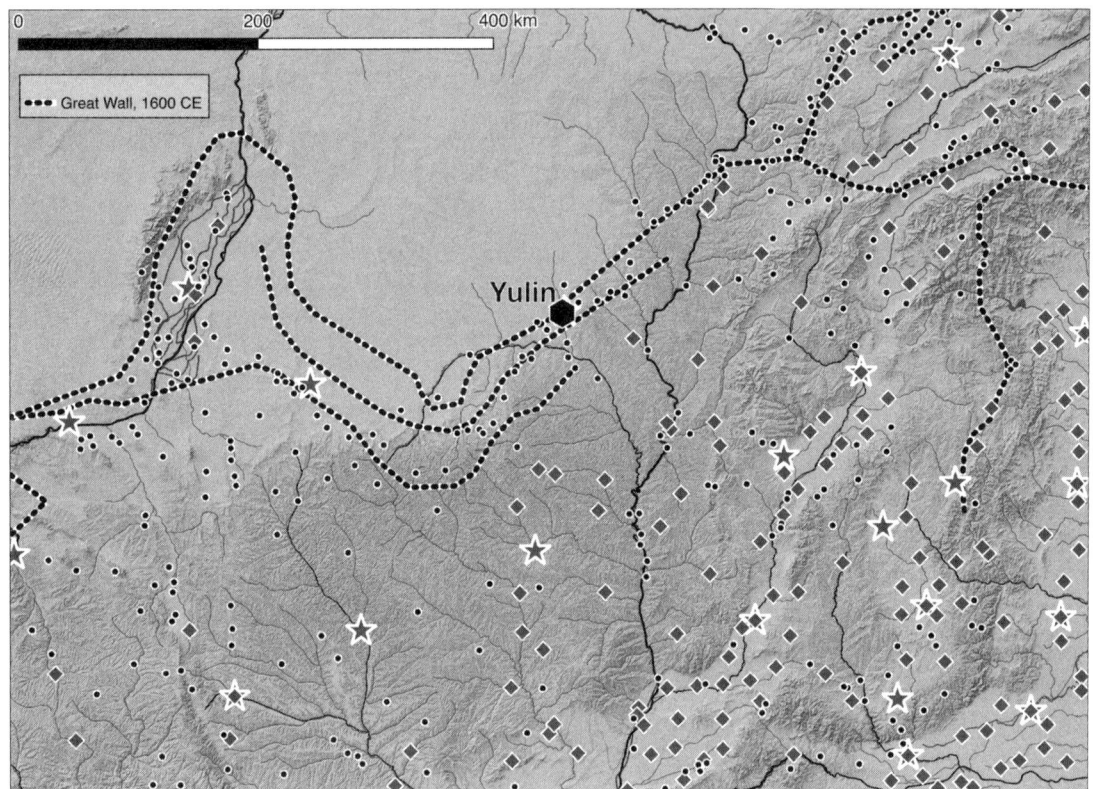

FIGURE 4.7
Yulin and its Hinterland During the Ming. The city of Yulin, north of the Wuding River at the edge of the Mu Us Desert, was a strategic site during the Great Wall era. The prefecture of Ningxia was located in the Yinchuan plain, and the prefecture of Datong lies at the northeastern perimeter of the Loess Plateau. The line of Ming fortifications lies north of the twelfth-century line. It is immediately at the desert edge, in a location that was feasible to settle only during the unusually moist sixteenth century, and in a region where vegetation would have particular difficulty rebounding after disturbance.

The Ming military occupation of the northern perimeter of the empire was known as the Nine Defense Areas (Jiubian). As figure 4.6 depicts, it was divided into nine sections along the borderline that stretched for over 2,400 kilometers, or 1,500 miles, from the sea to the deserts, each one commanded from a major fort (*bianzhen*) headed by a regional military commander.[14] The Yulin command, situated near the banks of the Wuding River, occupied the central Ordos. It was flanked by the Ningxia command to the west and the Datong command to the east. Although the Yulin command occupied most of the Loess Plateau, the Ningxia and Datong commands incorporated some Ordos territory as well.

Defense priorities took precedence over environmental limits in determining state policy. Multiple parallel lines of fortified walls ultimately stretched across the grasslands and fragile soils of Yellow River's Great Bend: a "great border" wall (*dabian*) flanked by a "secondary border" (*erbian*) edifice to its south.[15] From its base near the

Wuding River at the edge of the Mu Us Desert, the Yulin command governed eight hundred strong points, sentry posts, beacon-fire towers, and other defenses. Figure 4.7 is a contemporary map of Yulin and its hinterland.[16] Plate 31 shows the Yulin Command today. As in the eleventh century, colonist farmers were compelled to farm the marginal lands of northern Shaanxi to feed the construction crews, artisans, troops, and horses.[17]

To a large extent, this infrastructure was less significant for preventing invasion from afar and had more to do with ethnically cleansing the Ordos and making it, for the first time, a region of exclusive Chinese political and cultural dominance that was devoted primarily to agriculture. Even though the defense policy would not ultimately prove sustainable, the expropriation of the multiethnic population of the Ordos was a success. By the seventeenth century, the Ordos was no longer a cultural frontier.

At first, even as the climate was deteriorating in the early 1400s, there was ample desirable land to establish farms and self-supporting garrison detachments. In 1435, the manager of frontier affairs reported that south of the Yulin garrison "there is a very broad area of land very suitable for farming." Elderly locals reported about salt ponds, lakes, and arable land.[18] There was lush grassland and fertile soil, and plenty of land not yet under cultivation.[19] Even during the arid period of 1473, one provincial governor asserted that "to reclaim wasteland is one of the important policies of frontier defense. South of Yulin City there is a large area of land that can be planted. In recent years, officers have advocated that people should grow crops."[20] Soon thereafter, a successor provincial governor oversaw the construction of 12 fortresses, 15 border towers, 78 small towers, and 719 cliff-top stockades.[21] By the final quarter of the sixteenth century, fortresses were spaced less than 100 *li*, about 50 kilometers, or 33 miles, apart.[22] The fifteenth- and sixteenth-century garrisons lay significantly to the north of the previous twelfth-century line.

In fact, the ambitions of the Nine Defense Areas policy almost immediately proved difficult to sustain at the desert edge in the central Ordos, where the demands of military colonization outstripped the carrying capacity of the land. In 1489, a year of average precipitation, a fortress east of Yulin had to be relocated because the water table at the original site had dropped too far and sand had encroached on its agricultural fields. The same year, another fortress near there also

FIGURE 4.8

Moisture Mean, 1500–1650.

The moisture record during this time period does not correlate to documentary reports of colonization or desertification, which means that desertification resulted from human activity.

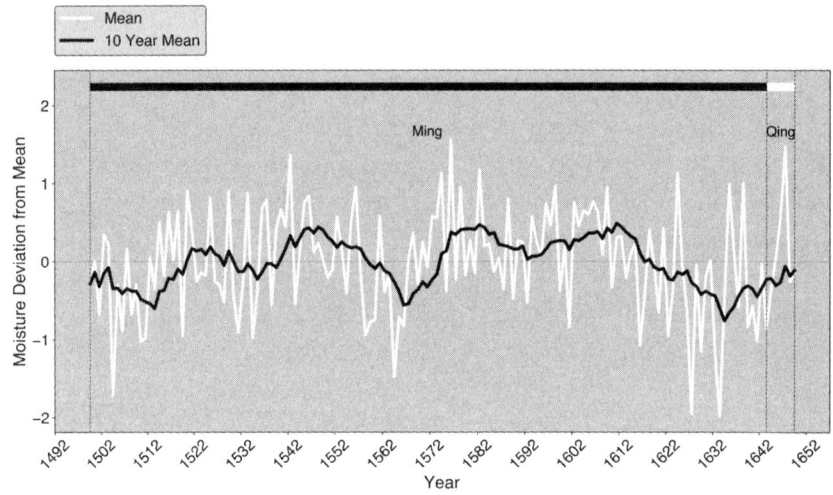

retreated because of desertification. In 1529, another year of moderate rainfall, supply trains carrying food to Yulin had to abandon wheeled vehicles in favor of donkey caravans because the roads had been engulfed by sand. In 1534, following several dry years, "the central city of Yulin Garrison was surrounded by the yellow sands, and there was no agriculture or commercial activity there. Provisions and funds for the troops [had] to be supported by other regions."[23]

The Mu Us Desert margin lay just north of the Yulin wall, its location fluctuating as the strength of the monsoon varied, as irrigated farming caused the water table to drop, and as forest barriers disappeared. The environmental scientist Jianxin Cui and his coauthors insist that climate played the primary role in desertification and the Ming retreat from the Ordos frontier.[24] However, although temperatures were low in northern China during the Little Ice Age between about 1300 and 1800, fluctuations in precipitation do not neatly match historical records that report on landscape deterioration or the history of civilian and military colonization. Although climate ultimately played a role in the Ming retreat from the desert edge, the predominant environmental impact during the earlier decades was intensive human exploitation of an agriculturally marginal landscape. Figure 4.8 depicts the moisture mean between 1500 and 1650.

By the middle of the sixteenth century, the defense system had clearly become untenable. In 1546, a year of average precipitation during a relatively moist decade, the governor-general of Shaanxi advocated abandoning the fortresses at the desert edge in favor of shoring up defenses further south. The commander of the Yulin garrison concurred. However, the court, far away in Beijing, did not follow their advice and instead continued to invest resources and personnel in the walls.[25] By 1547, even though precipitation remained above average, the region along the wall had deteriorated into desert. One observer attributed this to deforestation that resulted from farming, firewood collection, and brick making to repair the wall and to construct new buildings.[26] In some places, the long walls had been buried by sand and could no longer be used for defense because raiders could simply gallop across the dunes on horseback.[27]

The military farm system collapsed as well. Between 1566 and 1620, by which time severe drought was taking hold, yields on military farms dropped by 60–70 percent. As the loess environment deteriorated, soldiers began to flee or mutiny, and civilian farmers rebelled as well.[28] In 1608, a book about the history, geography, and military defense system of the whole Yulin garrison district reported that desertification had expanded across the whole region, that at least eight fortresses had been affected by severe desertification, and that the hillsides around the fortresses had also been deforested, had lost their topsoil, and had become sandy.[29] As precipitation plummeted at the beginning of the new century, documents from 1615 and 1616 reported on failing wells and desiccated rivers that could no longer support people and horses.[30]

Forest cover contracted more precipitously during the middle Ming period than it ever had before, particularly around the Great Wall. A wave of forest destruction on the northern Loess Plateau occurred in the sixteenth century before two generals halted it in 1580.[31] In addition to the military frontier, the large settler population harvested timber for construction and fuel, and cleared trees for farming and grazing. In the Heng Mountains south of Yulin, once the site of Song fortification against the Xi Xia, lands designated for military farms (*tuntian*) rapidly degraded into uncultivated "wasteland" (*huangdi*). Forests in the Lüliang Mountains vanished as well.[32]

Forest exploitation events tended to happen in sudden bursts in conjunction with military campaigns or commercial exploitation. The changes were rapid enough that commentators at the time realized that they were witness to an ecological transformation transpiring before their eyes. One individual wrote that "people are becoming numerous, a thousand people have come to the peaks, they cannot be expelled, and they cannot be interdicted.... [F]orest regions have been burned to ashes as far as the eye can see, with woodcutters swept away by the *li*."[33]

Civilian population growth and agricultural intensification far from the Mongol frontier played a significant role in the process of deforestation. In *The Shanxi Gazetteer* (*Shanxi tongzhi*), Yan Shengfang (1515–1565), a Shanxi native who passed the imperial civil service exams and became a writer, precisely and trenchantly connected each link between farming, forestry, erosion, flooding, and destitution in the Taihang Mountains, the easternmost range of hills on the Loess Plateau. His commentary begins by extolling the fertility of the region and explaining that people became prosperous by developing irrigation networks in order to expand agriculture there: "Before the Zhengde reign period (1505–1521), flourishing woods covered the southeastern slopes of the Shangzhi and Xiazhi mountains. They were not stripped because people gathered little fuel. Springs flowed . . . and it was never seen dry at any time of year. Hence villages from afar and in the north of the county all cut branch canals and ditches which irrigated several thousand *qing* of land. Thus Qi [county in east central Shanxi] became prosperous." Within decades, he continues, the situation had changed. The slopes were denuded, erosion was rampant, and wealth collapsed:

> At the beginning of the Jiajing reign period (1507–1567), people vied with each other in building houses and wood was cut from the southern mountains without a year's rest. Presently people took advantage of the barren mountain surface and converted it into farms. Small bushes and seedlings in every square foot of ground were uprooted. The result was that if the heavens send down torrential rain, there is nothing to obstruct the flow of the water. In the morning it falls on the southern mountains; in the evening, when it reaches the plains, its angry waves swell in volume and break through the embankments, frequently changing the course of the river. . . . Hence the county of Qi was deprived of seven-tenths of its wealth.[34]

As deforestation accelerated, a legal infrastructure for conservation came into place as well, but it was only sporadically effective. Around the town of Taihe in the eastern Qinling Mountains, for instance, the travel writer Xu Xiake (1587–1641) reported in 1623: "Mountains surround Taihe on all sides, with dense forests for at least 100 *li*, blotting out the sun and reaching to the sky. Within a distance of several tens of *li* of the mountain, there are strange firs and ancient cypress trees so big that three people can embrace them, and they spread out all over the valleys. This is due to their protection under state law." In contrast, he explained that farther east, on the slopes of Mount Song in Henan, "there is no trace [of forests] from the foothills to the very summits due to felling and firewood cutting."[35]

In Guanzhong, on the Wei River plain, fifteenth- and sixteenth-century engineers conducted large-scale work on the old Zhengguo Canal and offered ongoing technical innovation, but the system was capable of irrigating only between one thousand and eight thousand *qing* of farmland. By the seventeenth century, according to one report, "officials and people were exhausted, and obstruction by silt only grew worse by the day. Compared to the Song and Yuan, the benefits from water management were less than one tenth."[36]

During the early seventeenth-century drought, famine in Shaanxi was so severe that in 1628, one of the driest years on record, farmers resorted to eating weeds and tree bark for survival.[37] By that time, the government was struggling with fiscal collapse and the impacts of global cooling as well as environmental degradation, and a major epidemic struck around the same time.[38] As peasant rebellions swept northern China, the central Ordos—dry, eroded, and heavily burdened by supporting the still-numerous frontier garrisons—was an epicenter of unrest. Bands of desperate farmers left their lands and resisted tax collectors. Li Zicheng (1606–1645) and Zhang Xianzhong (1606–1647), both from the central Ordos, led rebellions that briefly seemed likely to topple the regime itself. Zhang Xianzhong ruled Sichuan in 1644 and 1645 until he was deposed and killed by the invading Manchu army, and Li Zicheng captured Beijing and declared a new dynasty that ruled briefly before its defeat by Manchu invaders and Ming loyalists.[39] The Shaanxi population declined for a number of decades as refugees fled the fighting. This would have offered some opportunity for vegetation to rebound, even though ecological conditions were not ideal.

The transition to a new imperial regime occupied several decades of the cold, dry, difficult mid-seventeenth century. On the Loess Plateau, drought and state retreat shuttered the garrisons, with their insatiable demand for timber, fodder, and drinking water. Numerous farming families died or migrated amid the earthquakes, famines, epidemics, and rebellions that marked the era. Political and environmental chaos blocked timber and other resources from commercial exploitation. Some forest regrowth appears to have occurred.

By the final decades of the seventeenth century, the Manchu Qing dynasty had secured its power throughout China. The new regime did not pursue the Ming fortification strategy. The Manchu rise to power depended in the beginning on alliances with the Mongols. The various Mongol clans declared fealty to the new empire in return for a great deal of autonomy in governing their own tribal affairs. The Manchu rulers were descendants of the Jurchen people who controlled northern China during the twelfth and thirteenth centuries and were long-term participants in the multiethnic federation politics of eastern Central Asia. Manchu leaders used diplomatic and military leverage to prevent the Mongol clans from reunifying and to persuade Mongol leaders to join the Qing enterprise. The Chahar Mongols, north of the Loess Plateau, declared fealty to the Manchus beginning in the 1580s.[40] Ming enclosure initiatives, an arid climate, Chinese population growth, Mongol political collapse, and Manchu politics all combined to ensure that, for the first time, the entire Loess Plateau was functionally an ethnically Chinese space. It became overwhelmingly agrarian as settler farmers completed the venture of dislodging pastoralists from their homelands. The story of the Loess Plateau in its final imperial centuries is primarily a civilian story, not a military one. Figure 4.9 depicts the settlement geography of the Loess Plateau in 1820. The line of settlements along the Ming Great Wall is still clearly visible on the map: they did not disappear even though imperial strategy toward the region had moved on. No place on the Loess Plateau is without towns and villages. New counties are exceptionally dense along the Yellow River tributaries, especially to the east of the Ordos.

Advantageous tax policies prompted a vast wave of migration to the Loess Plateau during the Qing dynasty. In the early eighteenth century, the emperor commanded provincial governors throughout the realm to intensify initiatives for land reclamation. The beginning of the eighteenth century was a time of ample rainfall, and he specifically identified the Loess Plateau as a region that should support

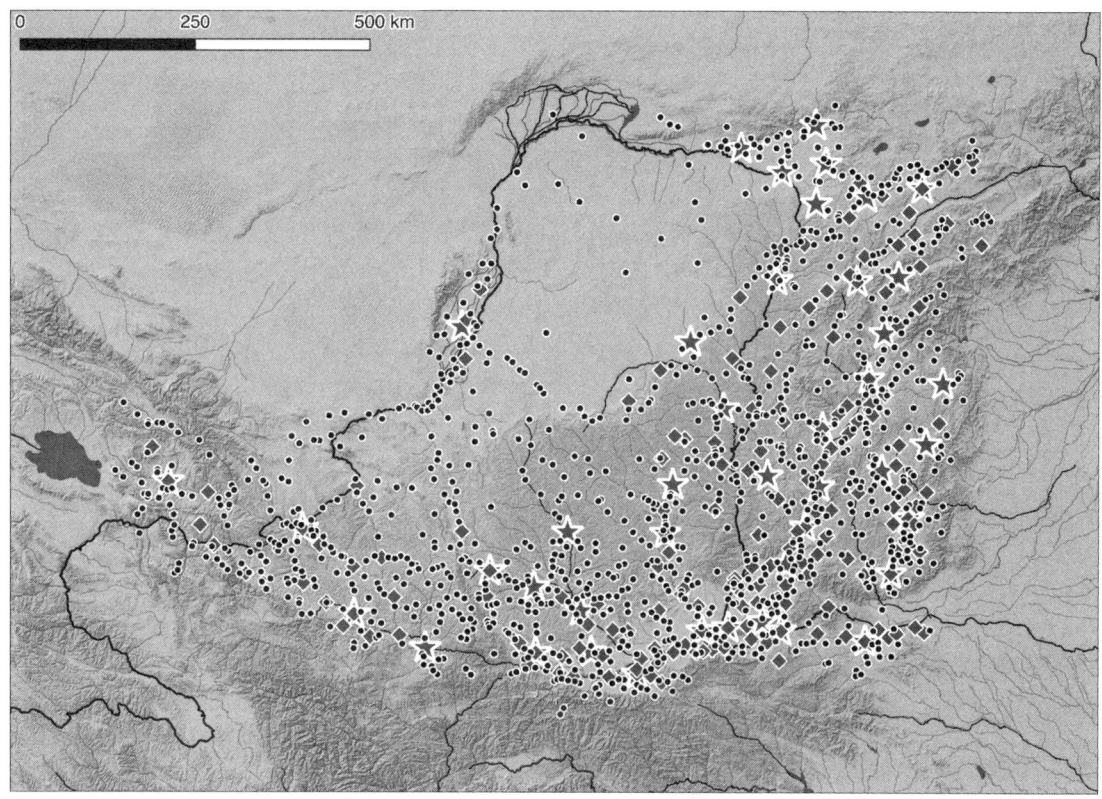

FIGURE 4.9
The Loess Plateau in 1820. By the nineteenth century, the dense population on the Loess Plateau was a result of the growth of population and commerce, and, for the first time, did not correlate closely with military strategy.

more agriculture.[41] From 1750 to 1860, the registered population on the Loess Plateau increased from around 24 million people to more than 40 million. The central and northern parts of the plateau were home to more than 31 million people by the 1780s. The population was more than six times larger than it had been in the late Ming, although the amount of cultivated land on the Loess Plateau merely doubled. For the first time, the machinery of the market economy, albeit with imperial hands on the levers, bore primary responsibility for the change. There was no coerced migration. New land came under cultivation, previously farmed land was exploited more intensively, and timber and other desert and forest commodities were exchanged on a commercial market.[42]

New commercial towns emerged along the eastern edge of the Ordos, on the banks of the southward-flowing reaches of the middle course, the navigable stretch of the river south of the Hukou waterfall. These included places like Qikou, on the eastern bank of the river

LEVIES AND LEVEES   197

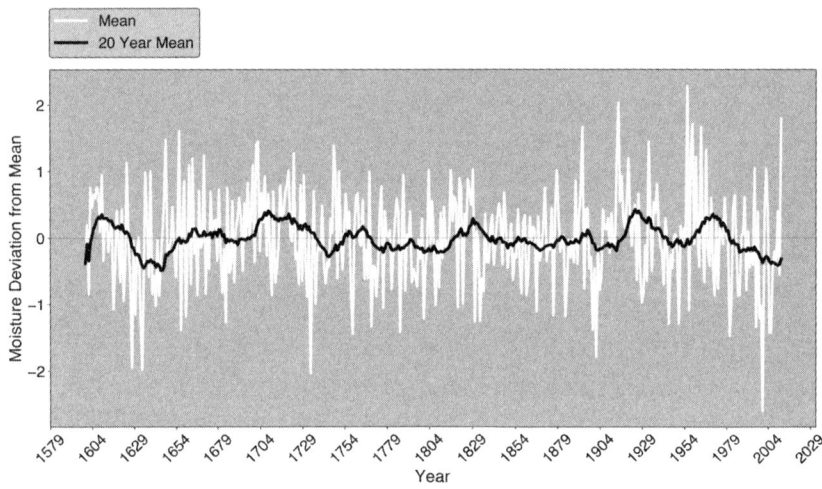

Figure 4.10
Moisture Mean, 1600–2012. Following the deep drought of the early seventeenth century and the moist decades of the early eighteenth century, average moisture levels hovered close to their historical mean for the remainder of the imperial era, although annual deviation could be quite considerable.

at the foot of the Lüliang Mountains. There, desert products like salt and alkali, and agricultural products like canola oil and, later, cotton, shipped out from the Wuding River and other tributaries, were gathered, traded, and sent downstream. Logging resumed and accelerated in the seventeenth and eighteenth centuries, forest coverage throughout the Ordos plummeted, and the rate of erosion accelerated, in particular as hillslope cultivation became more widespread. The Ordos was almost completely deforested by the nineteenth century.[43] Many people resided in cave houses (*yaodong*) carved into the sides of loess canyon walls. Plate 32 depicts Li Mountain Village, a Qing town built into the cliffs across the river from Qikou, with its cave houses and steeply terraced agricultural fields, none of which is currently farmed. As figure 4.10 depicts, most of the desertification visible in sediment cores occurred during the past three hundred years, a process that correlates better to agricultural intensification than to the climate record.[44]

To encourage potentially restive farmers to remain in place on the Loess Plateau even as conditions deteriorated, the Ming and Qing regimes offered lucrative salt monopoly contracts to well-connected merchants. In return, they were required to ship grain to the northwest to subsidize colonists there. The government granted land, seeds, loans, and information to settlers in sparsely populated regions, subsidized the digging of wells, allowed for new farmland to be left off the tax rolls, subsidized the state granary system, and supported handicraft textile production to supplement farming. The Qing state was willing to use policy tools and direct transfers to shift resources between regions of the empire in order to maintain some semblance of balance between subsistence on the poor peripheries and the rich center. The outcome succeeded, but only at the expense of the ecological stability of the northwest, which would not otherwise have been able to maintain such a high population.[45]

By the turn of the twentieth century, Shaanxi was one of the most precarious, impoverished regions of China. Once the Mongols and their pastoralist allies were consigned to the north or politically neutralized, most grazing land was converted to farmland. The farming-to-grazing transition zone shifted toward the north despite somewhat unfavorable climate conditions.[46] Marginal farmland came under cultivation, the commons was enclosed, and the herders were expropriated. Ordos forest cover declined from an estimated 15 percent of ground cover in the mid-seventeenth century to less than 6 percent today.[47] There were still two hundred springs on the Loess Plateau in 1667, and sixty of them were still used to irrigate farmland. None of them is to be found today.[48]

Maize and tubers, crops native to the Americas, are attested in China as early as the 1540s. They were widely cultivated in northern China by the late seventeenth century.[49] These crops were not suitable for cultivation on the entire Loess Plateau and were most heavily cultivated on its southern reaches. They flourished on hillslopes and in arid places that had until then maintained their native ground cover. The new crops transformed the terms of agrarian subsistence in an ecologically fragile region that had previously reached a stable population ceiling. The explosive migration into the region resulted from a combination of government incentives, subsidies, land commodification, and new crops. Population growth resounded even where the new crops were not being grown. The sudden demographic and ecological pressures were a source of social friction, endemic distress, and rebellion throughout Shaanxi.[50] The White Lotus Rebellion (1795–1804), a massive millenarian revolt against Qing rule that began as a tax rebellion, began in the Qinling Mountains bordering the Wei River valley to the south.

All of this resulted in landscape transformation that was of unprecedented speed and severity. An eighteenth-century gazetteer from Xing County in northwest Shanxi explained that during the mid-sixteenth century, the county had been a place of dense mountain forests, crisscrossed by streams, and it never flooded even with heavy rains. However, by the commentator's time, all the mountains were bare of vegetation. Every summer and fall, water rushed down from the peaks, collapsed the riverbanks in the valleys below, created standing pools of water, and forced people to flee their homes.[51]

People eradicated forests to meet household needs, to clear slopes for terraced agriculture, and to serve a commercial market. The rugged Qinling Mountains, their northern slopes draining to the Wei River, were a focus of the late-imperial timber market, and by the early eighteenth century, foresters needed to go 300 kilometers, almost 200 miles, deep into the mountains to find timber. By the late eighteenth century, whenever it rained, rocks and sand from the surrounding hills, which were denuded of soil as well as ground cover, tumbled into the Wei streambed and deluged the surrounding fields.[52] The Qinling forests were completely eradicated over the course of the Qing. By the early nineteenth century, all the forests in three Qinling counties had been replaced by farms.[53] There were "no old trees" in the early nineteenth-century Liupan Mountains in the Wei headwaters either.[54] Around Tianchi Lake in the Liupan Mountains, forest cover rebounded after the Song-Xia wars of the eleventh century, and the region lay far from the exploitation frontier of the Great Wall era. A new cycle of agricultural intensification began there around 1660, which ultimately resulted in the collapse of the entire forest ecosystem.[55]

At the northern perimeter of the Ordos, timber merchants hewed forests in Hetao, the saline wetlands north of the Yellow River's Great Bend.[56] Deforestation affected even remote mountains, with the most severe impacts on steep grades. Foresters cut trees for the paper and charcoal industries as well as to clear land for farming.[57] In the eighteenth and early nineteenth centuries, Shaanxi timber merchants sent trees all the way down the Yellow River to the floodplain, where they were used for waterworks construction. By the mid-nineteenth century, overcutting mostly ended this market.[58]

Close observers and hydrocrats on the Loess Plateau witnessed the erosive effects of the destruction of grassland and forest ecosystems. They understood that they were encountering unprecedented amounts of sediment. Ming and Qing local gazetteers from the counties and prefectures of the Wei River valley abound with descriptions of canals and irrigation works, with the comment that "all of the above have disappeared."[59]

Eighteenth- and nineteenth-century regional and local officials clearly connected erosion to deforestation. They recognized that reforestation initiatives were essential to ecosystem sustainability. On sporadic occasions throughout the eighteenth and nineteenth centuries, provincial and county officials stationed on the Loess Plateau sporadically issued orders to protect forests, to punish commercial loggers, and to reforest denuded slopes. They circulated books of forestry proverbs to residents and erected stone stelae carved with passages enjoining people to plant trees. Tao Mo (1835–1902), governor-general of Shaanxi and Gansu in the late nineteenth century, advocated for planting trees and grasses to prevent erosion. He knew that forests protected and improved water and soil quality and conservation, regulated the climate, and cleaned the air. However, although the Qing court, in faraway Beijing, recognized the severity of erosion on the Loess Plateau, they devoted no resources to ameliorating it.[60] Incentives for farmers and officials alike focused on agricultural intensification rather than ecological sustainability.

Qing personnel evaluation was based strictly on economic development. Successful land conversion for agriculture was a performance metric for local officials, and conservation played no role in the standards. Throughout the dynasty, emperor after emperor issued edicts to officials demanding that they open new land to farming. Sustained conservation initiatives could not find traction in the Qing political environment, and officials had no incentive to pursue it.[61]

Overwhelmed by sand, gravel, and boulders washed downstream from the mountains, the Zhengguo Canal had become effectively nonfunctional. Qing officials, under pressure to intensify agricultural technology, continued working on the Zhengguo system, but the scale of their attempts contracted, and their commentary focused increasingly on the environmental costs of their efforts and their unlikely success. As erosion in the Jing River watershed accelerated, as well as sediment accumulation on the Jing and the Wei, engineers developed increasingly risky and costly schemes to fetch clear water for the canal from higher up the gorge, expending funds and causing human suffering with increasingly smaller returns. In the mid-eighteenth century, one Wei valley county magistrate whose territory included key parts of the canal system noted that in ancient times, the river had carried silt that nourished agriculture, but by his time, all that was left was sand and gravel that did not contribute to soil fertility.[62]

Residents of the Loess Plateau attempted to sustain the rapidly changing environments in which they lived. They planted trees and built terraced fields and other earthworks to retain water and soil. They constructed small masonry check dams at the headwaters of gullies to create reservoirs, prevent flooding, and capture sediment. Townships issued regulations to control grazing and tree felling.[63] Even as ecological conditions declined, new technologies and ingenious methods allowed people to sustain themselves in changing environments.

In 1743, the Qing court dispatched an officer named Hu Ding (1709–1789) to the Loess Plateau. At the time, Hu Ding was an inspection official (*jiancha yushi*) whose home was in far-off Guangdong on the southeastern coast, but he had previously served in a military position in Shaanxi. Inspection officials were empowered to gather complaints from the people and review judicial cases; they were also authorized to submit recommendations to the court on changes to imperial policies.[64] On the basis of his investigations, Hu issued a report to the court asserting, correctly, that most of the sand on the Yellow River floodplain originated from two places upstream. One was in Shaanxi, upstream from Sanmenxia; the other was in the mountains of Shanxi. He urged the court to require local officials in those regions to lead efforts to build small check dams and other constructions on hillslopes near the headwaters of streams and creeks, which would catch sediment while permitting water to flow. This would, he believed, gradually fill gullies and create flat plains where winter wheat could be planted, solving several problems at once. One was food insecurity on the Loess Plateau, another was erosion on the Loess Plateau, and another was sediment accumulation in the Yellow River.

Hu Ding's effort was never adopted. Without visiting the Loess Plateau in person, Bai Zhongshan (d. 1761), the chief official in charge of the Grand Canal at its confluence with the Yellow River on the floodplain, contested Hu's assertion. From his vantage point downstream, Bai insisted that the Yellow River was naturally turbid. He believed that, so long as the river traversed the "loose and crumbling" (*shusan*) soils of Henan at the boundary between the Loess

Plateau and the floodplain, it was bound to carry a great deal of sediment no matter the intervention. He asserted that there was no use in creating infrastructure further upstream. In short, he argued for an approach that would preserve the budget and authority of his domain on the floodplain, rather than one that would transfer funds, influence, and imperial scrutiny to a distant, impoverished part of the river basin. The emperor adopted Bai Zhongshan's proposal to maintain the status quo rather than Hu Ding's intervention, which would have contravened two millennia of imperial policy toward the Yellow River. He opted to reinforce the waterworks on the floodplain, to accept the river's muddy character as a fact of nature, and to leave the Loess Plateau to its fate.[65]

In short, some residents and officials on the Loess Plateau during the final imperial life span of the Yellow River understood that, although they occupied a region that managed to support a very dense population, they did so at the cost of severe ecological degradation.[66] The situation was visible to outsiders, even if its history was not. European and American visitors, scientists, missionaries, and adventurers commented trenchantly on the state of affairs. In a fashion consistent with an emerging imperialist discourse of natural history that valued ecology over people, they blamed the denizens of the plateau for unsustainable practices; but they nevertheless described the late Qing environment of the Loess Plateau with reasonable accuracy and without being burdened by the bureaucratic infighting that troubled Hu Ding and Bai Zhongshan.

This tradition is the root of knowledge of the Yellow River in European languages. In 1875, the Vincentian priest Père David (Father Armand David; 1826–1900), who found his calling as a naturalist rather than a missionary, visited the upper reaches of the Yellow River. He decried its barrenness and deforestation and criticized the Chinese people for their destructive ways.[67] Western naturalists from the mid-nineteenth century through the 1930s remarked on the scarcity of forests in China and the "apparently relentless destruction" of what remained. They all identified deforestation as the cause of erosion on the Loess Plateau. Many of them linked upstream deforestation to flooding and misery far away on the alluvial plain.[68]

It is to the connection between the upstream and downstream components of the Yellow River basin that we now turn, circling back to the fourteenth century and onward from there.

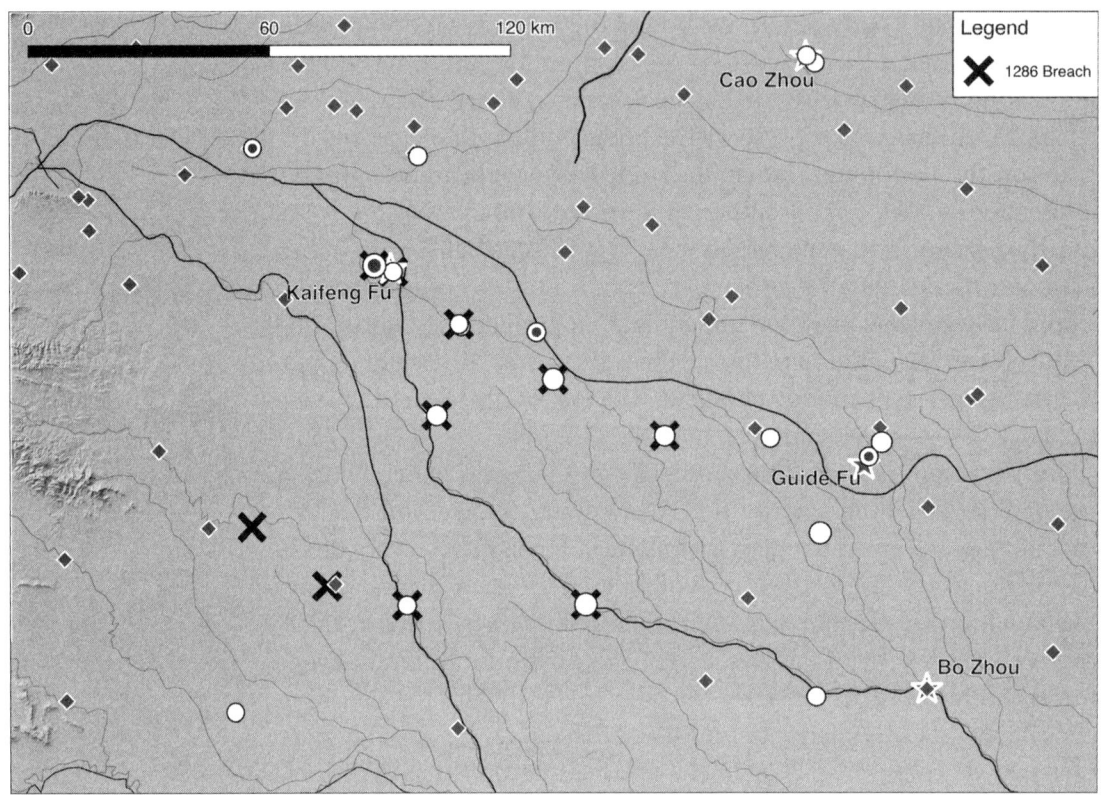

FIGURE 4.11
The 1286 Breach. The 1286 breach affected much of the territory between the Yellow River and the Huai.

THE MULTIPLE-CHANNEL ERA

A cataclysmic flood struck the Yellow River in the midst of Mongol Yuan rule over the basin, when the river breached its levees in fifteen places in 1286, the earthworks rupturing in succession as the flood crest moved downstream. After the flood, the course split in two, with both streams ultimately disgorging to the Huai River.[69] Figure 4.11 documents the region affected by this event. During the era of Mongol rule, the amount of sediment washing downstream from the Loess Plateau was low by historical standards. Nevertheless, the unmanaged river still flooded onto the plain and inundated farmland during the rainy season nearly every summer during this decade of above-average precipitation. The riverbed and the diversion channels filled with sediment. From this time forward, the canal system became unusable, as did the coastal saltworks. Another major levee breach, in Shandong in 1344, went unrepaired for eight years and filled the canals with additional sediment.[70]

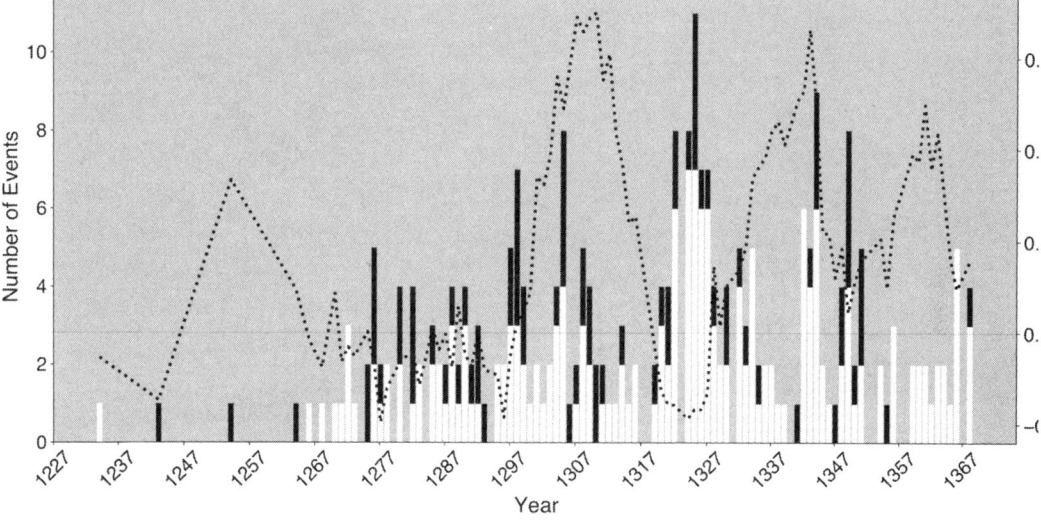

FIGURE 4.12
A Timeline of the Yuan Floodplain, 1234–1368. During the decades when control of the floodplain was split between the Yuan regime in the north and the Song regime in the south, Yuan sources attested few disasters and devoted little attention to events of infrastructure management. Immediately upon the unification of the realm in 1276, the regime redoubled attention to chronicling the behavior of the Yellow River on the alluvial plain and to managing its propensity to flood. This era is one in which political history had a tangible impact on the relationship between people and the river.

As Figure 4.12 depicts, the regime documented few floods and few infrastructure maintenance events until after they conquered the Southern Song and took control of the Yangtze basin after 1276. At that point, sources begin to attest multiple floods and breaches each year, and numerous events of floodplain management as well. Once the Yellow River and Yangtze River basins came under the control of a single regime, the floodplain and its canals could again serve as a transportation corridor between north and south. Figure 4.13 depicts the multiple channels and numerous events, spanning the entire floodplain, that were attested between 1234 and 1368. Disaster discourse and supervision of the landscape both followed from the political unification of the floodplain, a striking example of the ways that people and their political arrangements create the language for river history.

The fourteenth century dawned with decades of bitter cold, drought, and plague, which coincided with a collapse in legitimacy and effective governance at court. Earthquakes occurred almost annually between 1338 and 1352.[71] As the Yuan system of paper money broke down, so did the market economy and the tax system. By the 1350s, multiple rebel armies contended for control from bases of

LEVIES AND LEVEES

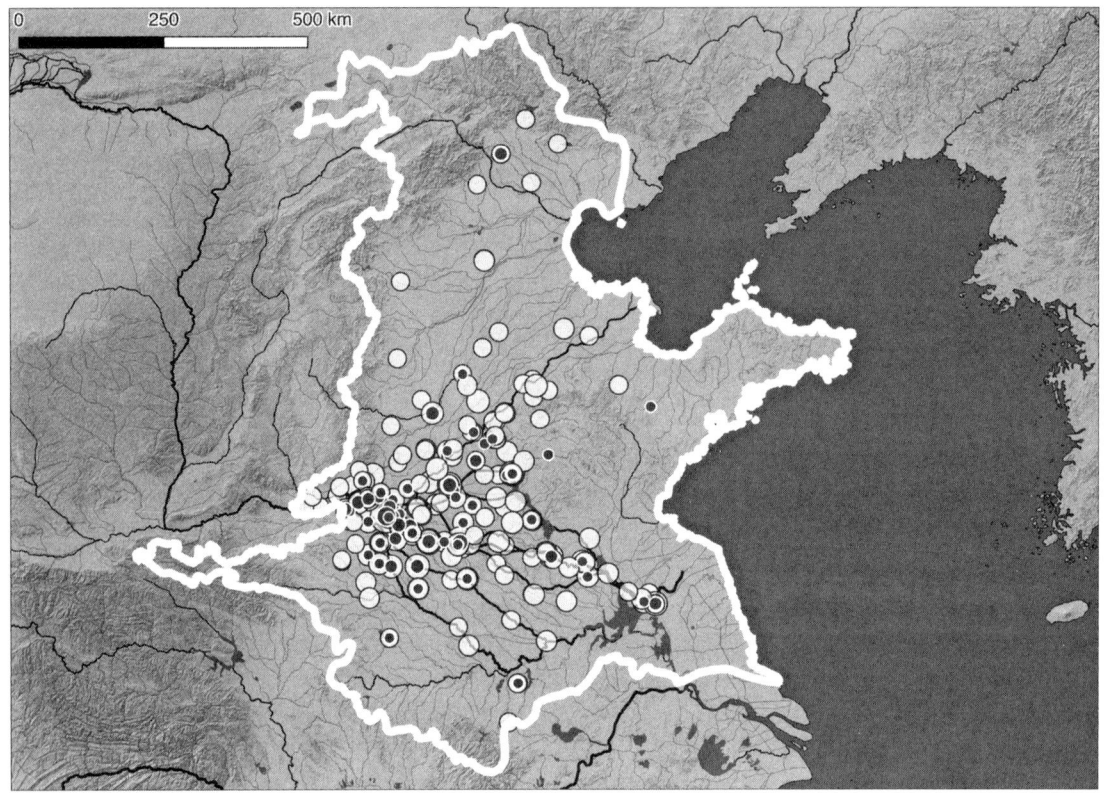

FIGURE 4.13
The Yuan Floodplain, 1234–1368. The river maintained multiple channels throughout most of the Yuan era. When the regime began recording significant numbers of disasters after 1276, they attested events that occurred throughout much of the floodplain. Hydrocrats sponsored waterworks activities on all the river courses.

power the size of provinces or smaller, warring with one another and with the Mongol regime. Displaced and hungry people were on the move. Canal workers joined the rebels as plague and famine devastated the Huai River plain. By the 1350s, "Chinese society was rent by disorder on a scale that had not been seen for centuries."[72]

This would seem an unlikely context for innovation in the history of river management. Nevertheless, even amid its disintegration, the regime prioritized the repair of the 1344 breach. Relieving disorder and distress on the floodplain was one objective. There was another goal as well. Pirates—former salt merchants and their followers among displaced coastal people—were attacking oceangoing grain boats. These were the crafts that shipped southern produce to the capital by sea after the Grand Canal failed by the 1330s. Under the circumstances, the court opted to secure and control the floodplain and to repair the canal. The regime faced an age-old dilemma: whether to permit the river to spread out between multiple channels

LEVIES AND LEVEES

and diversion canals or to lock it into a single course behind high embankments.

The monarch turned to Toqto'a (Tuo Tuo; fl. 1314–1356), a high-ranking Chinese-educated Mongol minister. As the lead editor of the official *Song History* and *Jin History*, compiled from archived government documents, he would have been well versed in river history. He had also been involved with pro-canal factions since the 1340s. Toqto'a advocated for restoring the river to a single southerly course. In 1351, Toqto'a hired polymath engineer Jia Lu (1297–1353) to direct a process for doing so. Jia was a member of the *Song History* editorial committee, an expert in river management, and a passionate writer on the human misery engendered by the inundated floodplain.

Under Jia Lu's guidance, 150,000 laborers, overseen by 20,000 sergeants, dredged the former course and widened the riverbed. They built levees along the north bank so that the river could not avulse toward the north, and they built sluice gates into the levees so that high waters would flow into drainage canals and ponds rather than rupturing the earthworks. They directed the bed of the Yellow River into a stable course through the beds of the Si and Huai Rivers.[73] Toqto'a funded Jia Lu's initiative by arranging for the printing of a new round of inflationary paper currency.[74] With some variations, the Jia Lu riverbed established the course that persisted until the nineteenth century.

From the vantage point of long-term river history, the significance of Jia Lu's work is primarily that he and his colleagues carefully documented their efforts, and less the sustainability of their work. The infrastructure they devised did not survive the travails of the fourteenth-century crisis. However, their texts created the road map for river and canal management that animated the remaining life spans of the imperial river. The text *Records of River Defense from the Zhizheng Era* (*Zhizheng hefang ji*), by an official and writer named Ouyang Xuan (1283–1358), collates information about the Jia Lu era work. One of the earliest river management handbooks still extant, it describes the innovations in strategy, techniques, soil analysis, and construction management that characterized the Jia Lu infrastructure. It classifies the functions and characteristics of different types of embankments and the stalk and wood materials used to construct them, it describes the steps that Jia Lu took to seal the breach, and it details various aspects of engineering and construction.[75]

Records of River Defense systematically describes how to integrate Grand Canal maintenance into Yellow River management. Ouyang Xuan explains that prior to the opening of the Grand Canal, river managers generally treated Yellow River infrastructure merely as a problem of defense against floods. They built dikes and diversion canals to protect specific fields and population centers. Now, river management needed a new and expanded repertoire of hydraulic systems, and methods to keep the river confined to its bed and simultaneously to channel it in ways helpful to canal transport. He knew that sediment from the Loess Plateau was a problem, so the system needed ways of dealing with silt: by creating drainage channels, keeping them dredged, and blocking floodwaters from entering the canal network.[76] Although the Jia Lu system was only sporadically maintained during the Yuan-Ming transition, the accomplishments of these hydrocrats demonstrated the viability of engineering floodplain hydrogeography around the Grand Canal and made sediment discharge into an aspect of Yellow River management. This canal-centered and floodplain-wide approach was the final conceptual innovation in the life of the hydrosocial river of the imperial era.

By 1368, rebels under the control of Zhu Yuanzhang (1328–1398), a millenarian sect leader from a family of destitute tenant farmers on the Huai River plain, had vanquished other claimants to imperial power and captured the Yuan capital in Beijing. He declared himself emperor, leader of a new dynasty that he named the Great Ming. The new regime took control over a floodplain region that had become notorious for poverty and banditry. The first Ming imperial capital was situated on the Yangtze River in Nanjing, rather than in the north, and the Grand Canal fell back into disrepair.

The river changed course in 1391, with two major branches primarily capturing the course of the Huai River. As figure 4.14 depicts, until the 1590s, the river again occupied multiple channels, each of which followed a muddy path that spread standing water and sediment throughout the low-lying wetlands, former canal courses, and Huai tributaries to the south of the Shandong Peninsula. At times, the river took multiple courses simultaneously. Figure 4.15 reveals an intriguing sequence of peaks and valleys in the ratio of documented infrastructure work to attested river disasters during these centuries. At times, half or more of river documents refer to management activity. About a third of the management activities involved repairs of

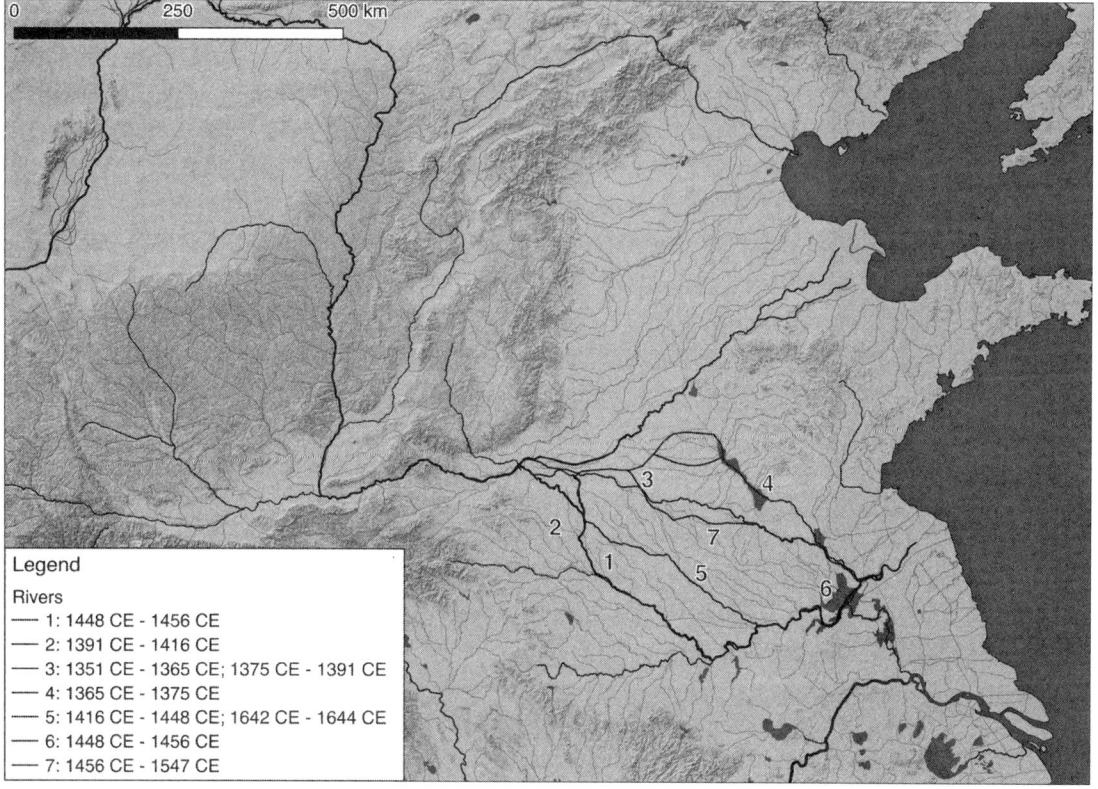

FIGURE 4.14
Early Ming River Courses. Between the mid-1300s and the mid-1600s, the river took a multiplicity of southern courses, sometimes occupying several of them at the same time.

existing structures, and about 10 percent entailed new construction. Decades of unmitigated flooding transpired between each activist episode.

Pointing to Yu the Great's legendary technique of "dispersing the nine rivers" (*shu jiuhe*), early Ming water managers diverted excess water into canals and tributaries rather than building levees. Good management was consistent with tolerating, or even promoting, multiple and shifting flows, even though this meant that commerce between north and south was more difficult and that populations and agricultural productivity on the floodplain were lower. With the Grand Canal out of commission, merchants and officials transferred grain shipments serially from one barge to the next, transiting multiple canals, streams, and lakes along complex and shifting routes to the north.[77] Water managers directed floodwater into one or more of the profusions of small streams that branched off the main channel and flowed into the Huai. River management policy focused on

FIGURE 4.15

Management to Disaster Ratio, 1368–1590. Attention to river management between the mid-1300s and the late-1500s came in bursts of policy initiatives followed by periods of less activism around infrastructure.

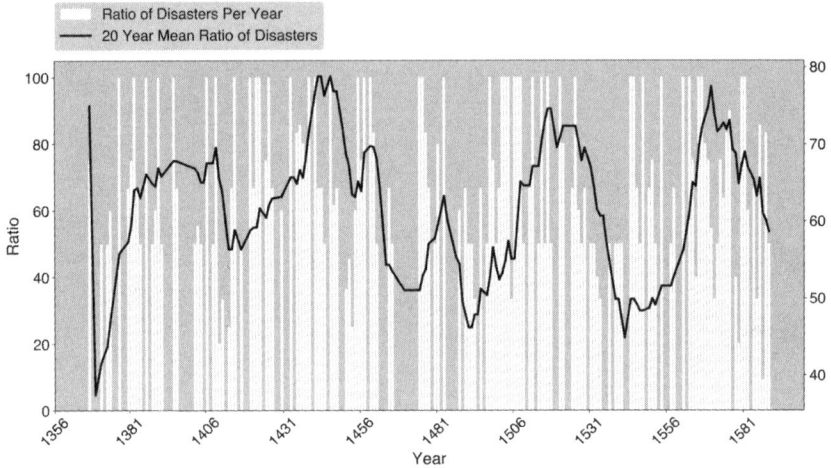

identifying the best diversion paths rather than anti-diversion strategies. Farmers and merchants subsisted in a mutable and marshy land.

Figure 4.16 is a population map of the Ming empire around 1550. The shift of the population center of the realm, which began before the ecological collapse of the north in the eleventh century, was now complete. The Huai basin had lost a great deal of its population. On the Loess Plateau, although population in the Ordos remained sparse relative to other regions, the Fen River valley in Shanxi was one of the most densely populated regions in Ming territory.

Only in the late sixteenth century—after two hundred years of design experiments and policy disputations, during a time of deteriorating conditions on the Loess Plateau—did the regime and its water managers unequivocally settle on a consensus conviction in favor of annually repaired tax-supported earthworks that could lock a stable single course into place at any cost during an era of rising sediment rates. The "modern" Yellow River was a conscious policy choice, and it dates only to that time.[78]

This is not to say that early and mid-Ming water managers ignored the floodplain. They closely observed rainstorms and improved their estimates for the timing and magnitude of floods. They developed new procedures for creating and maintaining embankments, building construction materials, and repairing breaches. Many innovations involved new means for directing silt and sand purposefully into drainage basins so that it would not clog the main river channel or the sluice gates and other mechanisms.[79] Between 1411 and

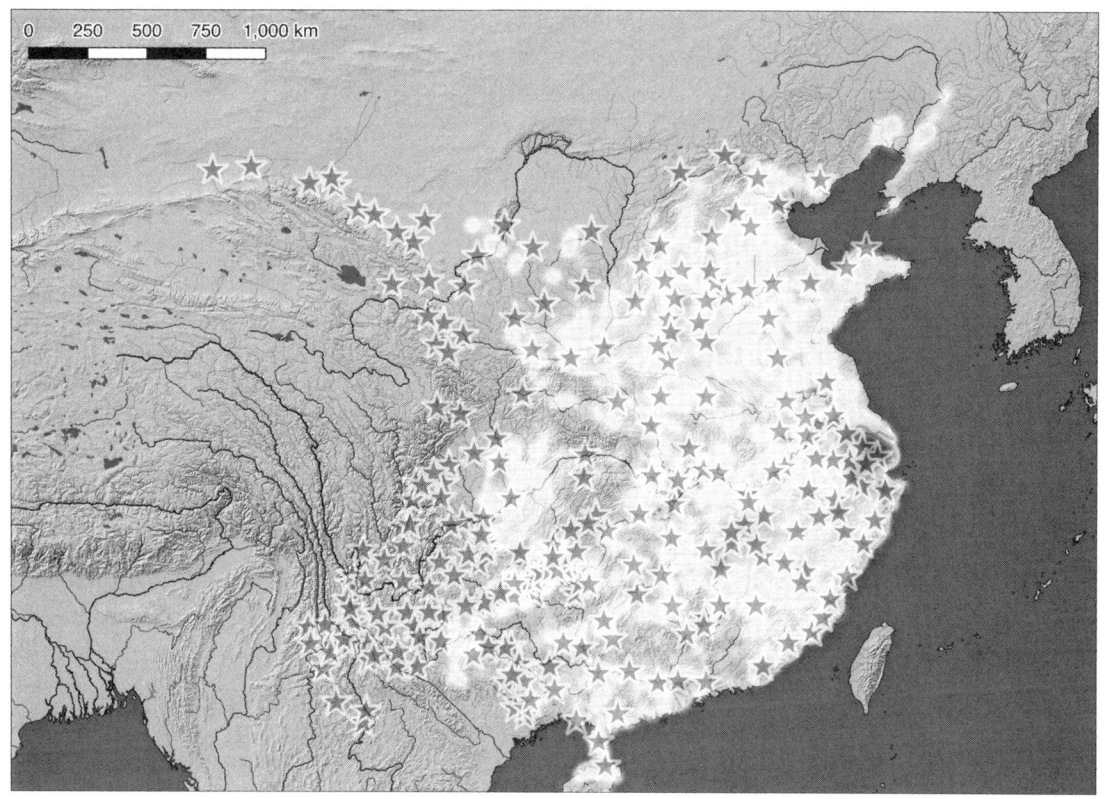

FIGURE 4.16
Population and Prefectures in 1550. By the middle of the sixteenth century, the center of dense population had moved to the Yangtze delta. The population on the floodplain was relatively sparse, and there were few prefectural seats..

1415, even as the river took multiple channels, the Grand Canal was reconstructed almost entirely. Engineers built dams, diverted rivers into the canal, and dug four large reservoirs to regulate water levels. A total of 165,000 laborers worked to dredge the canal bed of accumulated sediment and build new channels, embankments, and locks.[80] The capital moved to Beijing, where it was dependent on predictable and continuous Grand Canal transportation. Canal-side cities and cities connected with the coastal salt trade became wealthy metropolises occupied by politically connected merchants.[81] By the mid-fifteenth century, 121,500 soldiers and officers operated 11,775 government grain barges on the Grand Canal, and 47,004 full-time corvée laborers maintained the system.[82] The Grand Canal had fully displaced the sea route to the capital.[83]

However, the canal was useless without the right management of the Yellow River. Floods around the Grand Canal filled its bed with sediment while floods further upstream diverted the water away from the canal and left it inoperable. People living on the floodplain around the canal and upstream from it began to be protected from floods at all costs, but their well-being was not the point of the venture. They were compelled to contribute labor and building materials; their land became saline as tributaries were extinguished and drainage worsened; and when levees breached, or when water managers intentionally opened sluice gates onto their lands during times of high water, the results were catastrophic for them. Taxpayers in the Yangtze delta paid a surcharge on grain tribute, which covered the bill for the costs of transporting the grain and maintaining the Grand Canal and the Yellow River.

Nevertheless, these imperatives did not fully fix in place the river's muddy course across the floodplain. Between 1416 and 1448, the dominant southern course took the bed of the Guo River, except between 1448 and 1455, when a substantial northern course captured the channel of the Qin River. Following a series of floods in the late 1480s and 1490s, up to 70 percent of the water went north again. Although these events interfered with the south-to-north transportation of grain, and by extension the integrity of the empire itself, hydrocrats were not sure how to respond to the rising sediment burden and to the overwhelming challenges of hydroengineering.

There was never a consensus for the single-channel, high-intervention approach in the Ming: that is why the management-to-disaster ratio timeline depicts such vivid peaks and valleys of infrastructure maintenance activity (fig. 4.15). Consternation about the rising silt burden was a significant concern. In the late fifteenth century, policy makers even began to contemplate walking away from floodplain and Grand Canal management because so much sediment was accumulating that they could not figure out how to transport it from the river channels to the sea.

The river was beginning to break away from the Huai River plain and toward its former course north of Mount Tai. According to a 1492 account, "Today the old course has silted up and become shallow, so that [the river] has gradually shifted north, joining with the Qin River [northern Zhili], its situation becoming more and more unmanageable."[84] In 1493, Liu Daxia (1436–1516), head of the Ministry of War (*bingbu shangshu*), led a massive construction effort that resulted in the creation of a system of continuous levees that would assure an exclusive southern route.⁸⁵ However, this measure served only to move the problem downstream, transporting sediment to the river mouth, where it formed mounds that kept the river from draining out to sea. This, in turn, caused floods upstream, blocking the Grand Canal and disrupting grain transport.⁸⁶ The levees were abandoned. There was simply too much silt.

Some Ming commentators were entirely clear that their predicament stemmed from erosion on the Loess Plateau. Liu Tianhe (1479–1545), a famous minister who had served as a military officer on the Mongol front in Shaanxi, was also a polymath with an interest in engineering and medicine. In *A Compendium of Questions about Water* (*Wenshuiji*), he made all the connections between sediment, topography, water management engineering, and flooding. First he described the fact that the river was clogged with sediment: "The Yellow River is turbid. On the lower course it is blocked and obstructed, and thus it is silty. In the middle channel the water's flow is spread out and sluggish, and thus it is silty. The river's course is sinuous, and thus it is silty. In the hot summer and in autumn [the water level] rises dramatically and abruptly retreats, and thus it is silty." Next, he listed six origins of the silt. The first, unequivocal one was the condition of the Loess Plateau, a place where he had spent much of his career: "[Silt] comes down from the gullies that have come to exist in the high-altitude places in the northwest, from whence the water rushes with extreme violence that levees are unable to resist." He continues, detailing a chain of causality that extends from the gullies of Shaanxi to the breaches on the floodplain:

> Second, the silt raises the riverbed higher and higher, so today, if one surveys the middle of the stream as it flows through Kaifeng, [one discovers that] in winter and spring, it has a depth of only a bit more than one *zhang*, while even in summer and fall it does not exceed much more than two *zhang*. Its bed is high, unlike the abyss of the Changjiang [Yangtze] course. . . . Fourth, downstream, the land is extremely flat and there are no mountains to control and confine [the river]. Fifth, in all of the former northern and southern courses of the [river] through the central prefectures, the earth is mixed with mud and sand. Sixth, this perfectly causes [riverbank] collapse through susceptibility to breaches.⁸⁷

Liu Tianhe, with experience on the floodplain and on the Loess Plateau, was an ideal river interlocutor who offered trenchant expertise about rising erosion on the Loess Plateau and the existential challenges that sediment posed for river management downstream.

During the first three quarters of the sixteenth century, the Yellow River continued to take multiple simultaneous paths across the Huai floodplain, prompting significant commentary and disputation concerning possible solutions. In 1558, one observer explained that the power of the river was too weak to maintain a clear course or to entrain sediment, as the current was distributed among so many streams that sediment settled in riverbeds and river mouths rather than being pushed out to sea. Commentators noted sixteen different streams in 1565.[88]

Wu Guifang (1521–1578), director general of grain transport (*zongdu caoyun*), had authority over the Grand Canal and Yellow River. He was distressed about the numerous floods that inundated people's farms and homes and interrupted canal transportation. In 1577, he developed a radical proposal to build dams, reservoirs, and earthworks to create a narrow channel with a rapid current that would take advantage of the altitude gradient of the floodplain to scour sediment and push it out to sea. The following year, the court tasked Pan Jixun (1511–1595) with putting Wu Guifang's ideas into action, essentially reconstructing the entire lower course of the river. This was the end of the era of multiple courses and the beginning of the final life span of the imperial floodplain.

INTERLUDE: TECHNOLOGY, LABOR, AND HYDROGEOGRAPHY

Following a series of official posts at the court and in the southeast, Pan Jixun was appointed to the position of superintendent of the Grand Canal (*zongli hedao*) beginning in the 1560s. He directed a complete redesign of the Yellow River floodplain infrastructure. Hongze Lake, the reservoir that formed as an unintended consequence of human water management after a sixth-century flood, was the centerpiece of the new hydrosystem.

In different ways, figures 4.17 and 4.18 and plates 33 and 34 depict the fact that the lake was situated at a slightly higher altitude than the Yellow River. Plate 33 depicts a late eighteenth-century episode in the reconstruction of the Qingkou channel, "the clear passage" at the northeastern corner of the lake, designed to force lake water into the

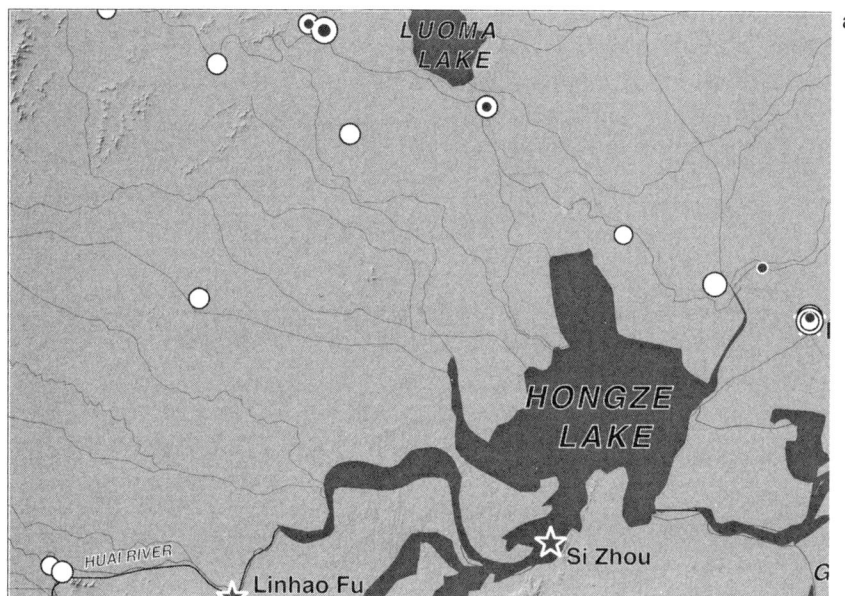

a FIGURE 4.17

Hongze Lake Terrain and Events, 1368–1911. Management and disaster events varied in intensity along the Yellow River north of Hongze Lake between (a) 1368 and 1468, (b) 1468 and 1568, (c) 1568 and 1668, (d) 1668 and 1768, (e) 1768 and 1868, and (f) the end of the Hongze Lake system between 1868 and 1911. The river runs just to the north of Hongze Lake. The Grand Canal skirts the eastern edge of Gaoyou Lake, passing through Huaian before entering the Qingkou infrastructure network at the northeastern perimeter of the lake and then paralleling the Yellow River. The Huai River enters Hongze Lake from the southwest.

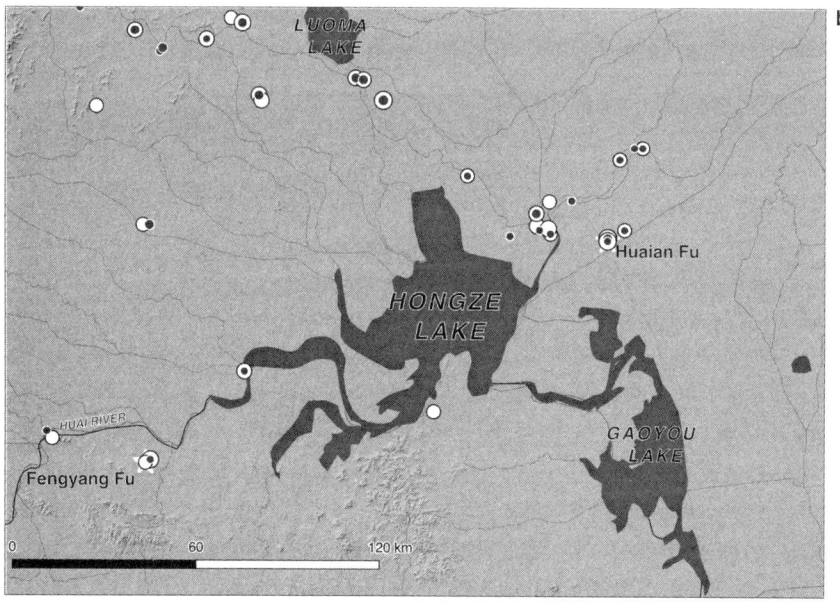

b

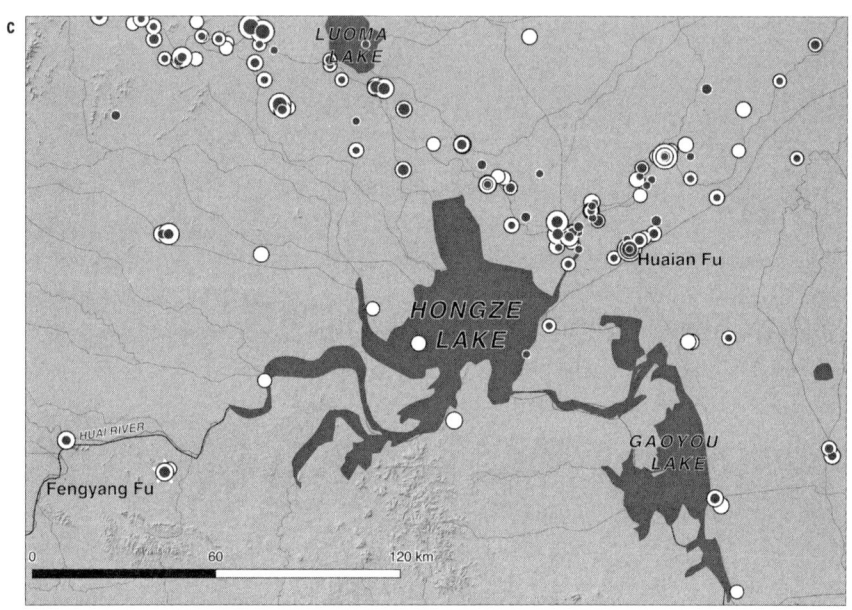
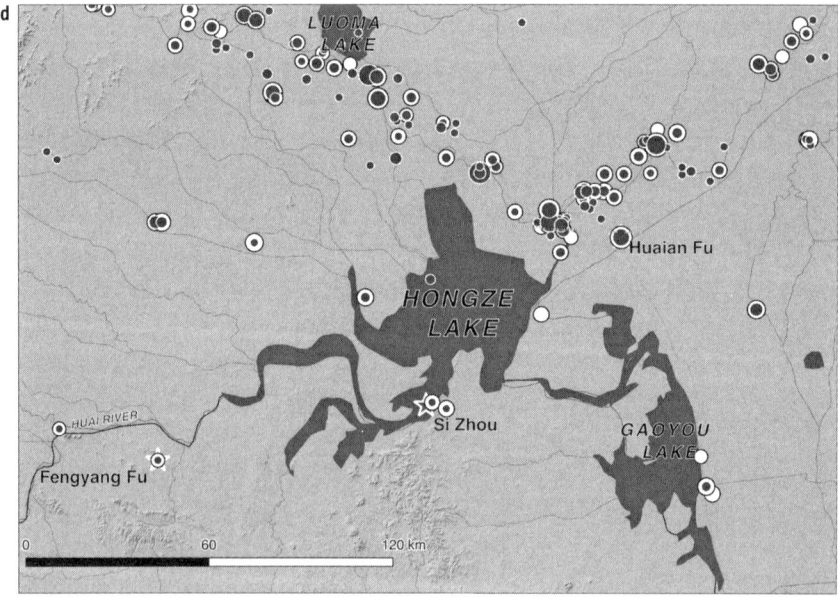

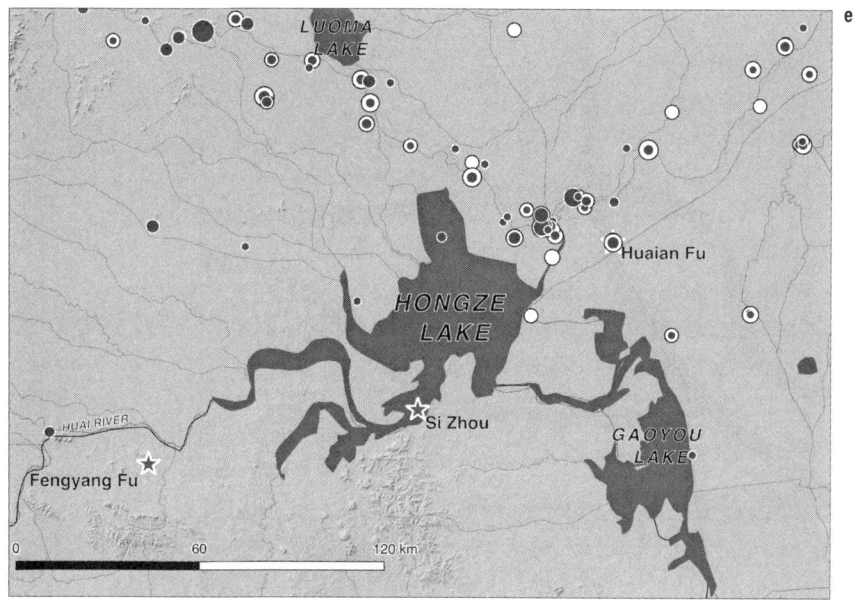
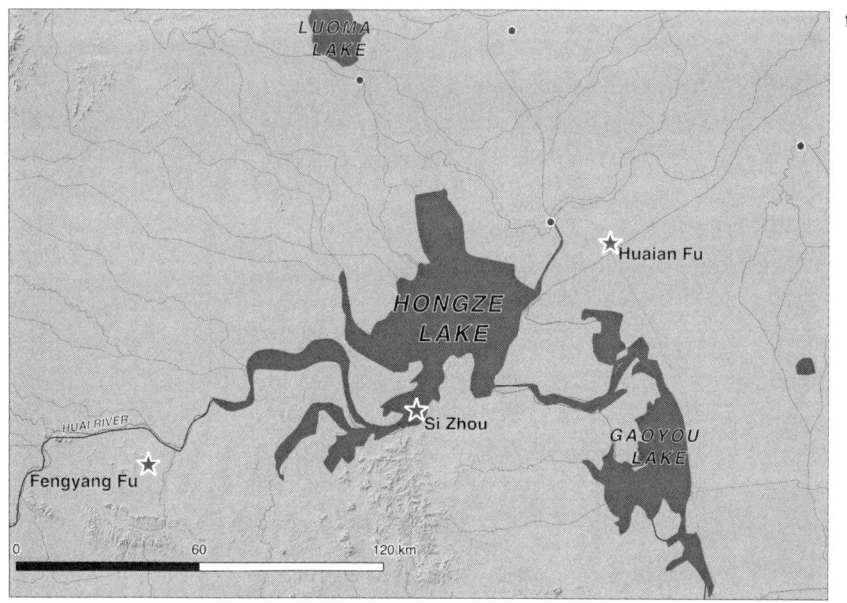

FIGURE 4.18

The Evolution of Hydrological Architecture around the Qingkou Confluence. The major features of the confluence of Hongze Lake, the Yellow River, and the Grand Canal were the Qingkou passage, Gaojiayan Dam, and various watercourses. Although construction of the Qingkou system began in the Ming era, it did not reach full buildout until the eighteenth century. These images depict the Qingkou confluence (a) prior to the fourteenth century, (b) in 1776, and (c) in 1854. Based on Zhang et al., "Qingkou Complex."

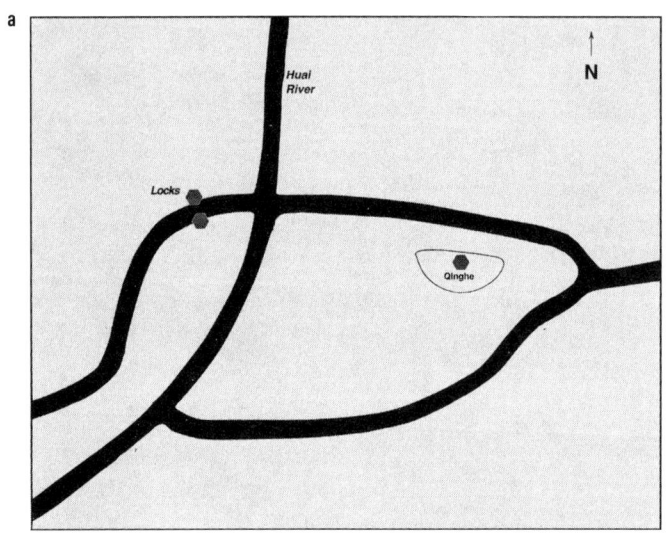
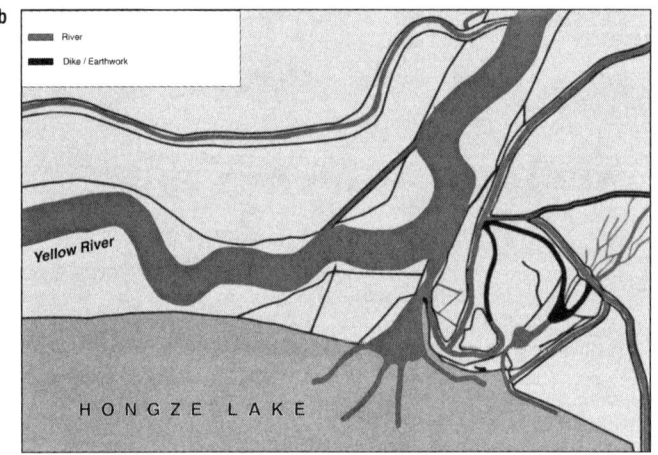
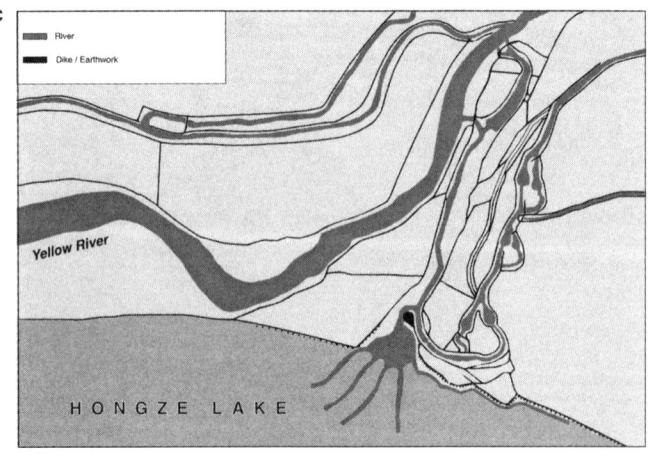

muddy Yellow River at high velocity. Plate 34 shows a spillway that disgorged high waters onto farmland at the lower-altitude southern side of the lake. The Huai River was slightly higher again than the lake itself. Late Ming and Qing engineers took advantage of this topography, constructing earthworks that directed Huai River water into the southwestern quadrant of Hongze Lake. A long dam, called Gaojiayan, constrained the lake on its eastern side and included floodgates to regulate water flow into the Yellow River and Grand Canal.[89]

The problem of managing rising quantities of sediment deposit was at the core of the system. During high-water seasons, lake water flowed rapidly through the narrow Qingkou channel. The torrent of lake water spilling into the river would cause the river current to accelerate as it flowed past its confluence with the Grand Canal, thus preventing sediment from settling on the riverbed. Close-set dikes reinforced with stone confined the river in a straight and narrow channel in all but the worst floods. Offset dikes up to a mile behind them contained the overflow in the highest flood stages. Additional embankments and sluice gates farther upstream helped ensure that the river would maintain a constant current: fast enough to reduce the risk of meandering and silting, and slow enough to reduce the current.[90]

During dry seasons, lake water flowed through the Gaojiayan floodgates and the Qingkou channel to raise water levels on the Grand Canal high enough to float the grain transport fleet to the Yellow River.[91] Barges traveling north moved from the Grand Canal to the Yellow River, then traveled on the main channel of the Yellow River before moving back to the Grand Canal for the rest of the journey to the capital.[92] As figure 4.17 and plate 33 depict, by the late eighteenth century, an intricate complex of locks and canals protected the canal water from the silty river and kept the boats almost entirely out of the river. Boats required a week of travel on a winding passage through a long series of water gates to traverse the few kilometers of the Qingkou complex.[93] An elaborate system of dikes, weirs, and gates on the river diverted floodwater into side channels. Plate 34 depicts one of these structures today.

Sedimentation caused the river to meander inside the outer dikes, making microscale course changes when silt forced the water into new channels. As water crashed against the levees in new locations, it undermined the earthworks that confined the riverbed. Even the restricted river was a dynamic stream, and keeping it curbed required an ever more extensive network of inner dikes, defensive barriers, locks, spillways, and drainage canals. The system was enhanced with dredging by human and mechanical means. Officials conducted extensive dredging during dry months to cut across meander loops and deepen the channel. In the end, there was simply too much dirt to move: every major flood deposited millions of tons of sediment.[94] Ultimately, the system evolved into an extensive series of locks and a complex network of dikes and channels that paralleled the river across the entire floodplain from Zhengzhou to the coast.[95]

The levees formed a barricade between farmers and their sources of irrigation water and fertile sediment. In earlier times, during seasons when the water was relatively calm, the heaviest particles of sand and gravel would settle to the bottom of the riverbed. Farmers could direct to their fields the fine granules of loess that remained suspended in the water column, where they would replenish inundated or exhausted soil. Once the levees were in place, this became impossible. Over time, the riverside farmlands, inundated only by the most turbulent floods that entrained heavy particles along with fertile loam, deteriorated into low-yield sand fields.[96] The late-imperial waterworks infrastructure was about protecting commerce and transportation, not improving the well-being of farmers.

Figure 4.19 is a dramatic depiction of the well-defined procedure for responding to levee breaches.[97] This dirty, freezing, and dangerous work required thousands of workers with a wide range of skills, as well as ancillary labor to build and maintain tools, feed people and animals, and prepare stalk bundles like those depicted in figure 4.20. Breach repair occurred during the dry winter season, which was also downtime for farmers. Laborers carried stalk bundles and wheelbarrows of dirt to the edge of the breach and stacked them on a special platform with ropes laid underneath, then other workers stamped on them to flatten them. Another crew tied bundles together with ropes and rolled them into the water, where workers on boats would bind them with more rope and tie them to a structure in the water before dumping earth onto it and tamping it down. If there was ice around the work site, ice-breaking boats and workers with poles to break the ice would be part of the crew as well. Proceeding this way, workers would construct two halves of a repair dike out into the current from either side of the breach, one stalk section at a time, pinching the

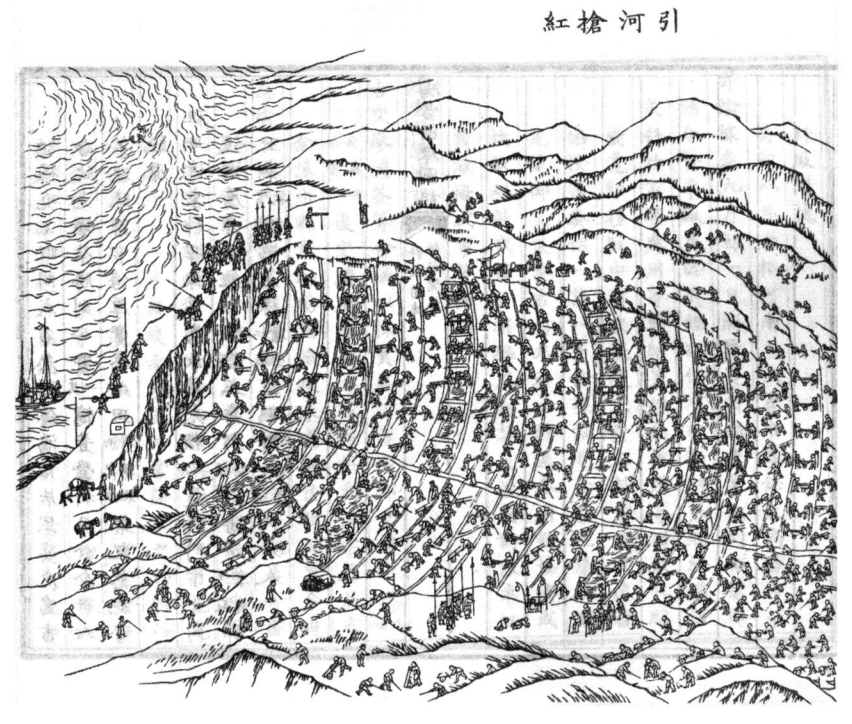

FIGURE 4.19
Responding to a Levee Breach. This image, entitled "Vying for the Red when Cutting a Canal," is accompanied by a text which explains that workers are racing to finish because of the time pressure of this kind of labor. The winners will have their names inscribed on a red silk lantern and will receive meat and wine, boots, and hats. Linqing, the author of the autobiographical work in which this image appears, explains that these incentives are preferable to threats and favoritism, which could cause workers to strike. This is a vivid image of the kind of labor that accompanied any major levee repair. From *Illustrated Records of the Tracks of Fate* (*Hongxue yinyuan tuji*). Courtesy of the East Asian Library and the Gest Collection, Princeton University Library.

flow at the breach, with the current speeding up. Meanwhile, another crew would excavate a diversion canal where the river turned toward the breach and a deflection dike to push the current away from the repair dikes and toward the diversion canal. When the two ends of the repair dikes drew close enough for a single bundle of stalks and earth to seal the breach, a crew would open the diversion canal to reroute the river while the workers at the breach put the final stalk plug in place and pounded it into the bottom. Once the emergency repair was complete, laborers would strengthen all the dikes, including new deflection dikes to keep the river at bay if it had shifted course.⁹⁸

South of Qingkou, drainage canals connected Hongze with the Gaobao lakes (fig. 4.17). These were what remained of the primeval Huainan wetlands landscape after centuries of water management. The lakes contracted during the fifteenth through seventeenth centuries, overwhelmed by sand and sediment deposited from the Yellow River. The massive deposits of sediment meant that flood resilience declined throughout the floodplain. Technical modifications improved sediment trapping at Gaojiayan in the eighteenth century,

FIGURE 4.20

Stalk Bundles. Bound sorghum stalks with attached hooks were the most common material used in earthworks construction and repair on the Yellow River floodplain. From an 1836 edition of *Illustrations and Explanations of Tools for River Work* (*Hegong qiju tushuo*). Courtesy of the East Asian Library and the Gest Collection, Princeton University Library.

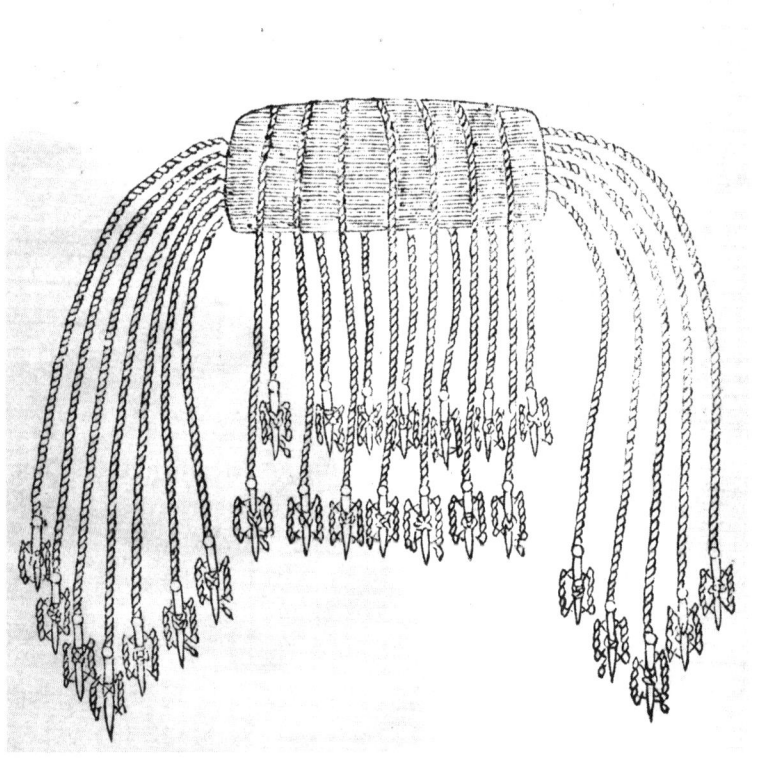

and the Gaobao lakes expanded again. Thereafter, they shrank once more as the system aged and deteriorated and the quantity of sediment in the system became overwhelming.[99] There was simply too much silt, its quantity rising at too rapid a pace, for even the best-designed preindustrial technology to manage.

In short, the system did not function exactly as planned, at least not in the long run. When the Yellow River flooded, Hongze's dams and embankments were not adequate to prevent its water and entrained sediment from flowing into the lake. Although the excess water could readily be drained out, sediment settled on the lakebed, raising it higher and higher, and negating the topographical advantages that animated the system. As the dam rose and the lake expanded, it intruded into the Huai River upstream. As the gradient of the Huai decreased, ultimately by 38 percent, water crossed the whole floodplain more sluggishly, sediment settling wherever water slowed and pooled. Below the land's surface, the groundwater level around the Huai rose, and the river's ability to discharge floods declined for that reason as well. Waterlogging and flooding became more frequent, and farmers' wells filled with sand and brackish water. The land around the river subsided, drowning cities and farms. A flood sluice on the Gaojiayan Dam collapsed in 1851.[100] Much of Hongze Lake drained out through the Gaobao Lakes and into the Yangtze, turning the Huai River into a Yangtze tributary.[101] The Yellow River levee system breached catastrophically four years later.

THE LAST IMPERIAL FLOODPLAIN: PAN JIXUN AND HIS LEGACY

As director general of the Grand Canal, Pan Jixun's innovation was to focus on silt rather than water. He observed, perhaps with some hyperbole, that the water of the Yellow River was 60–80 percent sediment and only 20–40 percent water.[102] His rubric was to "restrict the current to attack the sand" (*shushui gongsha*) and to "build levees to restrict the water, using the water to attack the sand" (*zhuti shushui, yishui gongsha*). In place of the channel diversion system that had existed until his time, he designed continuous close-set levees that would keep the current moving rapidly across the floodplain, and he designed the Hongze-Gaojiayan-Qingkou infrastructure that permitted water to be released periodically to scour the Yellow River bed.[103] The Pan Jixun system protected the Grand Canal from intrusion by the sands of the Loess Plateau. As the seventeenth-century historian and geographer Gu Yanwu (1613–1682) observed, "The Ming blocked the rivers in order to bring rice to the capital."[104]

The numbers are staggering. Pan's project built roughly 365 kilometers, or 1.2 million feet, of earthen embankment; and almost ten thousand meters, 30,000 feet, of stone embankment. Laborers under Pan's direction stopped 139 breaches, constructed four stone spillways of nearly a hundred meters each, dredged 35 kilometers of riverbed, planted 830,000 willow trees to stabilize the tops of the dikes, and

FIGURE 4.21

A Sixteenth-Century Map from *A Unified View of River Management*. Pan Jixun's accomplishments included documenting a waterworks vocabulary as well as engineering innovations themselves. This page from his text labels sluices, dams, canals, reservoirs, and levees. Courtesy of the East Asian Library and the Gest Collection, Princeton University Library.

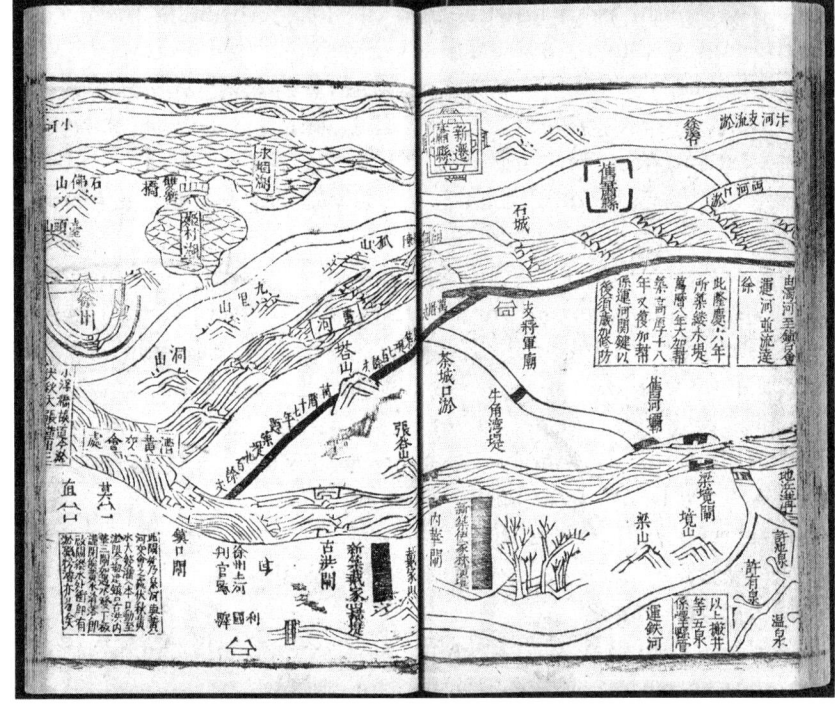

drove in a large but unrecorded number of tree trunks as piling under the embankments. The effort cost 500,000 ounces of silver and nearly 127,000 *dan* of rice, perhaps 7.5 million kilograms, or over 16 million pounds.[105] Figure 4.21 is a map from a first edition of Pan Jixun's book *A Unified View of River Management* (*Hefang yilan*), which spells out his accomplishments and engineering philosophy and pioneers a visual vocabulary for river management.[106] Pan's work occurred in two stages. After being criticized for interfering with navigation, Pan was demoted in 1571, before being reinstated to complete the task by 1580.[107]

The *Ming History* explains Pan's approach. It was explicitly designed for a time of rapid erosion on the Loess Plateau: "Although human power is incapable of dredging [sediment], the power [*shi*] of the water itself is capable of scouring it away by impact.... If we keep the dike defenses in good repair, on the other hand ... then the water will pass through the midst of the land and the sediment will be removed by being entrained in the water."[108] Another comment from the same time explains: "When waters are divided, their power is weakened, and when their power is weakened, sediments come to a standstill. . . . [But] when waters join, their power is fierce, and when their power is fierce, the sediments are scoured away. . . . One builds dikes to confine the water and one uses the water to attack the sediments. If the water does not precipitously overflow the two banks, then it has to scour in a straight line along the bottom of the river."[109] The discourse and debate about how to manage *shi*, the power of the river's current, a conversation that dates back to the days of the mythical Yu the Great, is now linked inextricably to the sediment crisis.

Pan Jixun's interventions finalized the hydrogeography of the highly engineered floodplain, which would remain stable for three hundred years. From then until the mid-nineteenth century, despite frequent floods, there were no major course changes or levee breaches.[110] The river of the late Ming and Qing loomed above its floodplain as hydrocrats built ever-higher levees above its rising bed; it flooded often, though rarely catastrophically; and it maintained its course until 1855. It flowed as much as twelve meters (forty feet) above ground on some occasions.[111] The levees ultimately extended over 1,400 kilometers—almost 900 miles. They required diligent maintenance and had to be rebuilt after every flood.[112] As the river ran above the surface of the plain, the floodplain bisected into two separate river systems: a Hai River system in the north and a Huai River system in the south. Tributaries reversed direction, flowing away from the Yellow River rather than into it. Rather than serving as a destination for tributaries, as it would have been in a less engineered system, the Yellow River became a source of tributaries, with seepage from its levees running off into the Huai and Hai watersheds and effectively sacrificing the Huai basin.[113]

In other words, the late-imperial floodplain infrastructure system had no solution to the accelerating quantity of sediment flowing from the Loess Plateau onto the floodplain: it simply moved it away from the Grand Canal confluence and into other locations that were of less strategic value to the imperial regime. As more sediment washed out to sea after the late sixteenth-century constructions, the rate of delta accretion increased from an average of 33 meters (just over a hundred feet) per year to 1,540 meters (almost a mile) per year, although that rate persisted only briefly. The river mouth became so long and gentle that backwash ultimately deposited more sediment in the river than had been present there before.[114] Silt stopped up the river when waters were high. Owing to sediment blockage at the sea mouth, a flood in 1591 inundated the city of Sizhou to a depth of a meter, and 90 percent of residents drowned. Pan's critics wanted to dismantle the new infrastructure and return to multiple channels.[115] Thereafter, levee maintenance relaxed somewhat, and new drainage canals were dug in 1595–1596 "to repress the power of the river." The rate of delta expansion became somewhat lower, though still much higher than it had been prior to the construction of the new waterworks.[116]

By the turn of the century, China was suffering from fiscal crisis and drought, and the aging Ming regime failed to resist the pressures of peasant rebellion and foreign invasion. In 1642, Ming armies, in a futile effort to resist a six-month siege by Shaanxi rebel Li Zicheng, destroyed the flood-control system that had successfully protected the city of Kaifeng since the mid-fifteenth century. The city was inundated, and most residents drowned or died from ensuing famine and plague. Li Zicheng's armies proceeded to Beijing; within two years, they too had been vanquished and the Manchu Qing dynasty was in power. Kaifeng was not rebuilt until the 1660s.[117]

The new regime maintained and extended the floodplain management techniques and principles of the late Ming. The hydrosocial system of the floodplain reached full administrative buildout by the eighteenth century. By the nineteenth century the lower course was an entirely engineered system of locks, spillways, dams, and dikes; lakes, wetlands, and streams; and tens of millions of people and other species who lived on the managed alluvial plain. The infrastructure continued evolving and complexifying until the early nineteenth century, becoming "more extensive, more technologically sophisticated, fiscally demanding and administratively challenging."[118] As figure 4.22 shows, there were two episodes, one in the early eighteenth century and one in the early nineteenth century, during which new construction rose far above historical precedent to become the predominant type of infrastructure activity.

The new regime instituted annual levee maintenance. Qing hydrocrats also significantly extended the Pan Jixun system. This entailed new engineering works, along the lines revealed in figures 4.17 and 4.18, and also new modes of administrative and fiscal organization.[119] Multiple Qing emperors toured the Yellow River and the Grand Canal and inspected the waterworks. In the tradition of Yu the Great, these monarchs considered the management of the floodplain

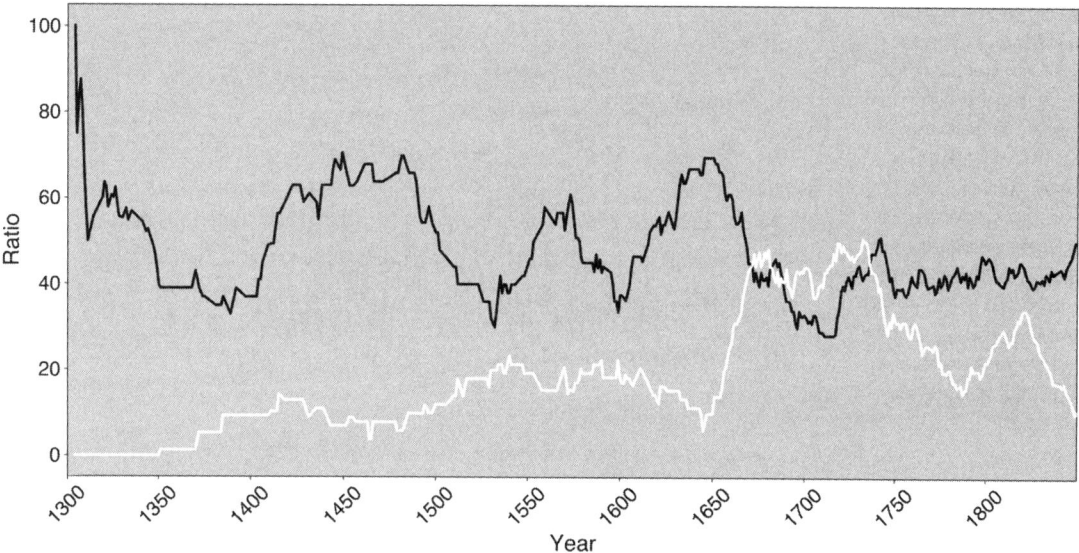

FIGURE 4.22
Ratio of Repairs to New Construction, 1300–1850. Watershed management not only changed in magnitude during the final centuries of river management; the characteristic activities evolved as well. During its first century in power, from the mid-seventeenth century to the mid-eighteenth century, the Qing regime promoted new construction more intensively than repairs to existing structures.

and the Grand Canal to be one of their core symbolic and pragmatic tasks.[120]

Most of the ecological decline on the floodplain occurred after the eighteenth century. In fact, Qing state water control effectively staved off ecological catastrophe in the Huai basin and offered advantages in the form of the economic benefits of the Grand Canal and the receipt of intensive state investment in the region.[121] However, flood crises became more frequent even as catastrophic levee breaches ceased, since the river carried a high and rising load of sediment and traversed a plain with no natural barriers to keep a channel in place. Plate 35 depicts the effects of frequent flooding on the city of Kaifeng, where the Qing-era city wall lies under meters of loess sediment, far below the current cityscape, and where floodwaters pooled into muddy lakes that are still visible in the historical landscape.[122]

The Qing population more than doubled during the eighteenth century. With settlements and cities built right up to levees' edges, dividing the flow of the river during flood season, which had been an option until the late 1500s, was no longer a viable policy choice. Even if displacing people were possible, the hydrology of the system

LEVIES AND LEVEES

FIGURE 4.23

Qing Dynasty Event Timeline. These timelines offer another perspective on the unprecedented focus on river management during the long eighteenth century. (a) Depicts all the events that occurred during the years in question. An intensive regime of management characterized the long eighteenth century during all rainfall conditions and ended abruptly by 1830. (b) Is a trendline that vividly depicts the intensive-management era as a period extending from about 1690 to 1830.

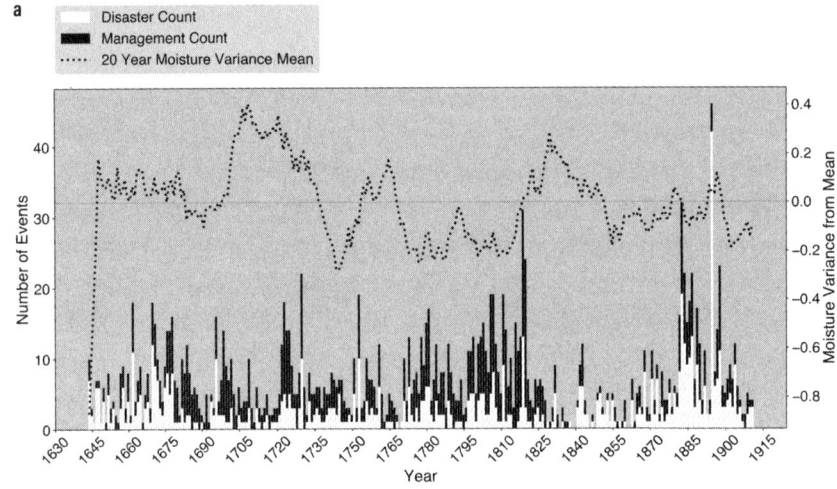

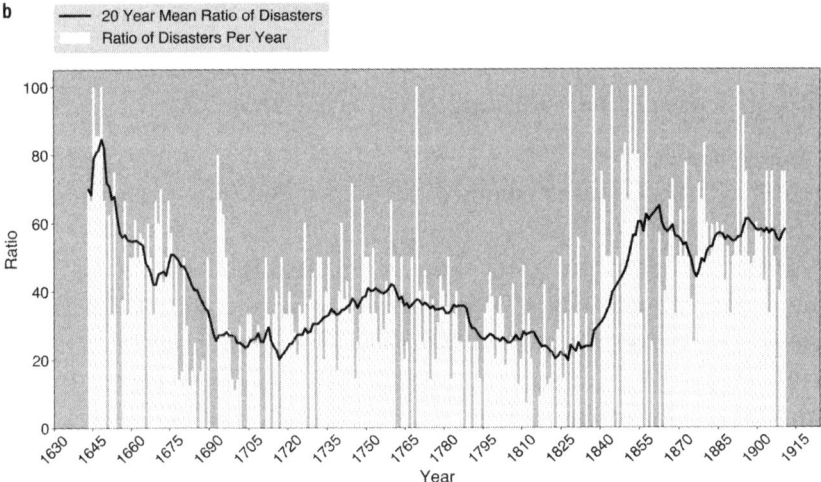

itself had become circular. There was no place to send the water and sand that did not lead back into another part of the hydrosystem. Diverting high waters into other channels only led them by circuitous routes back to the confluence at Qingkou or into the Grand Canal. There was only one option. As the historian Randall Dodgen puts it: "If the Yellow River was to be managed in such a way that its seasonal floods did not disrupt the schedule of tribute grain shipment, some method had to be found to keep it within its channel."[123] As the two timelines of figure 4.23 show, the long eighteenth century was an era of remarkably successful intensive floodplain management. Even in a time of unprecedented erosion and sediment deposit, rigorous attention to infrastructure construction and supervision kept the rate of floods extremely low.

Meanwhile, the Grand Canal had become a commercial artery. Grain merchants received huge government contracts, thousands of transport and river workers made their living on the canal, merchants had patronage networks, and so too did contractors and subcontractors for repairs, provisioning, labor contracting, and other services. These individuals were the political constituency for flood control management.[124] The survival of the realm depended on the transport of four million *shi* (nearly a hundred and ten million kilograms or about 240 million pounds) of tribute grain to the capital every year and on keeping the engineered river courses connected to one another rather than avulsing naturally as sediment rose and courses prepared to change. As the 1748 *Huaian Prefectural Gazetteer* reported: "Under these circumstances, the Yellow River must be diverted to feed the Grand Canal and cannot be permitted to flow northward. The Huai River must be used to scour the Yellow River and cannot be permitted to flow eastward. So, the reason for controlling the Huai is to control the Yellow River. The reason for controlling the Yellow River is to regulate the Grand Canal."[125]

Expenses rose fivefold between 1732 and 1821, and southern taxpayers bore almost the entire cost. By 1821, maintenance of the system consumed between 10 percent and 20 percent of all government spending, not including labor service requisitions.[126] For Dodgen, "two hundred years after the decision had been made to confine the Yellow River to its course, the hydraulic system was bedeviled by a myriad of interlocking problems and had developed a seemingly insatiable appetite for state resources. River control projects demanded specialized technical skills, a huge labor force, and substantial infusions of money."[127]

However, no amount of hydroengineering or funding requisition could halt sediment transport from the eroding Loess Plateau. Although the origin of sediment on the Loess Plateau was illegible to imperial policy making, it was not so to close observers of the situation. Chen Huang (1637–1688) was an influential commentator who clearly understood that the Loess Plateau was the root cause of sediment on the floodplain. He wrote two major works, *Commentaries on River Defense* (*Hefang shuyan*) and *River Defense Abstracts* (*Hefang zhaiyao*). Born and raised in Hangzhou, at the southern terminus of the Grand Canal, Chen studied agriculture and water conservancy as a youth before beginning an official career on the western and northern Loess Plateau. There he studied and surveyed the upper reaches of the Yellow River in person prior to returning to the floodplain. In *Commentaries on River Defense*, Chen wrote fatalistically about a rate of sediment discharge that he assumed had always stymied floodplain management: "Of China's many rivers, the [Yellow] River originates in the most distant place. It flows a long way; and since it flows so far, many tributaries enter it. Since the tributaries are so numerous, the power [*shi*] [of the river] cannot help but be turbulent and tempestuous. [Since] the nature of the earth in the northwest is loose and unstable, as the water rushes rapidly and forms waves, the river water becomes yellow. The people of Qin had a song: 'in one *shi* [about twenty-seven kilograms or sixty pounds] of Jing River water, several ladlesful are mud.'"[128]

Chen was very clear about the relationship of the character of the soil of the Loess Plateau to its erosion into the Yellow River and the impact of that process on the circumstance of the floodplain. Chen did not argue for abandoning the Grand Canal and the engineered floodplain. Based on his personal study and his upriver fieldwork, he advocated for combining diversion channels together with narrow-dike techniques. Chen was also the first Qing hydrocrat to explicitly propose that soil and water conservation were essential to flood control on the lower river.

Chen's recommendations did not have an effect on Qing river policy. By that time, the floodplain was too populous, its tax revenues were too important, and too many people were invested in the new high-intervention approach. A century later, Shen Menglan, a late eighteenth-century county magistrate (fl. 1766), used similar language and made a similar connection, again citing the Qin saying and invoking the character of the loess soil. He averred, "The nature of the earth in the northwest is to spread out flat and to be very dispersed, while the flow of the [Yellow] River is vigorous and very turbid. There is an ancient saying: 'in one *shi* of Yellow River water, six ladlesful are mud.' The Wei and Jing in Shaanxi, the Qin and Fen in Shanxi, and the

Hutuo and Yongding in Zhili are all like the Yellow in this way. When they flood, they are muddy and tempestuous, and when they breach it causes misfortune, since the river's mud is stagnant sediment, and the silted-up blockages cause a calamity."[129] As this commentary reveals, the gulf between expert knowledge on the one hand, and the universe of political possibility on the other, can sometimes be very wide indeed. During the age of environmental collapse on the Loess Plateau, sediment would always flow to the floodplain. The only question was where it would go from there, and when.

THE END OF THE IMPERIAL RIVER

By the turn of the nineteenth century, the floodplain infrastructure was aging, and the equilibrium of the hydrosocial system was facing pressure from many directions. Few tasks of infrastructure building and management took place after the 1820s. Sorghum and millet stalks, the building material that supported the floodplain earthworks, were becoming scarce and expensive. They rotted and had to be replaced every three or four years, and profiteers could hoard them and charge high prices.[130] The floodplain had no forests left by this time: visitors observed a total absence of trees outside of walled villages and family courtyards in Shandong and Henan. With the floodplain devoid of timber and brush, and the timber supply from deforested slopes upstream in decline as well, the stalks were a crucial source of fuel and building material for the local population. Diverting stalks into construction left few available for domestic fuel, fertilizer, and construction. This was one source of human misery on the floodplain, particularly for the residents of a narrow area close to the river itself. It also caused declining soil quality. Under other circumstances, the stalks would have been mulched back into the fields.[131]

Yellow River floodplain engineering during the long eighteenth century did not effectively support the livelihoods of most floodplain denizens, but it did serve the needs of state and elite power structures and prosperous urban life along the Grand Canal and in the capital at Beijing.[132] Although particular floods often came as a surprise, every resident and emperor knew that water and sediment periodically overwhelmed the river. Engineers built vast stone spillways and sluice gates that offered them a good deal of control over which people and places to intentionally impair or sacrifice, destining them to bear the brunt of long-term environmental degradation and short-term catastrophes at the expense of other more favored regions. Plate 34 vividly shows how floodwaters were directed straight onto disfavored low-lying lands.

Although certain places and peoples benefited, along with imperial state power broadly construed, the declining ecological stability of the Loess Plateau and the floodplain over the long term made many people poorer and rendered their lives more precarious. In locations where silt and sand covered the land, farmers had to dig wells ever deeper to reach the water table. Elsewhere, the amount of water was excessive rather than inadequate. After centuries of tributary blockage and wetlands extinction, waterlogging became a problem during the rainy season. Some floods were the result of local heavy rainfall, not high water upstream that overtopped banks and broke levees. The lowest-lying areas, former wetlands and lakebeds, became marshes and breeding grounds for the vectors of waterborne disease, and the soil became more saline. By the late-imperial era, this land was the last to be farmed, the least productive, and the first to be abandoned.[133] The region became "the heart of China's so-called flood and famine region," and the total amount of arable land fell between 1740 and 1930.[134] Throughout northern China, catastrophe coexisted with the routine decline of fertility in this heavily engineered landscape, and the disasters attested in the imperial record transpired in landscapes structured by routine slow violence.[135]

As erosion accelerated on the Loess Plateau, "a thickening blanket of silt" settled on the riverbed downstream.[136] The Hongze-Gaojiayan-Qingkou infrastructure could no longer take advantage of the differential between the higher-elevation lake and the lower-elevation Yellow River. This was the final imperial era of "silt crisis."[137] When hydroengineers opened the locks and dams on Gaojiayan, lake water no longer flowed downstream. Rather, the river poured "an oozing mass of mud" into the lake.[138] The quantity of silt was too great to dredge, and the changing elevation gradients meant that it was no longer possible to sluice the muddy water out to sea.

By that time, each major levee repair cost between 4.75 million and 12 million taels, but it was difficult to purchase materials even when budgets were approved. These are staggering numbers, amounts equivalent to the annual tax revenues of several entire provinces.[139] The money itself did not always arrive in time, and even when it did, the materials budget was often too low for local sellers, who ignored government-mandated prices and charged what the market would bear. When there was a flood, it destroyed many sources of local materials. Various government buyers—counties, prefectures, and regional water control agencies—competed with one another to acquire remaining supplies and often hoarded them to keep them from other agencies.[140]

In late 1824, storms ruptured the Hongze Lake dikes, unleashing floods that devastated the low-lying region around the Huai River confluence, destroyed the northern terminus of the Grand Canal, and discharged massive quantities of silt into the canal bed. By the beginning of the next year, if not earlier, canal managers realized that established techniques for seasonal and emergency repairs were no longer effective because there was too much silt to allow the system to function. As river managers began anticipating that the river could immanently shift course, the Daoguang emperor launched an investigation into sea-transport alternatives to the Grand Canal.[141]

In 1825, Wei Yuanyu (fl. 1793–1825), who had served as a provincial governor-general and as director general of grain transport on the Grand Canal, became the first official to declare that the system was no longer sustainable because of silt and that conditions had changed to make it that way. He pointed out that rising sediment rates had destabilized the riverbed and fundamentally changed its state. He asserted that "the canal course is different now than it was in the past. The Yellow River's fluctuations and meandering are uncontrollable, not like clear water, which can be controlled by human power."[142] The next year, another court official observed that "river conditions are now so dangerously unstable that it is necessary to supervise and defend the dikes at all time," and not to rely on the prior annual calendar with seasons of downtime built into seasonal fluctuations in water levels and precipitation.[143] In 1826, the emperor observed that with "the bed of the canal piled high with silt" and "the clear water from Hongze Lake not high enough to scour out the Yellow River," "conditions with the lake and river are not the same as they were" earlier in the century.[144]

Although the emperor and his hydrocrats recognized that nature and geography created almost-insurmountable conditions of instability in the floodplain system, the official report on the 1824 disaster emphasized the role of managerial failure over natural phenomena. It recommended that river managers double down on time-tested techniques and apply them more assiduously. As the historian Jane Kate Leonard puts it, "The emphasis on administrative practice suggests a rather optimistic belief that canal officials could sometimes temper the destructive forces of nature if they followed prescribed administrative routines—routines that were themselves shaped by natural cycles."[145]

The fiscal logic of the system failed along with its environmental premises. As the bed of the river rose above the countryside, as sediment inundated Hongze Lake, and as erosion ate away at the foundations of the existing levee systems, the engineering system became more "extensive, convoluted and complex." However, funding requisitions did not keep pace with the changes, nor did technical training for administrators. Without admitting that the system was becoming increasingly expensive to maintain, the court emphasized the need for fiscal rigor. The emperor expected costs to be fixed, and he did not trust his officials when they asked for progressively more funding for maintenance.[146]

Figure 4.24 depicts the end of the management era in the 1820s, the subsequent struggles to control the river as the late-imperial hydrosocial system collapsed, and the new crisis zone that emerged to the north of the Shandong Peninsula after 1855. A massive flood in 1841 initiated a chain of events that would end with the river avulsing to the north in 1855, the first time since the Song dynasty that it had occupied that bed. When the river overtopped the dike at its south bank near Kaifeng in 1841, seventy meters (about 230 feet) of the levee collapsed. Floodwaters spread 3 kilometers (5 miles) south to the city overnight. Many urban residents fled the city as refugees from the countryside, including poorly paid flood-control workers, came to Kaifeng, opting to try their luck in a new place. Some refugees joined roaming gangs of bandits. The government administered relief to encourage people to stay in their villages, but it was not enough. Between 1841 and 1855, emergency repairs rose to an unprecedented 25 percent of all documented management events.

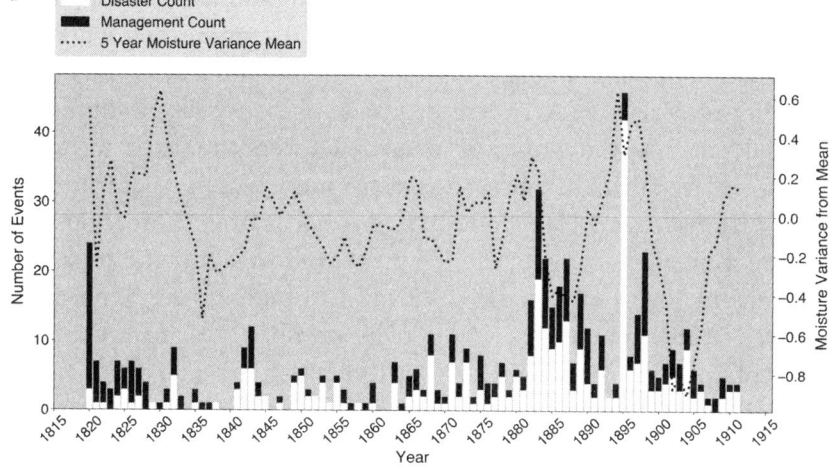

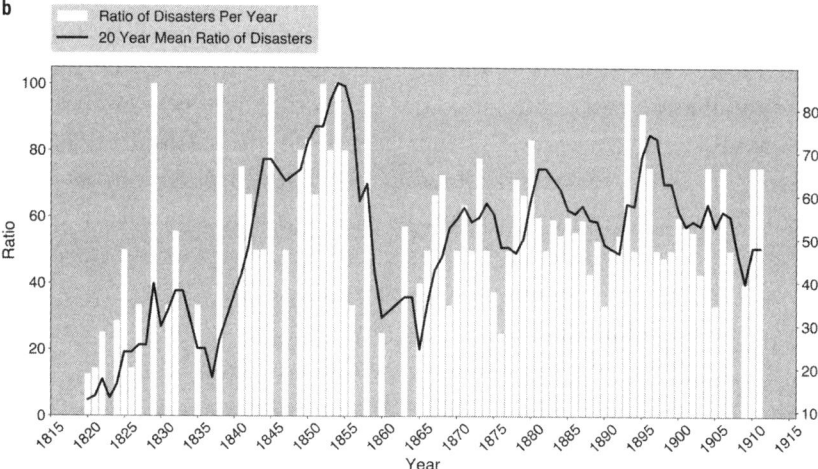

FIGURE 4.24

Floodplain Events, 1820–1911. By 1840, the ratio of disasters to all events on the river was significantly greater than it had been during the management era of the long eighteenth century. Figure (a) depicts the total number of events attested each year, along with moisture conditions. The high-rainfall year of 1895, transpiring after decades of neglect to earthworks, was especially disastrous. Figure (b) depicts the ratio of disasters to all events. After 1840, disasters were usually at least 50 percent of all attested events.

There was no repair material on hand in Kaifeng, as it had all been used for other emergency repairs, and there were not enough experienced officials, who would have known that the site was at risk. Amid the inaction, the breach widened to a kilometer, more than half a mile, and inundated much of the Huai basin with muddy water. Floodwater spread most of the way to Hongze Lake, which filled with silt, along with the rest of the nearby lakes, streams, and canal beds. It took eight months of work and several million taels of expenditure to return the river to its former course. The whole riverbed, from the breach site to Qingkou, had to be excavated, and the task required organizing, housing, transporting, and feeding troops and workers, along with special boats and skilled boatmen and special tools and building materials.[147] A violent storm at the very last minute destroyed much of the work and drowned several hundred workers. In the end, repairing the breach cost 6.5 million taels, requisitioned from various taxes on land, goods, and services from throughout the empire. Kaifeng became a "muddy pit" caked with salt and prone to subsequent recurrent flooding. Villages throughout the flood zone suffered production losses of 50 percent or more.[148] Many remaining residents of the counties in the path of the floodwater, their fields covered in sand, had to abandon agriculture. From then on, they subsisted by foraging for water plants, reeds, and medicinal herbs that they could consume or sell for cash.[149]

After the 1841 disaster, a new flood in 1842 interrupted Grand Canal transportation and deposited a meter or more of sediment. The silt blocked the river's outlet to the sea, and the water carved new small and turbid courses to the ocean through the sandbars. At this point, serious discussion began regarding the possibility of abandoning the old course and allowing the delta to persist in its new form, but commentators recognized that it would be futile to build defensive structures around the new course, which would itself silt up and meander soon enough.[150] Instead, in 1843–1844, there was a repair effort at a site between Kaifeng and Zhengzhou where tens of thousands of refugees had gathered. A frequently flooded place, inundated with sand, it was also a key site for labor and materials provision. There, engineers sought to refurbish the old course. This time, the repair cost a total of twelve million taels. The budget ran to twenty million taels including relief efforts and tax remissions. The undertaking was disorganized and inept, and a harbinger of things to come.[151]

The silt-clogged floodplain hydrosystem, which had been evolving since the time of Pan Jixun in the 1570s, never recovered from the crises of the 1840s. Gaojiayan, the dam at Hongze Lake, collapsed in 1851, and the levee system breached catastrophically four years later.[152] Under pressure from Loess Plateau silt and sand, the coupled Grand Canal and Yellow River floodplain system had outlived the logic of its possibilities. In 1855, the river breached about 20 kilometers northeast of Kaifeng at a site called Tongwaxiang. It divided into three streams, flooding several counties. The river breached to flow north of Mount Tai for the first time in hundreds of years, turning away from its shared course with the Huai, and returned to its medieval egress north of the Shandong Peninsula in the Bohai Gulf. Villages near the Tongwaxiang breach were half buried in sediment, and fifteen floods occurred around the breach site in quick succession.

The shift to the new course was not tidy. The river roamed across the northeast in an essentially unregulated multitude of pools and paths. It primarily occupied the channel of a Hai River tributary called the Daqing River, but it expanded to a width of almost 20 kilometers, or more than 12 miles, in some places. A massive swamp in Zhili, the province that was home to the capital at Beijing, gave "an inkling of how the entire region would revert to a mosaic of lakes and marshes in the absence of human engineering."[153] The river flooded frequently and did not follow a single fixed course. The emergent new course, initially lined with dikes built by local residents, was contained by government requisitioned embankments in a project that began in 1872 and ended in 1884.[154] The Grand Canal system was formally abandoned six years later.[155] As sediment continued to be transported downstream, river management frameworks endured as a value and as a source of legitimacy, even as the promise of industrial-era technology and hydraulic science emerged, along with new forms of bureaucratic organization and new political and ideological innovations.[156]

The 1855 course change reflected neither dynastic decline nor an irresistible natural cycle so much as the technological, economic, administrative, and most of all ecological limits of a complex system conceived and engineered during an era of vertiginously rising sedimentation rates. Although canal-side elites continued to advocate for returning the river to its southerly course, they could not prevail over the objections of powerful southern taxpayers who had subsidized the Grand Canal and the engineered floodplain for centuries. For them, canal-borne tribute was nothing but an expensive burden, and they were happy to see it end. With improved ocean shipping, and with Beijing importing grain from Manchuria, the importance of the Grand Canal had waned.[157] Only at this point did missionaries, revolutionaries, scientists, and other commentators throughout China and abroad begin to construe the North China Plain as a symbol of misery and backwardness that needed to be resolved with a modern state and modern technology. However, they read this stance back anachronistically into the more distant past.[158] In the end, it was the inexorable deposit of sediment from the eroded Loess Plateau, not any failures of engineering or administration, that doomed the late-imperial floodplain hydrosystem to collapse.

EPILOGUE

The Yellow River in the Anthropocene

If control is the problem, it must also, by the logic of the Anthropocene, be the solution.
—ELIZABETH KOLBERT, *New Yorker*, 2019

From 1855 through the 1940s, no entity in China had adequate political authority or military security to enact a long-term and comprehensive plan for river management: not for the unruly and multiple lower courses, and not for the dusty and impoverished Loess Plateau. The Qing dynasty persisted until 1911 amid growing calamities of civil war, imperialism, fiscal crisis, and sclerotic decision making. Following more than a decade of political turmoil after the fall of the Qing, the Guomindang party, emerging from a base in southern China, established control over most of the realm by 1927. The Guomindang aspired to effective governance. However, the regime was at war with invading Japanese armies from 1937 until 1945 and then fell to the Chinese Communist Party in 1949 after four years of civil war.

By the late nineteenth century, subsistence on the Yellow River floodplain demanded immense investment in water management infrastructure. Residents themselves, local officials, and the regimes that controlled the floodplain all built some defensive waterworks. However, flooding was endemic, as were the waterborne disease and crop failure that always accompanied it.

For its most vulnerable residents, northern China in the early twentieth century was a region of perpetual food insecurity. Both upstream and downstream, people in search of fuel destroyed what forests remained, and they gleaned grass and brush from hillslopes as well. Engineers and technocrats knew that there were large coal deposits throughout much of the Yellow River basin, and they understood that mining could alleviate the fuel crisis, but the supply of trees was not adequate to support mine construction.[1] Erosion and flooding rates continued to rise, ruining harvests and increasing soil salinity. Floodplain soil, mixed with sand and gravel from floods, underfertilized, and saline from poor drainage, had become deficient in nitrogen and phosphate.[2] Settlement along the river declined, transportation was difficult, and commercialization, investment, and infrastructure development were limited.[3] Poor residents sold foraged herbs and handicrafts when agriculture failed to meet their subsistence needs.

The Guomindang government, backed by American advisers from the Tennessee Valley Authority (TVA), founded the Yellow River Conservancy Commission (YRCC) in 1929. For the first time in history, a single agency was responsible for attending to both the lower course and the middle course of the river. The structural challenge of erosion on the Loess Plateau began playing a role in Yellow River management planning. On paper, the founding of the YRCC initiated a new era of basinwide administration and an unprecedented level of centralized control over the river basin. The YRCC of the 1930s, explicitly following a TVA road map, planned for modern civil engineering in support of a proposed Yellow River landscape of dams, irrigation, and hydroelectric power. TVA advisers recommended strategies for stabilizing erosion on the Loess Plateau to prevent sediment from building up behind the new dams or clogging the estuary. However, authority on the ground remained fragmented, and the crises of the day thwarted effective execution of the plans that the agency devised at its headquarters in Zhengzhou. In 1931, a massive levee breach at Hongze Lake resulted in a flood that affected 4.2 million households and lasted for fifty-one days. The river flooded in 1933, and once more in 1935.[4]

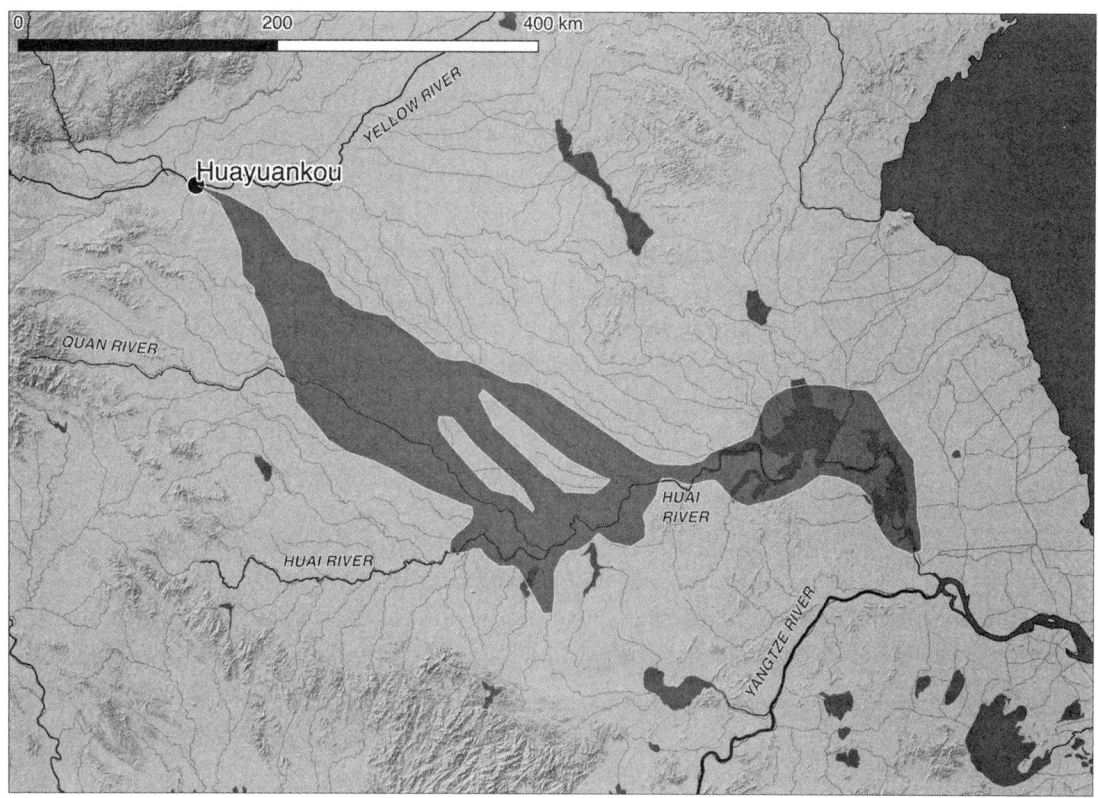

FIGURE E.1
The Huayuankou Breach. The breach emanated from near Zhengzhou and captured Huai tributaries in the lowest-lying part of the basin. Silty waters engulfed the entire Huai River and the whole of Hongze Lake and the Gaobao lake system, the low-lying and waterlogged region where drainage had been extinguished. Based on Stevenliuyi, "Huayuankou Flood." CC BY-SA 4.0.

By June 1938, the Japanese army controlled much of North China, and they too were planning for a postwar possibility of industrial-scale water control. Japanese forces captured Kaifeng and were marching upstream toward Zhengzhou, which had become an important railroad junction with direct lines to central and southern China. To prevent further Japanese advances, the Chinese defenders bombed the Yellow River at Huayuankou, near Zhengzhou.

The river avulsed southward, and the flood spread through three provinces. The river then resumed a version of its old southerly course, flowing into the Huai once again. Large regions were depopulated and buried in silt, fields and irrigation canals destroyed. Between three million and four million people evacuated the flood zone, many traveling west into Shaanxi. A postwar commission estimated that eight hundred thousand people died in the floods. Figure E.1 depicts the extent of the flood zone, which extended south through Huai tributaries and dumped sand, mud, and gravel into the bed of the Huai River, Hongze Lake, and the Gaobao Lake system south of there, having an impact even on the Yangtze delta.[5]

With transportation and communication lines disrupted, the Japanese failed to consolidate their hold over the flood region, which became a center of Chinese Communist Party (CCP) guerrilla activity. Displaced farmers moved into the drained northern bed of the river and turned it into productive farmland.[6] The Guomindang, in collaboration with the United Nations Relief and Rehabilitation Administration, repaired the Huayuankou breach in 1947 and returned the river to its northern course, where it has remained to this day.[7] Both the Guomindang and the CCP began planning large-scale dams and reservoirs. The central plains were highly contested during the civil war, but the levees there held.

After 1949, the victorious CCP, which took the river as "a cultural touchstone for collective identity" for the new nation, initiated a comprehensive reconstruction of the Yellow River basin. The regime sought to create a modern industrial sector and productive agriculture alongside a tame and predictable river.[8] Following a flood in 1950, the new government promulgated a five-year plan to build levees and reservoirs. They expanded the funding and authority of the YRCC and dug irrigation canals in Henan. Premier Zhou Enlai (1898–1976) invoked Yu the Great. In 1952, Chairman Mao Zedong (1893–1976) traveled to Tongwaxiang, the site of the 1855 course change. A stone stele there bears a proclamation in Mao's handwriting: "We must make things right on the Yellow River" (*yao ba Huanghe de shiqing banhao*). He visited the Mount Mang overlook near Huayuankou four times in the early 1950s, and he declaimed regularly about the importance of Yellow River management. Two distinctive kinds of activities simultaneously animated river work. High-technology initiatives favored elaborate modern dams, whereas labor-intensive mass mobilizations allowed the new regime to plant trees, terrace hillslopes, build earthworks, and dig canals, which channeled irrigation water to crops and conducted fertile silt to barren sand fields.[9]

In the mid-1950s, the CCP collaborated with engineers from the Soviet Union on "a great plan to conquer nature"—a massive hydroelectric dam on the Yellow River at Sanmenxia Gorge. The Soviet engineers brought blueprints from their own dam-building experience, none of which adequately accounted for the volume of sediment that washed into the river from the Loess Plateau. The dam opened in 1961, but the rate of sediment accumulation was much greater than projections. As sediment clogged the new dam, the Wei River began to back up. As the water table rose, so did salinity, threatening the potability of Xi'an's water. The turbines at Sanmenxia could not generate any electricity. A year after the dam opened, in 1962, the reservoir was half full of silt. A 1968 overhaul fixed the dam and permitted silty water to be sluiced downstream, but additional projects further downstream, which depended on clear water, had to be abandoned. The dam is still operational, but it has never fulfilled its hydroelectric potential, and sediment in its reservoir is again building up to alarming levels.[10]

Upstream, the new regime confronted a different kind of severely degraded landscape. A 1953 survey linked Loess Plateau erosion to flooding downstream. The CCP understood that environmental conditions throughout the floodplain were part of a single complex network. The 1953 report explained that each year, rainwater on the Loess Plateau "washes away the surface soil, causing cultivatable land to form barren, chaotic gullies and earthen hills." The report asserted that 90 percent of the hills on the Loess Plateau had become barren. The authors explained that this caused "direct, chronic and fatal disasters" in the northwest in addition to being the cause of the floods and course changes that had occurred for centuries downstream. The CCP needed to launch a "great struggle against nature" (*ziran huanjing douzheng*) that would include reforesting the Loess Plateau.[11]

Since the 1950s, a series of initiatives have begun to restore forests and grasslands. Erosion has diminished, although the gullied topography of recent centuries will persist into the future. On the floodplain, new concrete embankments with steel foundations, replacing earthworks built on bases of bundled sorghum stalks, have ended flooding. Around Hongze Lake, the new Grand Canal is dense with barge traffic. Along its historical route and amid the surrounding Qing waterworks, inscribed as a UNESCO World Heritage Site, tourists can survey preserved stoneworks, sample local seafood, and search for telltale mounds of yellow earth. Nevertheless, agriculture in the region still suffers today from poor drainage, waterlogging, and saline soil.[12] To this day, floodplain terrain reflects the history of the Yellow River just as much as the canyons of the Loess Plateau do.

Historical micro-highlands and flood splay fans with low groundwater and sandy and gravelly soil remain less agriculturally productive than surrounding lands, while the depressions behind historical levees and the depressions left by paleochannels are characterized by high water tables, fine silty clay, high salt content, and sometimes even salty marshes in the middle of depressions.[13]

Amid contemporary China's prosperity and its ambitious infrastructure development, many problems now are ones that Yu the Great and his successors could never have imagined. Climate change is leaving North China hotter and drier than ever before, and deserts are expanding in spite of forestation efforts. Although intensive irrigated agriculture feeds a large and increasingly affluent population, water diversion for farms means that the Yellow River has been experiencing episodes of desiccation since the early 1970s. The problem accelerated after the free-market reforms of 1987, which increased water consumption. By 1997, the river's desiccation had reached as far upstream as Zhengzhou. Dust storms whip up sediment from the dry riverbed that chokes the air downstream. Glacial volume is in steep decline on the Tibetan Plateau and in the Himalayas, and groundwater pumping from the North China aquifer is 200 percent more than rainfall replenishment. The water that remains in the river is polluted by fertilizers, pesticides, and industrial chemicals. Plate 36 depicts a beached ship on a sandbar in the middle of the muddy and desiccated Yellow River near Kaifeng in December 2019.

In the post-Mao reform era of the 1980s, two movies evoked a Yellow River imaginary. The 1984 movie *Yellow Earth* (*Huang tudi*) featured gorgeous cinematography that focused on the barren hills of the central Loess Plateau and a plot focused on the desperate poverty of the region in the 1930s. The movie is equivocal about whether the vanguard CCP would be able to create a new society in this region of dust and drought. Four years later, the six-part TV documentary *River Elegy* (*Heshang*) captured the imagination of Chinese intellectuals. The show lamented the fact that the regimes of the imperial era were so focused on the Yellow River that they turned their backs on the ocean and the world. After months of commentary, the movie was banned in a government crackdown.

The Yellow River in the Anthropocene is one of the world's many endangered waterways. Its big dams and middle-course irrigation works have functioned for a few decades, but their cost has been great. A new canal project is under way, a massive south-to-north water diversion venture in western China that seeks to move water from the Yangtze River of the subtropical south into the bed of the Yellow River in the semiarid north. The Loess Plateau is recovering as a result of an aggressive campaign to plant trees and stabilize hillsides, but it is not clear whether these monocropped grids of foliage will flourish as sustainable ecosystems. However, even today, much of the Yellow River basin is seriously affected by water and soil erosion, and the basin remains among the worst examples of erosion worldwide. The Loess Plateau is crisscrossed with terraces and check dams, canyons and gullies, sand deserts and alkali badlands, wide alluvial plains with farms and cities, muddy streams, gigantic open-pit coal mines from which exhaust-belching trucks stream onto narrow roads, and kilometers of young shrubs and saplings growing on hillslopes and retired farmland. At the subhumid southwestern edge of the Loess Plateau near the headwaters of the Jing River, bulldozers sculpt denuded clay hills into neatly terraced ziggurats for foresters to plant with trees. The future of the Yellow River and its denizens during an era of ecological collapse and climate catastrophe is uncertain, but its past offers lessons about both resilience and its limits. Plates 37 and 38 depict these efforts.

In a recent article about the Mississippi River, the environmental journalist Elizabeth Kolbert writes:

> At this point, it might be prudent to scale back our commitments and reduce our impacts. But there are so many of us—nearly eight billion—and we are stepped in so far, return seems impracticable. And so we face a no-analogue predicament. If there is to be an answer to the problem of control, it's going to be more control. Only now what's to be managed is not a nature that exists—or is imagined to exist—apart from the human. Instead, the new effort begins with a planet remade, and spirals back on itself—not so much the control of nature as the control of (the control of) nature.[14]

In Kolbert's imagination, the need to control nature in a densely populated environment is a contemporary conundrum. In China, however, people have debated and tinkered with the scope and character of intervention on the engineered Yellow River floodplain for more than two millennia. The Yellow River has coexisted with people for eons, mutating through a series of life spans and offering subsistence as well as peril to its ingenious and persevering denizens. May its example be a source of solace for our troubled present.

APPENDIX

Tracking Yu: Developing a Data System for Yellow River History

BY RUTH MOSTERN AND RYAN M. HORNE

When Ruth Mostern began conducting research into Yellow River history, she discovered that most publications about the Yellow River from China after the 1930s were claiming that, counting from the time of the legendary Yu the Great, the Yellow River had experienced 1,500 floods and twenty-six course changes. Those assertions were ubiquitous and consistent but seldom footnoted. When they were referenced, they routinely pointed toward a 1935 book called *The Yellow River Annals* (*Huanghe nianbiao*).[1] *The Yellow River Annals* is a chronological list of three thousand years of documented floodplain history. Each entry in the annals consists of a date, an event type, and a quotation from one or more primary sources. It seems that this work is the corpus upon which most assertions about the Yellow River have rested for the past eighty years. That makes it very influential, given the extent to which assumptions about the unstable and perilous character of the Yellow River have shaped warfare, large-scale engineering works, and even national identity during that time.

The Yellow River Annals focuses on the history of the alluvial plain, but it does not document erosion at its point of origin. Moreover, it has rarely been used as the basis for a database management system that takes advantage of an era of contemporary computing, the places named in it have never been georeferenced, and it has never been integrated with other annalistic lists of historical events or with other spatial and environmental information.[2] Mostern and Horne have therefore created their own chronicle of Yellow River history, with the conviction that the pursuit would inspire people interested in various fields, including the history of China, the world environment (because Yellow River records are the world's longest continuous collection of documentary observations about entanglements of people, water, and sediment), and the digital humanities. The remainder of this appendix describes the creation of the information system that forms the foundation of Mostern's research: the Tracks of Yu Digital Atlas (TYDA).

STEP 1: MAKING SPREADSHEETS AND SHAPEFILES

The TYDA aspires to be a comprehensive and interactive resource for identifying disruptive events, slow processes of change, and places of human activity throughout the entire Yellow River watershed and floodplain during all of recorded history. Its objective is to offer volume at the scale of the watershed and the Holocene rather than detail at the scale of the historical case study. For that reason, it is based on published sources with vast coverage rather than new primary source research. Inevitably, the TYDA inherits mistakes and biases from those sources. The project mitigates that problem by relying on sources that clearly reveal their own connection to the documentary and scientific record, and by cross-referencing multiple sources against one another. Moreover, we aim to make the data publicly available, thus allowing other people to improve its accuracy in the future.

THE MIDDLE COURSE

Four students at the University of California, Merced, and the University of Pittsburgh worked to create, merge, and clean information from print publications about settlements on the Loess Plateau, which is the erosion zone in the middle reaches of the river. Acknowledging them by name and in the order in which they worked on the project, they are Edward Lanfranco, Rocco Bowman, Zhifeng Shen, and Shaobai Xiong. Erin Mutch, the manager of the Spatial

Analysis and Research Center at the University of California, Merced, developed a work plan and helped to supervise Lanfranco and Bowman. Ryan Horne, digital history postdoctoral fellow at the University of Pittsburgh's World History Center, worked with Shen and Xiong.

The erosion zone component of TYDA is a database that is optimized for tracking changes to the distribution of settlements on the Loess Plateau, located primarily in Shaanxi province in northwestern China, which is the source of the sediment that periodically inundates the floodplain. Developing this content was time consuming but relatively straightforward. Mostern and her team georectified all of the maps in each volume of the *Historical Atlas of China* (*Zhongguo lishi ditu ji*) that overlapped the Loess Plateau.[3] Next they georeferenced all the units below county rank and created attribute tables that associated each unit with its parent prefecture and that also included the historical date referenced on the map in question. They supplemented the data from the *Historical Atlas* with information from a table of garrisons established during the Northern Song, because Mostern had identified that period as the most important turning point in the history of erosion on the Loess Plateau.[4] Next, they merged the attribute tables associated with each separate map to create a time series of establishment and disestablishment that spanned the whole imperial era. They identified continuous existence for any place that had the same name as a place that appeared on a previous map and that was within 5 kilometers of the identically named place.[5] Finally, they clipped the data to a shapefile that identified the extent of the Loess Plateau.[6]

THE FLOODPLAIN

The floodplain component of the TYDA database tracks waterworks constructions and repairs (e.g., levees, canals) and river disasters (e.g., floods, breaches). It comprises information from sixty-two lists and tables spread across ten Chinese publications about Yellow River history, agrarian history, and the history of natural disasters. *The Yellow River Annals* is one of those ten sources.[7] Every attested event in those lists and tables is referenced to one or more historical documents, and each one is also attributed to a single year and, where possible, at least one named location: 81 percent of entities in the database include an attested location.

Mostern and her graduate student Kaiqi Hua surveyed thirty-three published articles, books, and book series with structured information about Yellow River history. Hua made a data inventory spreadsheet based on these thirty-three publications. He assigned source codes, analyzed the content of each source, and classified each table in each source as either interesting, optional, important, or critical. In total, the sources comprised 1,017 pages and 8,711 entries. Mostern and Hua ultimately decided to include the important and critical tables: 3,407 entries. Hua digitized the contents primarily via hand typing after determining that optical character recognition for Chinese, including literary Chinese, formatted in ruled tables, was too prone to errors. His initial data entry parsed each of the sources into separate Microsoft Word documents with columns and rows that precisely reproduced the original form of each published table so as to maintain fidelity to the source material. He then converted each Word document into an Excel spreadsheet and assigned unique IDs for each published source table and each entity in each table. Next, maintaining those IDs, he reformatted each spreadsheet into a standardized column structure and then merged the spreadsheets into a first draft master list.

The next step, cleaning and parsing data, required significant historical expertise and semantic judgment. Under Mostern's direction, Hua removed records that were unrelated to the Yellow River or any other floodplain water system, excised events that occurred after the end of the imperial era in 1911, identified and merged records that were redundant across more than one of the ten core sources (taking care to maintain unique information and original ID numbers as he did so), and as necessary, parsed information across multiple spreadsheet columns.

Subsequently, he merged and split data to ensure that there was a single unique record for each event. The goal was a master table in which each event was associated with a specific date (at least a single year), at least one location, and a description in at least one primary source. This work relied primarily on Hua's exceptional skills of data selection and historical judgment. In some cases, historical sources conflated a series of events that transpired in the same year (e.g., a breach followed by flood followed by a repair), and we subdivided them accordingly. At the end of this task, we identified more duplicates and removed or reconciled them.

Next, we standardized event types. We began with the forty-nine event types in *The Yellow River Annals*. Some of these referred to dozens of events, and others to only a handful; some were distinctive, and others were nearly synonymous. With Mostern's input, Hua split, merged, and selected among them to create a vocabulary of twenty event types: ten items that refer to types of fluvial disasters and ten items that refer to river management activities. This set of categories allows for meaningful distinctions while still also permitting meaningful query and visualization. Where necessary, we have assigned multiple event types to individual events. The explanations of all the event types appear in table A.1, along with information about which synonymous terms from primary source and data tables are subsumed under each one. This table in itself is an accomplishment for river history research.

The final data development task for the floodplain tables in the TYDA was to identify all places named in the historical documents and event summaries. This was another task that required significant specialist research and aptitude in literary Chinese. First, Hua made a master list of the places named in river events. He followed the imperial spatial hierarchy. Sub-administrative places such as villages, levees, and buildings are all situated in counties, which nest into prefectures, then provinces. He assigned an ID to each unique place-name, and then linked event IDs and place IDs in many-to-many relationships (a single event may involve multiple places, and a single place may be the site of many events). He determined the latitude and longitude of each place with as much precision as possible, and then identified modern place-name matches for all historical place-names in the table. There are 1,642 unique places in the downstream place table in TYDA. These are categorized into four levels. When the sources name sub-administrative places, like towns or dams, these are Level 1 places, and there are 750 of these in the TYDA. The remainder of the places are counties or prefectures. A few sources name only provinces. In total, the TYDA floodplain data consist of 1,642 places and 3,754 events.

Next, Mostern and Horne imported all the locations in the place table into GIS along with China River Basins shapefiles.[8] The contemporary Yellow River floodplain is channelized between levees. To encompass the historical extent of the floodplain as it existed in all incarnations of the river course, we created one study area shapefile that includes the Hai River and Huai River basins as well as the modern Yellow River basin. That allowed us to identify places in the TYDA events database that lay outside of the study region and events that lacked a coordinate. Working with Mostern and Horne, undergraduate student Shaobai Xiong determined locations for all unreferenced places. In addition, for events located at places that fell outside the designated floodplain region, he determined whether the initially assigned place was mistaken and needed to be corrected, whether the original location was correct and should be maintained (we decided to retain a number of places at the edge of the floodplain that were technically outside the study area), or whether the event in question was outside the scope of this research and should be removed from the database. This activity—another task that required expert judgment and exceptional literary Chinese—also served to validate Hua's work.

CONTEXTUAL DATA

The previous section of this appendix describes two new databases created under Mostern and Horne's supervision. In addition, the full TYDA system also includes several previously published data sources, which support contextual queries and maps linking Yellow River history with enduring topics concerning Chinese historical geography and environmental history. Some of these resources are spatial data native to GIS systems, and some are data tables. Table A.2 is a table listing all these sources.

STEP 2: DESIGNING AND VALIDATING A SYSTEM

To finalize the TYDA and prepare it for eventual web publication, and to facilitate queries, visualizations, and other explorations of the data set, Horne integrated the various data sources into a PostGIS relational database. Several design considerations animated this stage of the project. One objective was to permit eventual interoperation with Linked Open Data projects like the World Historical Gazetteer, a digital spatial ecosystem for integrating specialist information about places from multiple contributors. Another was to create a union gazetteer, a merged list of all places relevant to the study, along with a list of all individual events and their types. A third was to be able to capture changes to the nomenclature, status, administration, and even location of individual places.

TABLE A.1

Contextual Data in the Tracks of Yu Digital Atlas

Category	Source
China Historical GIS	Historical Counties and Prefectures[a]
Lakes and Open Water	Harvard ChinaMap
Rivers	China Rivers Sorted by Basin[b]
Historic Courses of the Grand Canal	Harvard ChinaMap
Basemap	Ancient World Mapping Center, Natural Earth, NASA SRTM[c]
Vegetation	Harvard ChinaMap (2000)
Historical Courses of the Yellow River	Chen Yunzhen[d]
Dynasties and Reigns	Wikipedia
Mountain Peaks and Passes	Harvard ChinaMap
Ecosystems	World Wildlife Fund
Moisture Index	Monsoon Asia Drought Atlas[e]
Sedimentation Rate	Xu Jiongxin
Historical Census Data	Harvard ChinaMap

[a] Fairbank Center et al., "China Historical GIS."
[b] Berman, "China River Basins."
[c] Horne, "Map Tiles"; NASA, SRTM.
[d] Chen, "Course Change Shapefiles." Thank you to Chen Yunzhen for sharing this data with me and allowing me to use it.
[e] Cook et al., "Asian Monsoon Failure and Megadrought." Thank you to Amy Hessl for preparing this data and discussing it with me.

The initial design of the database remained largely unchanged throughout this process. The database was designed to have a Places table that would serve as a digital gazetteer for the project, which would include a single identifier, representative location, and source, and a Chinese-language title that originated from either the source material or from a modern place-name. Places associated with specialized or more verbose data would be linked to other tables that held that data, as the Places table was designed to focus on the smallest number of common elements between the various data sources.

The Events table was another key component of the database design. Just like entries in the Places table, each individual event needed to have its own unique identifier, a date, and a link to any number of types describing the event. Because multiple types could be assigned to any individual event, the types themselves need to reside in another table that expresses the relationship between an event and any number of event types. These types were themselves described in another table, as were the historical sources that describe each event. These are also in a many-to-many relationship, as any event may be described in multiple sources, and many events are listed in each source.

When Horne began his work, most of the data sets for this project were in the form of individual spreadsheets with their own internal schemas for generating IDs. This made a simple compilation of data impossible. Furthermore, the collected data, fields, and formats differed substantially between the upriver data (a collection of named places and dates at which they were attested) and the downriver data (a collection of historical events that transpired at particular places). Therefore, Horne created a series of staging tables to import data and align new standardized IDs with original ones while preserving the relationships specified in the tables themselves. He imported each data table into the system, disassembled it into its various relationships and components, and then entered it into the new database schema. This required significant trial and error, as some of the spreadsheets had conflicting formats within the same table, multiple character encodings, and potential data errors.

A particularly complex operation was sorting and normalizing the upstream data, as many places were present on more than one of the maps and other sources that constituted the data set. Shaobai Xiong disambiguated this part of the data, and Horne then placed it into a staging table where each place entry was assigned a unique ID. At that point he checked each place to see if it was present in multiple sources at different times. If so, the first chronological entry was treated as the "primary" place, with the information from other sources placed in another table and linked accordingly. The final list of upstream places served as the core table of places, to which downstream places were added later.

In addition to the Places table, Horne constructed a new version of the Events table and established a many-to-many relationship with the Places table, as any event can have an impact on multiple places, and any place can be the subject of multiple events. Once again, he used a staging table to ensure that each events entry was comparable with the new database schema, and that the old linkages from the individual spreadsheets were preserved with the appropriate new IDs. He also linked each event to the table housing the different primary sources describing it and linked each upstream place to a table containing all its historical information. All the linkages are treated as foreign keys in the database, which ensures that each linked value actually exists in its respective table.

The result is a stable system that allows for visualization via simple PostGIS query and in which all events are associated with geolocated places. All the data from the various spreadsheet inputs has been normalized with sanitized and stable IDs and can be linked to external Linked Open Data resources.

WHAT'S IN HERE?

By the numbers, the TYDA database includes 3,657 unique places, of which 1,642 are downstream and 2,015 are upstream. There are 2,533 total upstream attestations in the data, but many places occur more than once in the data set. There are 3,754 unique events. Two of these lack source connections, so there are actually 3,752 unique event IDs in the source connections. Mostern began this project in part to test and validate *The Yellow River Annals*, which seems to have remained unanalyzed since its publication in the 1930s despite its importance to the study of Yellow River history and even to public policy and cultural imagination about the river. We can now report that the *Annals* appears 2,139 times in the TYDA, and it serves as a source for 2,095 unique events of the 3,754 events recorded in the TYDA. (The difference between those numbers represents forty-four cases in which the TYDA merges multiple events from the *Annals* into a single event.) Among these, there are 482 instances in the TYDA in which the *Annals* and other sources describe the same event, among which are 453 distinct events. The *Annals* is therefore the sole source for 1,642 events in the TYDA.

There are also 2,068 instances in the TYDA for which the *Annals* is not a source, representing 1,657 unique events that exist in the TYDA but not in the *Annals*. This means that more than 44 percent of the events in the TYDA are not visible in the *Annals*. In short, although the *Annals* remains an important source for understanding flooding and flood control on the imperial Yellow River floodplain, it should no longer be considered comprehensive.

Specifically looking at data about floods and course changes, 667 instances in the TYDA appear in the *Annals*, representing 628 unique events. Of these, 419 instances (403 events) appear in no other sources and 248 instances (225 events) appear in at least one other source. Among the 1,645 (1,032 events) instances of floods and course changes that appear in other sources, 1,069 (807 events) do not appear in the *Annals*, and 576 (225 events, the same set as above) include the *Annals* among their sources. Here, too, the TYDA is a more comprehensive source than the *Annals*. It should also be clear from this discussion that it is always possible to query the TYDA for results from any particular historical compilation: for purposes of comparison like this one, or because a particular TYDA user may find a particular part of its source base particularly interesting or reliable. The TYDA retains all the information needed to make such selections possible.

CREATING A DIGITAL SYSTEM

Although the TYDA database is a comprehensive and powerful resource, there was still a need to visualize the data, to create printable maps and charts for this publication, and to enable nontechnical specialists to query and interact with the data. In the summer of 2019, Horne and Mostern decided that using the Python programming language through Jupyter Notebooks (jupyter.org) provided the best mix of power, flexibility, and usability for the TYDA project.

A Jupyter Notebook is an open-source application that allows a user to run and modify code, embed narrative text, and present data visualizations. By using software libraries like Matplotlib and Pandas/GeoPandas, Horne created new methodologies for generating consistently styled and visually appealing charts and maps. The basic functionality of the Jupyter Notebook for TYDA is provided by almost one thousand lines of code, which represent various function calls and database queries. These are designed to make extremely complex operations simple for any end user. For instance, Horne wrote a query to combine events, places, places to events, event types, floodplain geography, and an arbitrary point and buffer. This makes it possible to find all information about all places impacted by an event of a certain type, between an arbitrary number of years, and within a set distance from a point. A user simply needs to input particular years, distances, and point values into a function call. The final notebook contains more than five hundred such queries and calls. Each individual query can be styled and adjusted as needed. Box A.1 depicts an example of the query logic in Jupyter, and Figure A.1 depicts a query call.

Box A.1

A Query to the Tracks of Yu Digital Atlas

This generic query combines the events, places, places to events, event types, floodplain geography, and an arbitrary point and buffer. It generates all information about places that were affected by an event of a certain type, between an arbitrary number of years, and within a set distance from a point. Users need simply input the years, distance, and point values of interest into a function call. The final notebook contains more than five hundred such queries and calls, with each individual query capable of being styled and adjusted as needed.

```
def makeManInfoCount(startdate, enddate, category, area):
    sql ="""
SELECT
  c.events_id,
  c.w_date,
  c.events_type,
  d.type,
  c.events_old_name_c,
  c.events_old_name_p,
  c.places_c_title,
  c.places_en_tite,
  c.places_current_loc,
  c.places_hz_title,
  c.places_py_title,
  c.places_id,
  c.geom
FROM
  (
    SELECT
      a.events_id AS events_id,
      a.w_date,
      a.type_c AS events_type,
      a.old_name_c AS events_old_name_c,
      a.old_name_p AS events_old_name_p,
      b.c_title AS places_c_title,
      b.en_tite AS places_en_tite,
```

Box A.1 (continued)

```sql
      b.current_loc AS places_current_loc,
      b.hz_title AS places_hz_title,
      b.py_title AS places_py_title,
      b.geom AS geom,
      a.place_id AS places_id
    FROM
      (
       SELECT
         events.old_name_c,
         events.old_name_p,
         places_to_events.place_id,
         events.type_c,
         events.w_date,
         events.id AS events_id
       FROM
         events
         LEFT JOIN places_to_events ON events.id = places_to_events.event_id
       WHERE
         events.id IN (
           SELECT
             events_to_type.id
           FROM
             events_to_type
             JOIN event_types ON event_types.ch_title = events_to_type.type_c
           WHERE
             event_types.en_cat = {category}
           GROUP BY
             (events_to_type.id)
         )
      ) a
      LEFT JOIN (
       SELECT
         places_to_events.place_id,
         places.c_title,
```

```
      places.en_tite,
      places.current_loc,
      places.hz_title,
      places.py_title,
      places.geom
    FROM
    places
      LEFT JOIN places_to_events ON places.id = places_to_events.place_id
    WHERE
    st_intersects(
      places.geom,
      (
        SELECT
          ST_Union(
            ST_Buffer({area}.geom :: geography, 180000):: geometry
          )
        from
          {area}
      )
    ) = true
  ) b ON a.place_id = b.place_id
  WHERE
  a.w_date BETWEEN {startdate}
  AND {enddate}
) c
JOIN (
  SELECT
    event_types.type,
    events_to_type.id AS event_id
  FROM
    events_to_type
    JOIN event_types ON event_types.ch_title = events_to_type.type_c
  WHERE
    event_types.en_cat = {category}
```

> Box A.1 (continued)
> ```
>     ) d ON c.events_id = d.event_id
>   GROUP BY
>     c.events_id,
>     c.places_c_title,
>     c.places_en_tite,
>     c.places_current_loc,
>     c.places_hz_title,
>     c.places_py_title,
>     c.places_id,
>     c.w_date,
>     c.events_type,
>     c.events_old_name_c,
>     c.events_old_name_p,
>     d.type,
>     c.geom
>   ORDER BY
>     c.w_date ASC=""".format(startdate = startdate, enddate =
>       enddate, category=category, area=area)
>       dfManInfo = pd.read_sql_query(sql, cnx)
>       return dfManInfo
> ```

The TYDA Jupyter Notebook was a shared resource between Mostern and Horne. This allowed us to jointly track data requests and tweaks to the different visualizations and to make code modifications that could be run and analyzed in real time. Using this system, Mostern was able to change queries, request different permutations of visualizations, and ask new questions of the data, to which Horne could respond quickly by simply modifying the inputs of different Jupyter functions. Using the GeoPandas library, these queries could be exported into formats that were readable by the QGIS geographic information system application, where Horne could apply a "house style" and prepare maps for publication.

Although popular in the field of data science, the use of Jupyter Notebooks in humanities research and computing is relatively new, although the Programming Historian website added a lesson about

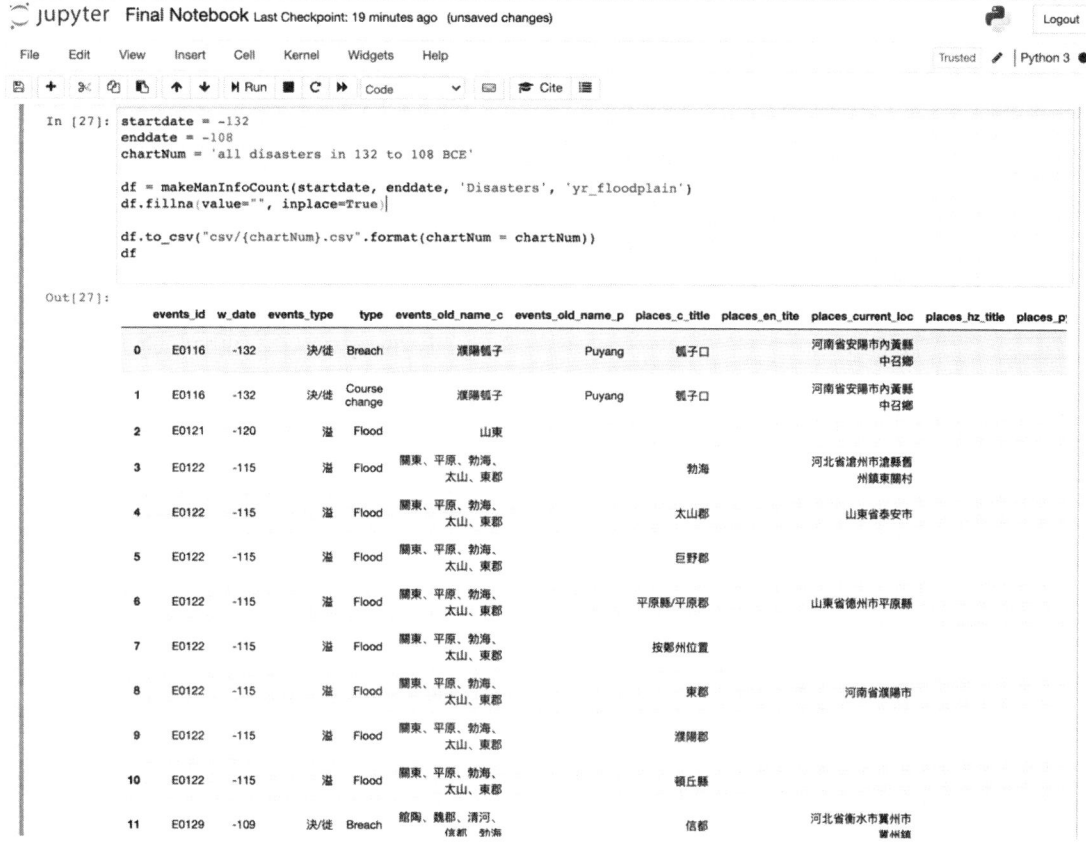

FIGURE A.1

Query Call in Jupyter. The Jupyter collaboration environment allows for collaborative editing of code. It is possible to generate tables and maps live in the application. This image also depicts the complexity of the queries that lie behind the images and assertions in this book.

Jupyter Notebooks at the end of 2019.[9] The power and flexibility of this system enabled the efficient creation of compelling images and opened up new queries that combined spatial reasoning, data points, and historical analysis. Although software installation was not trivial, it would simply not have been possible to manage the sheer volume of tables, charts, maps, and queries for this book without the use of this system. Even visualizations that did not make it into this volume proved their worth as insights into the structure and contents of the database.

CONCLUSION

In the end, it required almost a decade of effort to complete the TYDA, with support from multiple funding bodies, a great deal of

APPENDIX   261

TABLE A.2

Event Types in the TYDA Database and Their Most Common Synonyms in the Data Source

Chinese	Pinyin	Event type	Description
Disasters			
溢	*yi*	Flood	Floods may be signaled by the presence of terms such as *zhang* 漲 (raised water level) or *yan* 淹 (flood). *Daoguan* 倒灌 (water flowing against the current into a city or neighborhood) is another term that indicates the presence of a flood event.
決	*jue*	Breach	A synonym for *jue* is *beng* 崩 (a sudden opening). *Kui* 潰 is also a synonym for *jue* 決. A breach event (*jue* 決) occurs at a single location as a result of the rupture of a dike, whereas a flood event (*yi* 溢) involves a larger region.
旱	*han*	Drought	A drought may also be described by the synonyms *ganhe* 乾涸, *he* 涸, or *ku* 枯.
災	*zai*	Natural disaster	The natural disasters described as *zai* 災 in the sources are rainstorms, earthquakes, landslides, and locust swarms. If a rainstorm results in a flood, these are indicated in our database as two distinct events.
毀	*hui*	Anthropogenic disaster	This refers to an intentional dike breach, which may be performed by the government, usually during wartime, as a military tactic; by private individuals without approval for purposes of irrigation; or by mistake in a failed attempt to rectify an engineering error.
險	*xian*	Risky situation	The term *xian* 險 often co-occurs with the term *shixiu* 失修 (damage or disrepair), which is used to refer to poorly maintained waterworks. If a situation of disrepair becomes severe, it is referred to as a Risky situation.
兆	*zhao*	Omen	When the river is described as inexplicably running clear or, occasionally, dirty, this is referred to as an Omen.

Chinese	Pinyin	Event type	Description
徙	*xi*	Course change	This event type may also be signaled by the term *duo* 奪 (literally to seize), which indicates that one river has taken the course of another river or canal. Other specific types of course change are *fen* 分 (a river has divided in two) or *he* 合 (two rivers have come together as one). A Course change is the result of a Flood that occurs as a result of a Natural disaster or a disaster caused by humans.
絕	*jue*	River runs dry	This event type is also referred to by the synonym *duanliu* 斷流 (a break in the course). It is generally associated with Drought.
塞	*sai*	Blockage	The term *sai* 塞 often co-occurs with the word *yu* 淤 (silt), and *yusai* 淤塞 often appears as a compound term. *Yuzu* 淤阻 (silt obstruction) is another term for this event. *Se* 澀 is another synonym. In addition to silt, ice can also cause a blockage.

Management Events

Chinese	Pinyin	Event type	Description
議	*yi*	Proposal and discussion	This event type refers to several related activities: *yilun* 議論 (discussion), *jianyi* 建議 (a proposal), and *zouyi* 奏議 (a memorial that includes a proposal). Once a proposal is approved, it generally triggers a management event (*zhi* 治). Discussion and management are often co-occurring event types. The publication of a book or essay about the river is also described in the proposal and discussion event type.
遷	*qian*	Settlement relocation	A settlement relocation event occurs whenever people give up their residence. The term may refer to the relocation of a settlement or to the complete destruction or disappearance of a settlement. If a settlement ceases to exist altogether, it is generally referred to as *pi* 圮, or less commonly as *mo* 沒. Settlement relocation is often indicated by references to the presence of refugees (*zaimin* 災民).

TABLE A.2 (*CONTINUED*)

Management Events

放	*fang*	Dam or sluice opening	This event type is most commonly evoked in the Qing dynasty. The event is also referred to as *kaiba* 開壩 or *qiba* 啟壩.
疏	*shu*	Dredging	Dredging is the management event that rectifies the river event of blockage (*sai* 塞). It is synonymous with *jun* 浚 (dredging) or *fangyu* 放淤 (removal of silt). The term *shu* 疏 always refers to silt dredging. Other events associated with expanding water flow are covered by the dam or sluice opening event.
助	*zhu*	Using water for a benefit	This event type is the antonym of the disaster caused by humans in the river events. It refers to the use of river water for agriculture and transportation. In particular it encompasses sanctioned private actions that would generally be considered management (*zhi* 治) if they were government actions.
治	*zhi*	Management	Management is a catchall term that refers to events that cannot be specified more precisely from the source material or to complex acts of water conservancy with multiple components that cannot be separated into distinct events. The term generally refers to acts of policy implementation or policy enforcement. For instance, management of the river may mean that the emperor enforced a law, requisitioned a budget, appointed a supervisor, raised tax revenue, toured a site, or issued an edict (*zhao* 詔). Management may also refer to an engineering act intended to enhance the infrastructure system, such as opening a canal, building a dike, and repairing human damage and/or structures. However, it never refers to emergency repairs. Acts of management may be successful or unsuccessful. Management always refers to government activity, never to private activity.

Management Events

修	*xiu*	Repair of structures	This refers to any kind of repair to the river infrastructure, such as repairs to dikes or embankments, and any efforts intended to return the river to an old course (*gudao* 故道 or *jiudao* 舊道). A synonym for *xiu* 修 is *fujian* 復建 (rebuilding).
救	*jiu*	Emergency repairs	An emergency repair is an attempt to respond quickly to a flood (*yi* 溢), a breach (*jue* 決), a settlement relocation (*qian* 遷), a blockage (*sai* 塞), a human disaster (*hui* 毀), a natural disaster (*zai* 災), or a risky situation (*xian* 險). The presence of terms such as *ji* 即 (promptly) and *xuan* 旋 (immediately) distinguish an emergency repair from a repair of structures. *Sai* 塞 as a verb rather than a noun refers to an intentional emergency blockage of a breach and is a kind of emergency repair.
建	*jian*	New construction	New construction can refer to construction of the components of the river management infrastructure: dikes (*di* 堤), dams (*ba* 霸), or canals. A synonym for *jian* is *zhu* 筑.
探	*tan*	Scientific survey	This event type, attested beginning in the nineteenth century, refers to surveys and field trips conducted in the context of modern science and engineering techniques.

expert assistance, and untold hours of student and professional labor. Like many data-intensive digital history projects that are based on content of diverse formats from heterogeneous sources and compiled by many hands, this project evolved slowly from a wide assortment of poorly cross-referenced spreadsheets and shapefiles. It would have been ideal to have designed a database structure early on, to have linked every location in it with a controlled vocabulary of counties and prefectures from CHGIS, and to have created a system for data entry that would have generated consistent ID formats. However, that would have required Mostern to make binding semantic decisions early in the project, long before she was prepared to do so. Like many such endeavors, this one proceeded stepwise, with data development and historical insights continuously informing one another. In the end, this completed system allows for a unique approach to historical research, one that merges close readings of individual historical texts and scrutiny of particular events with epochal scale conclusions about landscape and environmental change.

GLOSSARY OF CHINESE CHARACTERS

An du 安都
Anjing 安靜
An Lushan zhi luan 安史之亂
Babaili Liangshan shuibo 八百里梁山水泊
Bayankala shan 巴顏喀拉山
Bai Juyi 白居易
Baima 白馬
Baiyin 白銀
Bai Zhongshan 白鐘山
Benxing 本性
Bianqu or Bianhe 汴渠 or 汴河
Bianzhen 邊鎮
Bingbu shangshu 兵部尚書
Bohai 渤海
Cen Zhongmian 岑仲勉
Chanzhou 澶州
Chang'an 長安
Changjiang 長江
Chao Cuo 晁錯
Chen Huang 陳潢
Chengshou zhanlue 城守戰略
Chongbo 崇伯

Chunqiu shidai 春秋時代
Dabian 大邊
Da he 大河
Dalu ze 大陸澤
Daqing he 大清河
Da sima 大司馬
Datong 大同
Dawen he 大汶河
Da Yu 大禹
Dayunhe 大運河
Daohe xingsheng shu 導河形勝書
Dongping hu 東平湖
Dongzhi yuan 董志原
Dou 斗
Du Chong 杜充
Dushuijian 都水監
Duwei 都尉
Erbian 二邊
Erlitou 二里頭
Fanren 番人
Fen he 汾河
Fen sha qi shi 分殺其勢
Fenshuishi 分水勢

Feng Qun 馮逡
Fengshan 封禪
Fudan daxue lishi dili yanjiu zhongxin 復旦大學歷史地理研究中心
Gansu 甘肅
Gaobao zhuhu 高寶諸湖
Gaojiayan 高家堰
Gu Yanwu 顧炎武
Guanzhong 關中
Guan Zhong 管仲
Guanzi 管子
Guangtong qu 廣通渠
Gun 鯀
Guo Chang 郭昌
Guo he 渦河
Guomindang 國民黨
Guo Shoujing 郭守敬
Guo Zhongyan 郭仲彥
Hai he 海河
Han 漢
Han shu 漢書
Hao 壕
Haojing 鎬京
Hefang shuyan 河防述言
Hefang yilan 河防一覽
Hefang zhaiyao 河防摘要
Hegong qiju tushuo 河工器具圖說
Hekou 河口
Henan 河南
Hequshu 河渠書
Heshang 河殤
Hetao 河套
Heyuan zhi 河源志
Heng shan 橫山
Honggou 鴻溝
Hong Mai 洪邁
Hongxue yinyuan tuji 鴻雪因緣圖記
Hongze hu 洪澤湖

Hou 堠
Hu Ding 胡定
Huzi 瓠子
Hukou pubu 壺口瀑布
Hua shan 華山
Huayuankou 花園口
Huazhou 滑州
Huai he 淮河
Huainanzi 淮南子
Huang Chao 黃巢
Huangdi 荒地
Huang he 黃河
Huanghe nianbiao 黃河年表
Huang tudi 黃土地
Ji An 汲黯
Ji Chaoding 冀朝鼎
Jiajing 嘉靖
Jia Lu 賈魯
Jia Rang 賈讓
Jiancha yushi 監察御史
Jiedushi 節度使
Jin 金
Jinshaan daxiagu 晉陝大峽谷
Jinshu 晉書
Jing he 涇河
Jiubian 九邊
Jiuzhou 九州
Juya 鋸齒
Juye 鉅野
Kaifeng 開封
Kunlun shan 崑崙山
Lankao xian 蘭考縣
Lanzhou 蘭州
Li Chui 李垂
Li Daoyuan 酈道元
Li Yizhi 李儀祉
Li Zicheng 李自成
Liang 梁 or 樑
Liao 遼

Lingzhou 靈州
Liu Daxia 劉大夏
Liupan shan 六盘山
Liusha 流沙
Liu Tianhe 劉天和
Longmen shiku 龍門石窟
Lüliang 吕梁
Lüliang shan 吕梁山
Luochuan yuan 洛川原
Luo he 洛河
Luoyang 洛陽
Luoyi 洛邑
Matou 馬頭
Mang shan 芒山
Mao 峁
Maowusu shamo 毛烏素沙漠
Mao Zedong 毛澤東
Mei shili li yi shuimen 每十里立一水門
Meng Ke 孟軻
Meng Tian 蒙恬
Mengzi 孟子
Ming 明
Mulong 木龍
Nihe 泥河
Ningxia 寧夏
Niushan 牛山
Ouyang Xiu 歐陽脩
Ouyang Xuan 歐陽玄
Pan Angxiao 潘昂霄
Pan Jixun 潘季馴
Pingliang 平凉
Qi 祁
Qidan 契丹
Qi Huan Gong 齊桓公
Qikou 磧口
Qin 秦
Qin he 沁河
Qinling shan 秦岭山

Qinzhou 秦州
Qing 清
Qinghai 青海
Qingkou 清口
Sanmenxia 三門峽
Sanmenxia difang shizhi bangongshi 三門峽地方史志辦公室
Sanyangzhuang 三楊莊
Sao 埽
Shaanxi 陝西
Shandongsheng Huanghe tu 山東省黃河圖
Shanhaijing 山海經
Shanxi 山西
Shanxi tongzhi 山西通志
Shanyang du 山陽瀆
Shang Yang 商鞅
Shen Dao 慎到
Shen Kua 沈括
Shen Menglan 沈夢蘭
Shi (power) 勢
Shi (twenty-seven kilograms or sixty pounds) 石
Shiji 史記
Shijing 詩經
Shi Nianhai 史念海
Shu jiuhe 疏九河
Shusan 疏散
Shushui gongsha 束水攻沙
Shuidao zili 水道自利
Shuiguan 水官
Shuihu zhuan 水滸傳
Shuijingzhu 水經注
Shui wang dichu liu, ren wang gaochu zou 水往低處流, 人往高處走
Shunshui 順水
Sima Qian 司馬遷
Song 宋

Song Jiang 宋江
Song shan 嵩山
Su Shi 蘇軾
Suzhou 蘇州
Sui 隋
Sui huo jumu zhi li 歲獲巨木之利
Taihang shan 太行山
Tai shan 泰山
Taiyuan 太原
Taizong 太宗
Taizu 太祖
Tan Qixiang 譚其驤
Tang 唐
Tao Mo 陶模
Taosi 陶寺
Titian 梯田
Tianchi 天池
Tie longzhua yangni che 鐵龍爪揚泥車
Tongguan 潼關
Tongji qu 通濟渠
Tongwaxiang 銅瓦廂
Tongwancheng 統萬城
Tuntian 屯田
Tuo Tuo 脫脫
Wang Anshi 王安石
Wang Jing 王景
Wang Mang 王莽
Wang Yanshi 王延世
Wei 衛
Wei he 渭河
Wei Yuanyu 魏元煜
Wenshuiji 問水集
Wudi 武帝
Wuding he 無定河
Wu Guifang 吳桂芳
Wuyue 五嶽
Xi'an 西安
Xihe jun 西河郡

Xi Xia 西夏
Xia 夏
Xiazhou 夏州
Xiaolangdi 小浪底
Xin Deyong 辛德勇
Xing 性
Xing xian 興縣
Xiongnu 匈奴
Xu Xiake 徐霞客
Xuanfang 宣房
Xunheguan 巡河官
Xunzi 荀子
Yan'an 延安
Yan Shengfang 閻繩芳
Yao 堯
Yao ba Huanghe de shiqing banhao 要把黃河的事情辦好
Yaodi 謠堤
Yaodong 窰洞
Yi he 伊河
Yijian dongxi ba tu 移建東西壩圖
Yinchuan 銀川
Yinchuan pingyuan 銀川平原
Yinzhang shier qu 引漳十二渠
Yongji qu 永濟渠
Yugong 禹貢
Yugongtu 禹貢圖
Yujitu 禹跡圖
Yulin 榆林
Yuan (regime) 元
Yuan (tablelands) 原 or 塬
Yuanhe junxian zhi 元和郡縣志
Zhanguo shidai 戰國時代
Zhang Rong 張戎
Zhangshui shier qu 漳水十二渠
Zhang Xianzhong 張獻忠
Zhenzong 真宗
Zhengde 正德
Zhengguo qu 鄭國渠

Zhengzhou 鄭州
Zhihe sance 治河三策
Zhizheng hefang ji 至正河防記
Zhongguo lishi ditu ji 中國歷史地圖集
Zhongyue 中嶽
Zhou 周
Zhou Enlai 周恩來
Zhouli 周禮
Zhuti shushui, yishui gongsha 筑提束水，以水攻沙

Zhu Yuanzhang 朱元璋
Ziran huanjing douzheng 自然環境鬥爭
Zongdu caoyun 總督漕運
Zongli hedao 總理河道
Zou Yilin 鄒逸麟
Zuowan 坐灣
Zuozhuan 左傳

NOTES

INTRODUCTION

1 I am grateful to the Facebook Sinology group for responding to my query about this saying on May 16, 2017, and especially to Ralph Litzinger, Megan Cai, and Benjamin Ridgeway. The gloss on the saying is based on an entry in the Baidu Baike online encyclopedia, at https://baike.baidu.com.. It is also available at the Sohu online encyclopedia, at https://www.sohu.com/a/278943339_674517. Two earlier versions of the phrase appear in works from the fifteenth and seventeenth centuries.
2 *The Mencius*: Gaozi I. Lightly edited translation based on Legge, *Chinese Classics*. Bilingual edition is available online at Sturgeon *The Chinese Text Project* (https://thechinesetextproject.org).
3 *Shangshu*: The Tribute of Yu. Bilingual edition online at Sturgeon *The Chinese Text Project* (https://thechinesetextproject.org); Allan, *Way of Water*, esp. 41–46.
4 Jullien, *Treatise on Efficacy*.
5 *Shangshu* (*The Classic of Documents*): "Yugong" ("The Tribute of Yu"). Bilingual edition online at Sturgeon, *The Chinese Text Project*; Lewis, *Flood Myths*, 39–43, on Gun's failure, and generally on the history of the Yu story and its variant forms.
6 Da Cunha, *Invention of Rivers*.
7 Acciaviati, "Great Sediment Sorting Machine."
8 Linton and Budds, "Hydrosocial Cycle."
9 This is the "seeing" that James Scott describes in his essential *Seeing Like a State*.

10 Josh Viers quoted in Pottinger, "Why We Need Working Floodplains."
11 White, *Human Adjustment to Floods*, 2.
12 See Zhang, *The River, The Plain, and the State* for the term *hydrocrat*, which I adopt gratefully throughout this book.
13 White, *Organic Machine*, 108.
14 White, *Organic Machine*, 112.
15 Cheng and Brown, *Berkshire Encyclopedia*, 1125.
16 Elvin, "Three Thousand Years of Unsustainable Growth."
17 Chen et al., "Socio-Economic Impacts on Flooding."
18 Dodgen, *Controlling the Dragon*, 27; Will, "On State Management of Water Conservancy."
19 Pomeranz, "Transformation of China's Environment," 125.
20 *Technology complex* is a key term throughout Pietz, *Yellow River*.
21 Mukerji, *Impossible Engineering*, 216.
22 Pomeranz uses this terminology throughout his work, but especially in "Transformation of China's Environment."
23 Elvin and Su, "Action at a Distance."
24 Mostern and Johnson, "From Named Place to Naming Event."
25 Sewell, *Logics of History*.
26 Klein, *The Shock Doctrine*.
27 Nixon, *Slow Violence*.
28 Richards, *The Unending Frontier*. As Richards points out, states and settlers can also invest in overseas colonies, but that option is not germane to this book and plays a relatively small role in Chinese history in general.
29 Zhang, *The River, The Plain, and the State*, 177–80.
30 Pomeranz, *Making of a Hinterland*; Chi, *Key Economic Areas*.
31 Hance, "Proving the 'Shifting Baselines' Theory."
32 Unless otherwise noted, translations of official titles are from Hucker, *Dictionary of Official Titles*.
33 Ban, *Han shu* 29.1697.
34 Zhang, *The River, the Plain, and the State*, 109, citing Ouyang Xiu, "Lun xiuhe," chap. 109.
35 Li Ronghua, "Qingdai huangtu gaoyuan," 87, citing Li, *Xu xing shui jinjian*, 11.255.
36 Quoted in Chen, *Huanghe fanlan shi*.
37 These paragraphs draw gratefully on Chen Yunzhen's doctoral dissertation, "Huanghe fanlanshi," and upon personal communications with her.
38 This Chinese tradition is also inspired by German and American writing about erosion, including Lowdermilk, *Forestry in Denuded China*. Xin, "You Yuanguang hejue" reviews that tradition. I am grateful for personal correspondence with David Pietz for the reminder about the role that

debates over the Sanmenxia Dam played in this discourse, and with Brian Lander, who urged me to include this paragraph. I am honored to have spent time at the Historical Geography Institute, where Han Zhaoqing has been my host.

CHAPTER ONE. THE NATURAL AND UNNATURAL HISTORY OF THE YELLOW RIVER

1. Based on Parara-Caryannis, "Historical Earthquakes in China."
2. Greer, *Water Management*, 7.
3. Based on Yi, "Holocene Vegetation."
4. Gong, *Huanghe shihua*, 153–56.
5. Greer, *Water Management*, 15–17.
6. Kidder and Zhuang, "Anthropocene Archaeology of the Yellow River."
7. Cook, "Monsoon Asia Drought Atlas." Thank you to Edward Cook for agreeing to share this data and to Amy Hessl for providing a subset of the Atlas focused on the study area of this book and for helping me to interpret it.
8. Greer, *Water Management*, 14. Wang and Su, "Geo-pattern of Course Shifts," includes significant discussion of sunspot data.
9. Holmes, Cook, and Yang, "Climate Change over the Past 2000 Years in Western China."
10. Xu, "Naturally and Anthropogenically Accelerated Sedimentation." Recent studies of California project that global warming will be associated with conditions that are hotter and drier overall but that there will also be more frequent intense "atmospheric river" winter storms. See Ingram and Roam, *The West without Water*.
11. Zhou et al., "Environmental Variability."
12. Gong, *Huanghe shihua*.
13. For Teh's photographs "Traces II: The Source" and "Traces II: The Middle" (2014), see ianteh.com/traces-ii and ianteh.com/traces-ii-the-source.
14. Lin et al., "How and When Did the Yellow River Develop Its Square Bend?"
15. Gong Li, *Huanghe shihua*, 126–30.
16. Gong Li, *Huanghe shihua*.
17. Gong Li, *Huanghe shihua*.
18. Greer, *Water Management*, 17–22.
19. Lander, *The Nature of Political Power*.
20. Size comparisons throughout the book are from Bluebulb Projects, *The Measure of Things*.
21. Chi, *Key Economic Areas*.
22. Needham, "Hydraulic Engineering," 275.
23. Wang et al., "Precipitation Gradient."
24. Rosen et al., "The Anthropocene and the Landscape of Confucius," 1642.
25. Xu, "Naturally and Anthropogenically Accelerated Sedimentation."

26 Shi, *Huangtu gaoyuan senlin*.
27 Rosen et al., "The Anthropocene and the Landscape of Confucius," 1641–42.
28 Menzies, "Forestry," 558.
29 Ren and Zhu, "Anthropogenic Influences."
30 Ren and Zhu, "Anthropogenic Influences."
31 Liu, *Loess in China*, 73–77; Lander, *The Nature of Political Power*.
32 Menzies, *Forest and Land Management*, 21.
33 Shi, *Huangtu gaoyuan senlin*.
34 Miller, *Fir and Empire* and personal communication. Thank you to Ian Miller for providing me with prepublication access to relevant portions of this book.
35 Shi, *Huangtu gaoyuan senlin*, 139–68.
36 Greer, *Water Management*, 17.
37 Liu, *Loess in China*, esp. xvi–xvii, 4–5, 27, 132, 139–40.
38 See, e.g., Tata, "Dust from Chinese Storm Reaches California."
39 Wang et al., *The Loess Plateau*, 11.
40 Ren and Zhu, "Anthropogenic Influences."
41 Ren and Zhu, "Anthropogenic Influences."
42 Laflen, *Soil Erosion and Dryland Farming*.
43 Wang et al., *The Loess Plateau*.
44 Richter et al., "Soil in the Anthropocene."
45 Fang and Xie, "Deforestation in Preindustrial China."
46 Wang et al., *The Loess Plateau*.
47 Ren and Zhu, "Anthropogenic Influences."
48 Zhang et al., "Late Holocene Vegetation Dynamics."
49 Zhang et al., "Late Holocene Vegetation Dymanics."
50 Xu, "Wind-Water Two-Phase Erosion."
51 Xu, "Wind-Water Two-Phase Erosion."
52 Xiang et al., "East Asian Winter Monsoon (EAWM) Changes."
53 Greer, *Water Management*, 17–22.
54 Qiao et al., Sedimentary Records."
55 Xu, "A Study of the Accumulation Rate of the Yellow River."
56 Shi et al., "Changes in the Sediment Yield."
57 Ren and Zhu, "Anthropogenic Influences."
58 Shi et al., "Changes in Sediment Yield."
59 Xu, "A Study of the Accumulation Rate of the Yellow River."
60 He et al., "Soil Erosion Response to Climatic Change and Human Activity."
61 Saito, Yang and Hori, "The Huanghe and Changjiang."
62 Han, *Songdai nongye dili*. Shi, *Huangtu gaoyuan senlin* also ascribes erosion to a mix of human and natural causes. The soil itself was prone to ero-

sion, and climate change exacerbated its tendency to erode. Zhang, *Liang Song shengtai*, addresses this issue as well.

63 Ren and Zhu, "Anthropogenic Influences."
64 He et al., "Soil Erosion Response to Climatic Change and Human Activity."
65 Richter et al., "Soil in the Anthropocene." Also see Montgomery, *Dirt*.
66 Klein, "Ritual Cleansing."
67 Shen, *Jingguan, shengji, liebian*.
68 Wang and Su, "The Geo-Pattern of Course Shifts."
69 One *li* is approximately one-third of a mile.
70 Shi, *Zhongyi shuihu quanshu* 2.17.
71 Wang and Su, "The Geo-pattern of Course Shifts."
72 Needham, "Hydraulic Engineering," 237.
73 Department of Regional Development, *Primer on Natural Hazard Management*.
74 Dodgen, *Controlling the Dragon*, 12.
75 Zhang, *Huanghe liuyu dituji*, 192.
76 Hummel, *Shandongsheng Huanghetu*.
77 Gong, *Huanghe shihua*, sec. 3.3.
78 Xu, "A Study of Long Term Environmental Effects."
79 Gong, *Huanghe shihua*, 151
80 Shen, *Jingguan, shengji, liebian*.
81 Xu, "A Study of Long-term Environmental Effects."
82 Ping et al., "Time Constraints for the Yellow River Traversing the Sanmen Gorge."
83 Elvin and Su, "Action at a Distance."
84 Rosen et al., "The Anthropocene and the Landscape of Confucius."
85 Rosen, "The Impact of Environmental Change and Human Land Use"; He et al., "Soil Erosion Response to Climatic Change and Human Activity."
86 Rosen et al., "The Anthropocene and the Landscape of Confucius."
87 Holmes et al., "Climate Change over the Past 2000 Years in Western China."
88 Multiple-proxy climate reconstructions for this region are based on Arabian Sea marine sediments that reveal variations in the Indian summer monsoon; South China Sea sediments that reveal variations in the Asian summer monsoon; and ice core records with oxygen isotopes and dust signatures that reveal temperature changes that tend to correlate with precipitation, tree rings, lake and peat sediments, glacier fluctuations, and historical documents. Holmes et al., "Climate Change over the Past 2000 Years," provides a comprehensive analysis based on all these data sources.

89 Ren and Zhu, "Anthropogenic Influences on Changes in the Sediment Load."
90 Huang et al., "Charcoal Records of Fire History."

CHAPTER TWO. BEFORE IT WAS YELLOW
1 Zhuang and Kidder, "Archaeology of the Anthropocene."
2 Kidder and Liu, "Bridging Theoretical Gaps."
3 Rosen et al., "The Anthropocene and the Landscape of Confucius," 1648–49; Kidder and Liu, "Bridging Theoretical Gaps."
4 Kidder and Zhuang, "Anthropocene Archaeology."
5 All the timelines of Loess Plateau settlements are based on two data sources. For counties, continuous data for every year comes from the China Historical GIS, and for settlements that did not have an administrative designation, digitized information is from Tan, *Zhongguo lishi ditu ji*.
6 Tan, "Heyi Huanghe." Thank you to David Pietz for pointing out the timing of this article relative to the completion of the Sanmenxia Dam.
7 Xin, "You Yuanguang hejue."
8 Kidder and Zhuang, "Anthropocene Archaeology of the Yellow River," 1627.
9 Kidder and Zhuang, "Anthropocene Archaeology of the Yellow River," 1627. Various tables of sedimentation rates assert different numbers, which are based on different methods and collection locations but follow the same broad trends.
10 Shi et al., "Changes in Sediment Yield of the Yellow River."
11 Scott, *Against the Grain*.
12 Based on Li, "The Products of Minds as Well as Hands." This map focuses on major sites yielding jade artifacts.
13 Zhou et al., "Environmental Variability within the Chinese Desert-Loess Transition Zone."
14 Zhou et al., "Environmental Variability within the Chinese Desert-Loess Transition Zone."
15 Zhuang and Kidder, "Archaeology of the Anthropocene."
16 Rosen et al., "The Anthropocene and the Landscape of Confucius."
17 Rosen, "The Impact of Environmental Change."
18 Kidder and Zhuang, "Anthropocene Archaeology of the Yellow River."
19 Huang et al., "Holocene Environmental Change Inferred from the Loess-Palaeosol Sequences"; Huang et al., "Extraordinary Floods of 4100–4000 BP."
20 Zhuang and Kidder, "Archaeology of the Anthropocene."
21 Kidder and Zhuang, "Anthropocene Archaeology of the Yellow River."
22 Huang and Su, "Climate Change and Zhou Relocations."
23 Rosen et al., "The Anthropocene and the Landscape of Confucius."

24 Kidder and Zhuang, "Anthropocene Archaeology"; see Kidder et al., "New Perspectives on the Collapse and Regeneration" on increasing flood frequency.
25 Kidder and Zhuang, "Anthropocene Archaeology," 2015; Zhuang and Kidder, "Archaeology of the Anthropocene" for great acceleration. The term is from McNeill and Engelke, *Great Acceleration*, which refers to the accelerating environmental impacts of human activity beginning in the middle of the twentieth century.
26 Lander and Brunson, "The Sumatran Rhinoceros," R245–53.
27 Zhuang and Kidder, "Archaeology of the Anthropocene."
28 Zhuang and Kidder, "Archaeology of the Anthropocene."
29 Rosen et al., "The Anthropocene and the Landscape of Confucius."
30 Rosen, "The Impact of Environmental Change."
31 Elvin, "Three Thousand Years"; Scott, *Against the Grain*.
32 Rosen et al., "The Anthropocene and the Landscape of Confucius," 1642.
33 Kidder and Liu, "Bridging Theoretical Gaps."
34 Rosen et al., "The Anthropocene and the Landscape of Confucius."
35 Huang et al., "Charcoal Records of Fire History."
36 Huang et al., "Charcoal Records of Fire History."
37 Shi, *Huangtu gaoyuan senlin*, 140–45.
38 Zhou et al., "Environmental Variability."
39 Zhang et al., "Late Holocene Vegetation Dynamic."
40 Di Cosmo, "The Northern Frontier in Pre-Imperial China"; Li, *Landscape and Power in Early China*.
41 Hsu, "The Spring and Autumn Period," 567.
42 Lewis, *Flood Myths*.
43 Rosen, "The Impact of Environmental Change."
44 Kidder and Liu, "Bridging Theoretical Gaps."
45 Xu, "A Study of Long-Term Environmental Effects," esp. fig. 3.
46 Rosen et al., "The Anthropocene and the Landscape of Confucius," 1646.
47 Rosen et al., "The Anthropocene and the Landscape of Confucius," 1647.
48 Rosen, "The Impact of Environmental Change."
49 Loewe, *Cambridge History of Ancient China*.
50 Needham, "Hydraulic Engineering," 232.
51 Guan, *Guanzi*, chap. 57 (*dudi*). See also Needham, "Hydraulic Engineering," 223.
52 Huanghe shuili weiyuanhui, *Huanghe dashi ji*.
53 Sima, *Shiji*, chap. 29, *shiqu shu* (Rivers and Canals Monograph). Translation modified from Watson, *Records of the Grand Historian*, 231.
54 Saito et al., "The Huanghe and Changjiang"; Fang and Xie, "Deforestation in Preindustrial China."

55 Gong, *Huanghe shihua*, 157–63.
56 Cited in Ren and Zhu, "Anthropogenic Influences."
57 Shi, *Huangtu gaoyuan senlin*, 52–57, citing Sima, *Shiji* and Ban, *Han shu*.
58 Menzies, *Forest and Land Management*, 15–19.
59 Cited in Fang and Xie, "Deforestation in Preindustrial China."
60 Fang and Xie, "Deforestation in Preindustrial China."
61 Allan, *Shape of the Turtle*, 29–62.
62 Jullien, *Treatise on Efficacy*, 16; Mengzi, *Mengzi, gongsun zhouyi* chapter. See also Legge, trans., *The Chinese Classics*, 183.
63 Jullien, *Treatise on Efficacy*; Mostern, "Mapping the Tracks of Yu."
64 *Zhouli* "Dong Gong kaogongji" section. Translation based in part on Needham, "Hydraulic Engineering."
65 *Shenzi* cited in Needham, "Hydraulic Engineering," 343. On Shen Dao see Yang, "Shen Dao's Theory of *Fa*."
66 *Shangshu*; see also Mostern, *Dividing the Realm in Order to Govern*, 62. Lewis, in *Flood Myths*, describes many versions of the Yu story and situates it in a written tradition that stretches back to the early Bronze Age.
67 Sima, *Shiji*; Watson, *Records of the Grand Historian*.
68 Animals, much like water was, were a key metaphorical concept in the early Chinese philosophical tradition. See Sterckx, *The Animal and the Daemon*.
69 Mencius, *Mencius*. See also Legge, *The Chinese Classics*.
70 Mencius, "Gaozi I" section. Translation informed by Needham, "Hydraulic Engineering," and Legge, *The Chinese Classics*.
71 Shi, *Huangtu gaoyuan senlin*, 140–45.
72 Chang, *The Rise of the Chinese Empire*, 17; Grousset, *Empire of the Steppes*, 26–27.
73 Zhang, *The River, the Plain, and the State*, 31.
74 Will, "Clear Waters versus Muddy Waters."
75 Will, "Clear Waters versus Muddy Waters."
76 Will, "Clear Waters versus Muddy Waters."
77 *Han shu* 29.1685. The translation is my own.
78 Kidder et al., "New Perspectives on the Collapse and Regeneration of the Han Dynasty."
79 Kidder et al., "New Perspectives on the Collapse and Regeneration of the Han Dynasty," citing Donald Wagner, *The State and the Iron Industry in Han China*.
80 Zhuang and Kidder, "Archaeology of the Anthropocene," 1612.
81 Hou, "Ancient City Ruins."
82 Kidder et al., "New Perspectives on the Collapse and Regeneration of the Han Dynasty," citing Ban, *Han shu*, "Shihuoshi" section. See Tan, "Heyi Huanghe," on the colonization campaigns on the Ordos.

83 Zhang, *The River the Plain, and the State,* 31.
84 Chen et al., "Socio-Economic Impacts on Flooding."
85 Dorofeeva-Lichtmann, "A History of a Spatial Relationship," citing the "Treatise on the Western Regions" (*Xiyu juan*) in the *History of the Han.*
86 Chang, *The Rise of the Chinese Empire,* 20.
87 Hou, "Ancient City Ruins."
88 Hou, "Ancient City Ruins."
89 Chang, *The Rise of the Chinese Empire,* 17–20.
90 Chang, *The Rise of the Chinese Empire,* chap. 1.
91 Miller, "The Southern Xiongnu."
92 Based on a figure in Miller, "The Southern Xiongnu."
93 Gong, *Huanghe shihua,* sec. 2.2.
94 Shi, *Huangtu gaoyuan senlin,* 145–50.
95 Shi, *Huangtu gaoyuan senlin,* 151.
96 Fang and Xie, "Deforestation in Preindustrial China."
97 Fang and Xie, "Deforestation in Preindustrial China."
98 *Huainanzi* 8.11 (*benjingxun*). Translation based on Needham, "Hydraulic Engineering," 245. The location of the forests in question is unclear from the text itself, although the passage vividly reflects the fact of deforestation.
99 Shi, *Huangtu gaoyuan senlin,* 64–66.
100 Shi, *Huangtu gaoyuan senlin,* 76–88.
101 Rosen et al. "The Anthropocene and the Landscape of Confucius," 1647.
102 Kidder and Zhuang, "Anthropocene Archaeology."
103 Kidder and Liu, "Bridging Theoretical Gaps."
104 Kidder et al., "New Perspectives."
105 Kidder et al., "New Perspectives."
106 Gong, *Huanghe shihua,* 161–63.
107 Huanghe shuilishi weiyuanhui, *Lidai zhi Huang wenxuan,* "Ji An" entry.
108 Sima, *Shiji,* chap. 29.
109 Watson, *Records of the Grand Historian,* 235–38; Sima, *Shiji,* chap. 29.
110 Greer, *Water Management,* 29–36.
111 Gong, *Huanghe shihua,* sec. 2.3.
112 Gong, *Huanghe shihua,* sec. 2.3.
113 Cited in Xu, "A Study of Long-Term Environmental Effects."
114 Ban, *Han shu,* chap. 29.
115 Huanghe shuilishi weiyuanhui, *Lidai zhi Huang wenxuan,* citing Sima, *Shiji.*
116 Needham, "Hydraulic Engineering," 329–31.
117 Needham, "Hydraulic Engineering," 232; original in Ban, *Han shu,* chap. 29.
118 Ban, *Han shu,* chap. 29.

119 Ban, *Han shu*, chap. 29; translation based on Needham, "Hydraulic Engineering," 232.
120 Kidder et al., "New Perspectives."
121 Bielenstein, "Wang Mang."
122 Kidder et al., "New Perspectives," 84.
123 Bielenstein, "Wang Mang."
124 Paul Wheatley, *Pivot of the Four Quarters*.
125 Kidder and Zhuang, "Anthropocene Archaeology."
126 Bielenstein, "Wang Mang."
127 Bielenstein, "Wang Mang," 240–41.
128 Miller, "The Southern Xiongnu."
129 Miller, "The Southern Xiongnu."
130 Miller, "The Southern Xiongnu."
131 Miller, "The Southern Xiongnu."
132 Bielenstein, "Wang Mang.
133 Gong, *Huanghe shihua*, sec. 4.2.
134 Fan, *Hou Han shu* 76.2458.
135 Gong, *Huanghe shihua*, sec. 2.3.
136 Xin, "You Yuanguang hejue" offers a detailed reading of sources about the Wang Jing initiatives. My thinking about agency in preindustrial waterworks initiatives owes a great deal to Mukerji, *Impossible Engineering*.
137 Huanghe shuili weiyuanhui, *Lidai zhi Huang wenxuan*, citing Fan, *Hou Han shu* 76.2464–66 (Wang Jing biography); Xin, "You Yuanguang hejue," 21–43.
138 Fan, *Hou Han shu* 76.2464–66 (Wang Jing biography). A *li* is approximately one-third of a mile.
139 See Xin, "You Yuanguang hejue," 34–43, on the debates over the Wang Jing water gates and their subsequent deployment during the imperial era.
140 Fang et al., *Jin shu*.
141 Chen et al., "Socio-Economic Impacts on Flooding."
142 Chin, "Climate Change and the Migrations of People."
143 Shi, *Huangtu gaoyuan senlin*, 89–103.
144 Yue, *Taiping huanyu ji*, "Shaanzhou" section.
145 Yue, *Taiping huanyu ji*, chap. 33, cited in Shi, *Huangtu gaoyuan senlin*, 186.
146 Shi, *Huangtu gaoyuan senlin*, 155.
147 Li, *Shuijing zhu*, cited in Fang and Xie, "Deforestation in Preindustrial China."
148 Huanghe shuili weiyuanhui, *Lidai zhi Huang wenxuan*.
149 Obrusanszky, "Tongwancheng."

CHAPTER THREE. LOESS IS MORE

1. Saito, Yang, and Hori, "The Huanghe (Yellow River) and Changjiang (Yangtze River) Deltas."
2. Ren and Zhu, "Anthropogenic Influences."
3. Gong, *Huanghe shihua*, 9.
4. Huanghe shuili weiyuanhui, *Huanghe dashi ji*. The text, *Tufan Huanghe lü*, is lost. It is referenced in the biography of the geographer Jia Dan in Ouyang, *Xin Tangshu*.
5. Zhang, "Changing with the Yellow River," 6.
6. See, e.g., Hymes and Schirokauer, *Ordering the World*.
7. Robert Hartwell's classic article "Demographic, Political, and Social Transformations of China" combines these two threads in Tang-Song transition theory.
8. Cao, *Shaanxi shengzhi renkou zhi*, 331–32. These numbers should be taken at heuristic value rather than as precise counts because of problems with conducing and comparing historical censuses.
9. Shi, *Huangtu gaoyuan senlin*, 151–55, citing Toqto'a, *Song shi*.
10. Shi, *Huangtu gaoyuan senlin*, 158–59, citing Ouyang, *Xin Tangshu* chap. 167.
11. Skaff, *Sui-Tang China*, 50.
12. Huanghe shuili weiyuan hui, *Lidai zhi Huang wenxuan*, citing Liu, *Jiu Tang shu*, Jiangshidu biographies.
13. Shi, *Huangtu gaoyuan senlin*, 188.
14. Skaff, *Sui-Tang China*, 27–28, 3–4, and app. C.
15. Skaff, *Sui-Tang China*, 29–30.
16. Zhang et al., "Late Holocene Vegetation Dynamics."
17. Zhang et al., "Late Holocene Vegetation Dynamics."
18. Chen et al., "Socio-Economic Impacts on Flooding."
19. Zhang et al., "Late Holocene Vegetation Dynamics."
20. Durand, "The Population Statistics of China."
21. Di Cosmo et al., "Environmental Stress and Steppe Nomads."
22. Skaff, *Sui-Tang China*, 44–48.
23. Chen et al., "Socio-Economic Impacts on Flooding."
24. Ren and Zhu, "Anthropogenic Influences."
25. Obrusanszky, "Tongwancheng, the City of Southern Huns."
26. Shi, *Huangtu gaoyuan senlin*, 103–12.
27. Zhou et al., "Environmental Variability."
28. Obrusanszky, "Tongwancheng, the City of Southern Huns."
29. Zhang et al., "Late Holocene Vegetation Dynamics."
30. Zhang et al., "Late Holocene Vegetation Dynamics."
31. Huanghe shuili weiyuanhui, *Huanghe dashi ji*.
32. Xu, "Naturally and Anthropogenically Accelerated Sedimentation."

33 Hartwell, "Demographic, Political, and Social Transformations of China," 365–83.
34 Hartwell, "Demographic, Political, and Social Transformations of China," 386–87.
35 Huanghe shuili weiyuan hui, *Huanghe dashi ji*.
36 Twitchett, *Cambridge History 3.1*, 134–35; Needham, *Hydraulic Engineering*, 269–70.
37 Twitchett, *Cambridge History 3.1*, 177.
38 Huanghe shuili weiyuan hui, *Lidai zhi Huang wenxuan*, citing Ouyang, *Xin Tangshu* Commodities Monograph and Pei Yaoqing Biography.
39 Cited in Pietz, *Engineering the State*, 8.
40 Huanghe shuili weiyuan hui, *Lidai zhi Huang wenxuan*, citing Ouyang, *Xin Tangshu* Commodities Monograph; Adshead, *T'ang China*, 50.
41 Huanghe shuili weiyuan hui, *Lidai zhi Huang wenxuan*.
42 Huanghe shuili weiyuan hui, *Huanghe dashi ji*.
43 Huanghe shuili weiyuan hui, *Huanghe dashi ji*, citing Ouyang, *Xinwudai shi* Biography of Duan Ning.
44 Zhang, *The River, the Plain, and the State,* 30.
45 Huanghe shuili weiyuan hui, *Lidai zhi Huang wenxuan*.
46 Hartwell, "Demographic, Political, and Social Transformations of China"; Shiba, *Commerce and Society*; Ho, "Early Ripening Rice."
47 Hartwell, "A Revolution in the Chinese Iron and Coal Industries."
48 Chen et al., "Socio-Economic Impacts on Flooding."
49 Needham, "Hydraulic Engineering," sec. 41, 245–62.
50 This figure from my database represents events within 50 kilometers of a modern Kaifeng centroid location.
51 Liu, "Picturing Yu Controlling the Flood," 91–92.
52 Shi, *Huangtu gaoyuan senlin*, 157.
53 Liu, "Picturing Yu Controlling the Flood," 102–3; Hartwell, "A Revolution in the Chinese Iron and Coal Industries."
54 Liu, "Picturing Yu Controlling the Flood," 104.
55 Zhang, *The River, the Plain, and the State*, chap. 5, esp. 184.
56 Shi, *Huangtu gaoyuan senlin*, 167.
57 Zhang, *The River, the Plain, and the State*, 268 and 275.
58 Toqto'a, *Song shi* 266.9183; Shi, *Huangtu gaoyuan senlin*, 157.
59 Miller, *Fir and Empire*, chap. 1.
60 Menzies, "Forestry," 560.
61 Elvin, *Retreat of the Elephants*, 110–12.
62 Mostern, "Cartography on the Song Frontier."
63 Shi, *Huangtu gaoyuan senlin*, 122–23.
64 Zhang, *The River, the Plain, and the State*.

65 Shi, *Huangtu gaoyuan senlin*, 157.
66 Li, *Changbian*, 186.7a; Tackett, "The Great Wall and Conceptualization of the Border."
67 Ma, "Song Zhezong qinzhengshi."
68 Dunnell, "The Hsi Hsia," 170.
69 McGrath, "Frustrated Empires," 153.
70 Li, *Song-Xia guanxi shi*, citing Li, *Changbian*, chap. 139.
71 Li, *Song-Xia guanxi shi*, 179.
72 Li, *Changbian*, 172.17a–20a.
73 Li, *Changbian*, 134.3b–4a.
74 Li, *Song-Xia guanxi shi*.
75 Mostern, *Dividing the Realm in Order to Govern*.
76 Xu, *Song huiyao*, bing 8.41–42.
77 Alyagon, "Soldier Mutinies and Resistance."
78 Alyagon, "Solider Mutinies and Resistance, 282, 286.
79 Li, *Changbian*, chap. 139
80 Li, *Changbian*, 204.4a–b.
81 Li, *Changbian*, 507.
82 Li, *Song-Xia guanxi shi*, 171, citing *Mengxi bitan* chap. 13.
83 Li, *Changbian*, chap. 132.
84 Li, *Changbian*, chap. 506.
85 Li, *Song-Xia guanxi shi*, citing Toqto'a, *Song shi*, chap. 264.
86 Xu, *Song huiyao*, bing 27.
87 Xu, *Song huiyao*, fangyu 19.
88 Shi, *Huangtu gaoyuan senlin*, 157.
89 Li, *Song-Xia guanxi shi*, 171–79.
90 Cao, *Shaanxi shengzhi renkou zhi*, 331.
91 Shi, *Huangtu gaoyuan senlin*, 122–23.
92 Cao, *Shaanxi shengzhi renkou zhi*, 331–32.
93 Will, "Clear Waters versus Muddy Waters," 300; Zhang, "Changing with the Yellow River," 6.
94 Toqto'a, *Song shi*, chap. 94.
95 Will, "Clear Waters versus Muddy Waters," 300.
96 Liu, "Picturing Yu Controlling the Flood," 106.
97 Liu, "Picturing Yu Controlling the Flood," 96–99.
98 Zhang, *The River, the Plain, and the State*.
99 Zhang, *The River, the Plain, and the State*, 58–80.
100 Lamouroux, "From the Yellow River to the Huai."
101 Liu, "Picturing Yu Controlling the Flood," 104, 109.
102 Alyagon, "Soldier Mutinies and Resistance," 284.
103 Lamouroux, "From the Yellow River to the Huai."

104 Liu, "Picturing Yu Controlling the Flood," 100–101.
105 Lamouroux, "From the Yellow River to the Huai," 549, citing Toqto'a, *Song shi*, chap. 91.
106 Xu, *Song huiyao: fangyu*, 14.4.
107 Lamouroux, "From the Yellow River to the Huai."
108 Lamouroux, "From the Yellow River to the Huai."
109 Zhang, *The River, the Plain, and the State*, 118–28.
110 Zhang, "Changing with the Yellow River," 35.
111 Dodgen, *Controlling the Dragon*, 13.
112 Gong, *Huanghe shihua*.
113 Zhang, "Changing with the Yellow River," 12–16.
114 Zhang, *The River, the Plain, and the State*, 191.
115 Zhang, "Changing with the Yellow River."
116 Zhang, "Changing with the Yellow River."
117 Zhang, "Changing with the Yellow River"; Meeks and Mostern, "The Politics of Territory in Song Dynasty China."
118 Zhang, *The River, the Plain, and the State*, 152–53.
119 Lamouroux, "From the Yellow River to the Huai."
120 Ouyang, "Lun xiuhe," 38.1b–4b.
121 Ouyang, "Lun xiuhe dierzhuang," 38.4b–8b.
122 Zhang, *The River, the Plain, and the State*, chap. 5.
123 Elvin and Su, "Action at a Distance."
124 Needham, "Hydraulic Engineering," 266; Huanghe shuilishi weiyuanhui, *Huanghe dashi ji*.
125 Toqto'a, *Jin shi* Rivers and Canals Monograph 27.669.
126 Mostern, "From Battlefields to Counties," 227.
127 Mostern, "From Battlefields to Counties," 241.
128 Mostern, "From Battlefields to Counties."
129 Perry, *Rebels and Revolutionaries in North China*, 12.
130 Toqto'a, *Jin shi* Rivers and Canals Monograph, cited in Huanghe shuilishi weiyuanhui, *Lidai zhi Huang wenxuan*.
131 Toqto'a, *Jin shi* Rivers and Canals Monograph, cited in Huanghe shuilishi weiyuanhui, *Lidai zhi Huang wenxuan*.
132 Pietz, *Engineering the State*, 8.
133 Huanghe shuili shi weiyuanhui, *Huanghe dashi ji*.
134 Information from my database.
135 Dodgen, *Controlling the Dragon*, 1–4.
136 Huanghe shuilishi weiyuanhui, *Lidai zhi Huang wenxuan*; Taubes, "Some Early Chinese Descriptions of the Myer Ma Chen Range."
137 Dorofeeva-Lichtman, "A History of a Spatial Relationship."
138 Huanghe shuili shi weiyuanhui, *Lidai zhi Huang wenxuan*.

CHAPTER FOUR. LEVIES AND LEVEES

1. Pomeranz, "The Transformation of China's Environment."
2. Richards, *Unending Frontier*.
3. I am indebted to Pomeranz's introduction, 5, for the elegant characterization of the era.
4. Pomeranz, "The Transformation of China's Environment," 119.
5. Pomeranz, "The Transformation of China's Environment," 122.
6. Cao, *Shaanxi shengzhi renkou zhi*, 331–32. This should be taken as a heuristic rather than a price count given problems with conducting and comparing historical censuses. Consistent with the population numbers for Shaanxi alone, another source estimates a total Loess Plateau population of 14,732,742 in the mid-Ming and 40,315,284 in 1820. Zhongguo kexueyuan, *Huangtu gaoyuan diqu*, 11–12.
7. Du et al., "Late Quaternary Activity of the Huashan Piedmont Fault."
8. Shi, *Huangtu gaoyuan senlin*.
9. Mote, *Imperial China*, 688–89, 691–92; see also Rossabi, "The Ming and Inner Asia"; Waldron, *The Great Wall*.
10. Waldron, *The Great Wall*, 77–78.
11. Waldron, *The Great Wall*, 96–102.
12. Perdue, *China Marches West*, 60–62; Mote, *Imperial China*, 693–97; Waldron, *The Great Wall*, 105.
13. Waldron, *The Great Wall*, chap. 9, 140–64.
14. Based on Mote, *Imperial China*.
15. Perdue, *China Marches West*, 60–62; Mote, *Imperial China*, 693–97; Waldron, *The Great Wall*, 105. See also Liu, *Mingdai jiubian shidi yanjiu*.
16. Wei, *Huang Ming jiubian kao*.
17. Waldron, *The Great Wall*, 107.
18. Cui et al., "Climate Change, Desertification and Societal Responses," citing *Ming Xianzong shilu* chap. 27 and *Ming jingshi wenbian* chap. 43.
19. Cui et al., "Climate Change, Desertification and Societal Responses," citing *Ming jingshi wenbian* chap. 61 and *Shuyu zhouzilu* chap. 18.
20. Cui et al., "Climate Change, Desertification and Societal Responses," citing *Ming Xianzong shilu* chap. 122.
21. Cui et al., "Climate Change, Desertification and Societal Responses," 91.
22. Cited in Cui et al., "Climate Change, Desertification and Societal Responses," 86.
23. Cui et al., "Climate Change, Desertification and Societal Responses," 91; citation for the sand-covered roads from *Huang Ming jingji wenlu*, chap. 40; for Yulin from *Jiubian tu lun*.
24. Cui et al., "Climate Change, Desertification and Societal Responses," 81.

25 Cui et al., "Climate Change, Desertification and Societal Responses," citing *Ming jingshi wenbian* chap. 238.
26 Gu, *Dushi fangyu jiyao*, chap. 61.
27 Cui et al., "Climate Change, Desertification and Societal Responses," 91.
28 Cui et al., "Climate Change, Desertification and Societal Responses," 91.
29 Cui et al., "Climate Change, Desertification and Societal Responses," citing Zheng Rubi et al., *Yansui zhenzhi*.
30 Cui et al., "Climate Change, Desertification and Societal Responses," *Yulin fuzhi* 21 and *Shaanxi sizhen tushuo* 1.
31 Cited in Needham, "Hydraulic Engineering," 240.
32 Shi, *Huangtu gaoyuan senlin*, 160–66, citing *Da Ming huidian*.
33 Shi, *Huangtu gaoyuan senlin*, 160–66, citing *Ming jingshi wenbian* chap. 63.
34 Needham, "Hydraulic Engineering," 245, citing *Shanxi tongzhi* chap. 66.
35 Cited in Menzies, *Forest and Land Management in Imperial China*, 71.
36 Will, "Clear Waters versus Muddy Waters," 305.
37 Cui et al., "Climate Change, Desertification and Societal Responses," 92.
38 On the plausible, if circumstantial, links between Ming collapse and the Little Ice Age, see Fan, "Climatic Change and Dynastic Cycles."
39 Brook, *The Troubled Empire*.
40 Mote, *Imperial China*, 868–73.
41 Li, "Qingdai huangtu gaoyuan shuitu liushi," 83.
42 Li, "Qingdai huangtu gaoyuan shuitu liushi."
43 Li et al., "Historical Changes in the Environment of the Chinese Loess Plateau"; Needham, "Hydraulic Engineering," 240.
44 Zhou et al., "Environmental Variability Within the Chinese Desert-Loess Transition Zone."
45 Pomeranz, "The Transformation of China's Environment," 124.
46 Chen et al., "Socio-Economic Impacts on Flooding."
47 Li et al., "Historical Changes in the Environment of the Chinese Loess Plateau."
48 Gu, *Dushi fangyu jiyao*, chap. 61, cited in Fang and Xie, "Deforestation in Preindustrial China."
49 Huanghe shuili weiyuanhui, *Lidai zhi Huang wenxuan*.
50 McMahon, *Rethinking the Decline of China's Qing Dynasty*, 133. On the impact of New World crops in the Yangtze region, see Osborne, "Highlands and Lowlands."
51 Li, "Qingdai huangtu gaoyuan shuitu liushi," citing *Xingxian zhi*, 17.12.
52 Li, "Qingdai huangtu gaoyuan shuitu liushi," citing *Qinjiang zhilue*, 65–66.
53 Li, "Qingdai huangtu gaoyuan shuitu liushi," citing *Sansheng bianfang luelan*, 6.
54 Li, "Qingdai huangtu gaoyuan shuitu liushi," citing *Sansheng bianfang luelan*.

55 Zhang et al., "Late Holocene Vegetation Dynamic."
56 Shi, *Huangtu gaoyuan senlin*, 160–68.
57 Chen et al., "Socio-Economic Impacts on Flooding."
58 Pomeranz, *Making of a Hinterland*, 134.
59 Will, "Clear Waters versus Muddy Waters."
60 Li, "Qingdai huangtu gaoyuan shuitu liushi," citing *Chongxiu chongxin xianzhi* 4.318–9.
61 Li, "Qingdai huangtu gaoyuan shuitu liushi," 84–86.
62 Will, "Clear Waters versus Muddy Waters."
63 Li, "Qingdai huangtu gaoyuan shuitu liushi."
64 Hucker, *A Dictionary of Official Titles*, 146.
65 Li, "Qingdai huangtu gaoyuan shuitu liushi," citing Xu, *Xing shui jin jian*, 255.
66 Will, "Clear Waters versus Muddy Waters."
67 Cited in Menzies, "Forestry."
68 Menzies, "Forestry." Key works in this genre are Fortune, *A Residence among the Chinese*; Wilson, *A Naturalist in Western China*; David, *Abbe David's Diary*; Garrison, "Some Effects of Deforestation in China"; Lowdermilk, "Forest Destruction and Slope Denudation"; *Forestry in Denuded China*; Mallory, *China Land of Famine*.
69 Gong, *Huanghe shihua*, 143–46.
70 Huanghe shuili weiyuanhui, *Huanghe dashi ji*, 51.
71 Brook, *The Troubled Empire*, 62.
72 Mote, *Imperial China*, 517–21 at 521.
73 Huanghe shuili weiyuanhui, *Lidai zhi Huang wenxuan*.
74 *Huanghe dashi ji*, 51.
75 Huanghe shuili weiyuanhui, *Lidai zhi Huang wenxuan*, citing *Zhizheng hefang ji*.
76 Dodgen, *Controlling the Dragon*, 1–4, 13.
77 Brook, *Confusions of Pleasure*, 46–47.
78 Dodgen, *Controlling the Dragon*, 13–14.
79 Huanghe shuili weiyuanhui, *Lidai zhi Huang wenxuan*.
80 Brook, *Confusions of Pleasure*, 47.
81 Sun Jinghao personal communication and Sun, "City, State and the Grand Canal."
82 Brook, *Confusions of Pleasure*, 48.
83 Farmer, *Early Ming Government*, 156–62, on the displacement of the sea route by the canal.
84 Cited in Elvin and Su, "Action at a Distance."
85 Xu, "A Study of Long-Term Environmental Effects"; Pietz, *Engineering the State*, 9.
86 Elvin and Su, "Action at a Distance."

87 Liu, *Wenshui ji*.
88 Elvin and Su, "Action at a Distance"; Mostern, "Mapping the Tracks of Yu."
89 Figure 4.18 based on Zhang et al., "Qingkou Complex."
90 Chang, "Fathoming Qianlong," 58–62.
91 Xu, "Long Term Environmental Effects."
92 Dodgen, *Controlling the Dragon*, 17.
93 Li, *Ming Qing Huang Yun diqu*, and personal communication.
94 Dodgen, *Controlling the Dragon*, 107.
95 Pietz, *Engineering the State*, 8–17.
96 Shen, *Jingguan, shengji, liebian*, and personal communication.
97 Linqing, *Hegong qiju tushuo*; Needham, "Hydraulic Engineering," 262, summarizes the text and explains the meaning of the image.
98 Dodgen, *Controlling the Dragon*, 96–103.
99 Yang and Han, "The Change of the Gaobao Lakes," and personal communication with Yang Xiao and Han Zhaoqing.
100 Xu, "Long Term Environmental Effects."
101 Xu, "Long Term Environmental Effects."
102 Pan, *Hefang yilan* (1970).
103 Dodgen, *Controlling the Dragon*.
104 Cited in Pomeranz, *Making of a Hinterland*, 128.
105 Vermeer, "P'an Chi-hsun's Solutions."
106 Theobald, "*Hefang yilan*"; Cheng, "Hefang yilan," 299.
107 For Pan's life and career, see Goodrich and Fang, *Dictionary of Ming Biography*.
108 Zhang, *Ming shi* 84.2047–74.
109 Cited in Elvin and Su, "Action at a Distance," 401.
110 Gong, *Huanghe shihua*, 143–46.
111 Gong, *Huanghe shihua*, sec. 2.3.
112 Gong, *Huanghe shihua*, 96.
113 Xu, "Long Term Environmental Effects."
114 Xu, "A Study of the Accumulation Rate."
115 Vermeer, "P'an Chi-Hsun."
116 Xu, "Long Term Environmental Effects."
117 Lorge, *War, Politics and Society*, 147.
118 Dodgen, *Controlling the Dragon*, 4.
119 Leonard, *Controlling from Afar*, 86; Dodgen, *Controlling the Dragon*, 18; Guy, *Qing Governors and Their Provinces*, 246.
120 Chang, "Fathoming Qianlong."
121 Pomeranz, *Making of a Hinterland*; Sun, "City, State, and the Grand Canal."
122 Thank you to Henan University archaeology professor Qin Zhen for arranging a tour of this site.
123 Dodgen, *Controlling the Dragon*, 21.

124 Dodgen, *Controlling the Dragon*, 17.
125 Cited in Chang, "Fathoming Qianlong," 57.
126 Chen et al., "Socio-Economic Impacts on Flooding"; Pomeranz, "The Transformation of China's Environment," 130; Pomeranz, *Making of a Hinterland*.
127 Dodgen, *Taming the Dragon*, 6.
128 Chen, *Hefang shuyan*, cited in Huanghe shuili weiyuanhui, *Lidai zhi Huang wenxuan*.
129 Li, "Qingdai huangtu gaoyuan," citing *Wusheng gouxu tushuo*. Also see Shi, *Huangtu gaoyuan senlin*, 190, citing *Xu xingshui jinjian*, chap. 13.
130 Dodgen, *Controlling the Dragon*, 42–68.
131 Pomeranz, *Making of a Hinterland*, 135–37.
132 Sun, "City, State, and the Grand Canal."
133 Gong, *Huanghe shihua*, 151.
134 Perry, *Rebels and Revolutionaries*, 10; Pomeranz, "Transformation of China's Environment," 130.
135 These concepts are explored in Nixon, *Slow Violence and the Environmentalism of the Poor*.
136 Leonard, *Controlling from Afar*, 109, and chap. 4.
137 Leonard, *Controlling from Afar*, 140.
138 Leonard, *Controlling from Afar*, 150.
139 Lawrence Zhang, personal communication to the Facebook Sinology group, July 12, 2017. According to his calculations, 1 million taels was the approximate annual tax revenue of one small province.
140 Dodgen, *Controlling the Dragon*, 89–95.
141 Leonard, *Controlling from Afar*, 109.
142 Leonard, *Controlling from Afar*, 142, citing Secret Edict [*Shangyudan fangben*] 4.12.18b held in Qing archives, Palace Museum, Taibei.
143 Leonard, *Controlling from Afar*, 220.
144 Leonard, *Controlling from Afar*, 241, citing P5.11.28 (public edict, *Shangyudang fangben* in Qing archives).
145 Leonard, *Controlling from Afar*, 125.
146 Dodgen, *Controlling the Dragon*, 27–34.
147 Dodgen, *Controlling the Dragon*, 84–89.
148 Dodgen, *Controlling the Dragon*, 97–107.
149 Shen, *Jingguan, shengji, liebian*.
150 Dodgen, *Controlling the Dragon*, 108–12.
151 Dodgen, *Controlling the Dragon*.
152 Xu, "Long Term Environmental Effects."
153 Pietz, *The Yellow River*, 72.
154 Pomeranz, *Making of a Hinterland*.
155 Dodgen, *Controlling the Dragon*, 112–45.
156 Pietz, *The Yellow River*, chaps. 2–3.

157 Pomeranz, *Making of a Hinterland*, 155–57.
158 Pietz, *The Yellow River*, chap. 3.

EPILOGUE

1 Pomerantz, *Making of a Hinterland*, 126.
2 Perry, *Rebels and Revolutionaries*, 20.
3 Perry, *Rebels and Revolutionaries*, 20, 33.
4 Pietz, *The Yellow River*.
5 Map based on Stevenliuyi, "Huayuankou Flood," CC BY-SA 4.0, https://commons.wikimedia.org/w/index.php?curid=59053473.
6 Lary, "Drowned Earth"; Muscolino, *Ecology of War in China*.
7 Muscolino, *Ecology of War in China*; Pietz, *The Yellow River*.
8 Pietz, *The Yellow River*, 133.
9 Pietz, *The Yellow River*. For the process of soil restoration on sand fields, see Shen, *Jingguan, shengji, liebian*.
10 Pietz, *The Yellow River*.
11 Muscolino, "Soil and Society on Shaanxi's Loess Plateau."
12 Pomeranz, *Making of a Hinterland*, 128.
13 Xu, "A Study of Long Term Environmental Effects."
14 Kolbert, "Under Water."

APPENDIX

1 Shen, *Huanghe nianbiao*. Thank you to Wang Yingjie and Su Yanjun for sharing with me a digitized version of this source.
2 The work of Chen Yunzhen and her coauthors is an exception, as is the work of Wang Yingjie and Su Yanjun. I am grateful to all of these authors for sharing their findings with me.
3 Tan, *Zhongguo lishi ditu ji*.
4 Sun, "Bei Song shiqi huangtu gaoyuan."
5 We did not require that identically named places needed to maintain the same suffix or place-name type across periods and regimes, because changing suffixes reflect changes to civil and military government systems over time and not the persistent existence of a particular settlement.
6 Wang, "Boundary Data of Loess Plateau Region."
7 The other sources are Huanghe shuilishi weiyuanhui, *Huanghe dashi ji*; Huanghe fanghongzhi bianzuan weiyuanhui, *Huanghe fanghong zhi*; Huanghe shuilishi weiyuanhui, *Huanghe liuyu zongshu*; Tan, *Huanghe shi luncong*; Huanghe shuilishi shuyao bianxiezu, *Huanghe shuilishi shuyao*; *Huanghe shuilishi yanjiu*; *Zhongguo gudai zhongda zaihai*; and Yuan, *Zhongguo zaihai tongshi*.
8 Berman, "China River Basins."
9 Dombrowski et al., "Introduction to Jupyter Notebooks."

BIBLIOGRAPHY

Acciaviati, Anthony, "The Great Sediment Sorting Machine," unpublished paper, Conference on Digital Methods for the Study of Environmental History, Fudan University (June 2019).

Adshead, Samuel Adrian M., *T'ang China: The Rise of the East in World History* (New York: Palgrave Macmillan, 2004).

Allan, Sarah, *Shape of the Turtle: Myth, Art, and Cosmos in Early China* (Albany: State University of New York Press, 1991).

Allan, Sarah, *The Way of Water and Sprouts of Virtue* (Albany: State University of New York Press, 1997).

Alyagon, Elad, "Soldier Mutinies and Resistance," in Patricia Buckley Ebrey and Paul Jacov Smith, eds., *State Power in China, 900–1325* (Seattle: University of Washington Press, 2016), 277–306.

Ban Gu 班固, *Han shu* 漢書 (Beijing: Zhonghua shuju; online at Scripta Sinica, http://hanchi.ihp.sinica.edu.tw).

Berman, Merrick Lex, ed., "China River Basins," Harvard WorldMap, 2011. worldmap.harvard.edu.

Bielenstein, Hans, "Wang Mang, the Restoration of the Han Dynasty, and Later Han," in Denis Twitchett and Michael Loewe, eds., *The Cambridge History of China*, vol. 1, *The Ch'in and Han Empires, 221 BC–AD 220* (Cambridge: Cambridge University Press, 1986).

Bluebulb Projects, *The Measure of Things* (2020). www.bluebulbprojects.com/measureofthings/default.php.

Bray, Francesca, and Joseph Needham, *Science and Civilization in China*, vol. 6, *Biology and Biological Technology, Part 2, Agriculture* (Cambridge: Cambridge University Press, 1984).

Brook, Timothy, *The Confusions of Pleasure: Commerce and Society in Ming China* (Berkeley: University of California Press, 2011).

Brook, Timothy, *The Troubled Empire: China in the Yuan and Ming Dynasties* (Cambridge, MA: Harvard University Press, 2010).

Cao Zhanquan 曹占泉, *Shaanxi shengzhi renkou zhi* 陝西省志人口志 (Xi'an: Sanqin chubanshe, 1986).

Cen Zhongmian 岑仲勉, *Huanghe bianqian shi* 黃河變遷史 (1957; Beijing: Zhonghua shuju, 2004).

Chang, Chun-shu, *The Rise of the Chinese Empire*, vol. 2, *Frontier, Immigration, and Empire in Han China, 130 BC–AD 157* (Ann Arbor: University of Michigan Press, 2007).

Chang, Michael, "Fathoming Qianlong: Imperial Activism, the Southern Tours, and the Politics of Water Control, 1736–1765," *Late Imperial China* 24.2 (2003), 51–108.

Chen Yunzhen 陳蘊真, "Huanghe fanlan shi: Cong lishi wenxian fenxi dao jisuaji moni" 黃河氾濫史: 從歷史文獻分析到計算機模擬, PhD dissertation, Nanjing University, 2013.

Chen Yunzhen, James Syvitski, Shu Gao, Irina Overnee, and Albert Kettner, "Socio-Economic Impacts on Flooding: A 4000-Year History of the Yellow River, China," *Ambio* 41 (2012), 682–98.

Cheng, Linsun, and Kerry Brown, *Berkshire Encyclopedia of China* (Great Barrington, MA: Berkshire Publishing, 2009).

Cheng Pengju 程鵬舉, "Hefang yilan" 河防一覽, in Zhou Gucheng 周谷城, ed., *Zhongguo xueshu mingzhu tiyao* 中國學術名著提要, *Jingji* 經濟 (Shanghai: Fudan daxue chubanshe, 1994).

Chi Ch'ao-ting, *Key Economic Areas in Chinese History, as Revealed in the Development of Public Works for Water-Control* (1936; New York: A. M. Kelley, 1970).

Chin, Connie, "Climate Change and the Migrations of People During the Jin Dynasty," *Early Medieval China* 13–14.2 (2008), 49–78.

Cook, E. R., K. J. Anchukaitis, B. M. Buckley, R. D. D'Arrigo, G. C. Jacoby, and W. E. Wright, "Asian Monsoon Failure and Megadrought during the Last Millennium," *Science* 328 (2010), 486–89 (Monsoon Asia Drought Atlas, http://drought.memphis.edu/MADA/Default.aspx).

Cui Jianxin, Hong Chang, Kaiyue Cheng, and George Burr, "Climate Change, Desertification and Societal Responses along the Mu Us Desert Margin during the Ming Dynasty," *Weather, Climate, and Society* 9.1 (2017), 81–94.

Da Cunha, Dilip, *The Invention of Rivers: Alexander's Eye and Ganga's Descent* (Philadelphia: University of Pennsylvania Press, 2018).

David, Armand, and Helen Fox, *Abbe David's Diary* (1949; Cambridge, MA: Harvard University Press, 2013).

Department of Regional Development and Environment Executive Secretariat for Economic and Social Affairs, Organization of American States, *Primer on*

Natural Hazard Management in Integrated Regional Development Planning, "Floodplain Definition and Flood Hazard Assessment" (1991). http://www.oas.org/dsd/publications/unit/oea66e/ch08.htm.

Di Cosmo, Nicola, "The Northern Frontier in Pre-Imperial China," in Michael Loewe and Edward Shaughnessy, eds., *The Cambridge History of Ancient China* (Cambridge: Cambridge University Press, 2009), 885–966.

Di Cosmo, Nicola, Amy Hessl, Caroline Leland, and Oyunsanaa Byambasuren, "Environmental Stress and Steppe Nomads: Rethinking the History of the Uyghur Empire (744–840) with Paleoclimate Data," *Journal of Interdisciplinary History* 48.4 (2018), 439–63.

Dodgen, Randall A., *Controlling the Dragon: Confucian Engineers and the Yellow River in Late Imperial China* (Honolulu: University of Hawai'i Press, 2001).

Dombrowski, Quinn, Tassie Gniady, and David Kloster, "Introduction to Jupyter Notebooks," *Programming Historian* 8 (2019). https://doi.org/10.46430/phen0087.

Dorofeeva-Lichtmann, Vera, "A History of a Spatial Relationship: Kunlun Mountain and the Yellow River Source from Chinese Cosmography through to Western Cartography," *Circumscribere* 11 (2012): 1–31.

Du Jianjun, Li Dunpeng, Wang Yufang, and Ma Yinsheng, "Late Quaternary Activity of the Huashan Piedmont Fault and Associated Hazards in the Southeastern Weihe Graben, Central China," *Acta Geologica Sinica* 91.1 (2017), 76–92.

Dunnell, Ruth, "The Hsi Hsia," in Herbert Franke and Denis Twitchett, eds., *Cambridge History of China*, vol. 6, *Alien Regimes and Border States*, 154–214 (Cambridge: Cambridge University Press, 1998).

Durand, John, "The Population Statistics of China, A.D. 2–1953," *Population Studies* 13.3 (1960), 209–256.

Elvin, Mark, *The Retreat of the Elephants: An Environmental History of China* (New Haven, CT: Yale University Press, 2008).

Elvin, Mark, "Three Thousand Years of Unsustainable Growth: China's Environment from Archaic Times to the Present," *East Asian History* 6 (1993), 7–46.

Elvin, Mark, and Su Ninghu. "Action at a Distance: The Influence of the Yellow River on Hangzhou Bay Since A.D. 1000," in *Sediments of Time: Environment and Society in Chinese History*, ed. Mark Elvin and Liu Ts'ui-jung (Cambridge: Cambridge University Press, 1998), 344–410.

Fairbank Center for Chinese Studies and the Center for Historical Geographical Studies at Fudan University, "China Historical GIS, Version 6" (2016), https://dataverse.harvard.edu/dataverse/chgis_v6.

Fan Ka-wen, "Climatic Change and Dynastic Cycles in Chinese History: A Review Essay," *Climatic Change* 101 (2010), 565–73.

Fan Ye 范曄 et al., *Hou Han Shu* 後漢書 (Beijing: Zhonghua shuju; online at Scripta Sinica).

Fang Jin-qi and Zhifen Xie, "Deforestation in Preindustrial China: The Loess Plateau as an Example," *Chemosphere* 29.5 (1994), 983–99.

Fang Xuanling 房玄齡 et al., *Jin shu* 晉書 (Beijing: Zhonghua shuju; online at Scripta Sinica).

Farmer, Edward, *Early Ming Government* (Cambridge, MA: Harvard University Press, 1976).

Fortune, Robert, *A Residence among the Chinese: Being a Narrative of Scences and Adventures during a Third Visit to China, from 1853 to 1856* (Cambridge: Cambridge University Press, 1857).

Garrison, F. Lynwood, "Some Effects of Deforestation in China," *Journal of the Franklin Institute* 152.2 (1901), 141–52.

Gong Li 龔莉 et al., eds., 黃河史話 *Huanghe shihua* (Beijing: Zhongguo dabaike quanshu chubanshe, 2007).

Goodrich, L. Carrington, and Chaoying Fang, *Dictionary of Ming Biography, 1368–1644* (New York: Columbia University Press, 1976).

Greer, Charles, *Water Management in the Yellow River Basin of China* (Austin: University of Texas Press, 1979).

Grousset, Rene, *The Empire of the Steppes: A History of Central Asia*, Naomi Walford, trans. (1939; New York: Barnes and Noble, 1996).

Gu Zuyu 顧祖禹, *Dushi fangyu jiyao* 讀史方輿紀要 (ca. seventeenth century; Taibei: Taiwan shangwu, 1968).

Guan Zhong 管仲, *Guanzi* 管子 (Shanghai: Commercial Press, 1923; Chinese Text Project digital version of the Sibu congkan edition).

Guy, Kent, *Qing Governors and Their Provinces: The Evolution of Territorial Administration in China, 1644–1796* (Seattle: University of Washington Press, 2013).

Han Maoli 韩茂莉, *Songdai nongye dili* 宋代農業地理 (Taiyuan: Shanxi guji chubanshe, 1993).

Hance, Jeremy, "Proving the 'Shifting Baselines' Theory: How Humans Consistently Misperceive Nature," *Mongabay* (2009). https://news.mongabay.com/2009/06/proving-the-shifting-baselines-theory-how-humans-consistently-misperceive-nature/.

Hartwell, Robert M., "Demographic, Political, and Social Transformations of China, 705–1550," *Harvard Journal of Asiatic Studies* 42.2 (1982), 365–442.

Hartwell, Robert M., "A Revolution in the Chinese Iron and Coal Industries during the Northern Sung, 960–1126 AD," *Journal of Asian Studies* 21.2 (1962), 153–62.

He Xiubin et al., "Soil Erosion Response to Climatic Change and Human Activity during the Quaternary on the Loess Plateau, China," *Regional Environmental Change* 6 (2006), 62–70.

Ho Ping-Ti, "Early Ripening Rice in Chinese History," *Economic History Review*, n.s., 9.2 (1956), 200–218.

Holmes, Jonathan, Edward Cook, and Bao Yang, "Climate Change over the Past 2000 Years in Western China," *Quaternary International* 194.1–2 (2009), 91–107.

Hong Mai 洪邁, *Rongzhai suibi* 容齋隨筆 (Taibei: Taiwan shangwu yinshuguan, 1965).

Horne, Ryan, "Map Tiles" (Chapel Hill, NC: Ancient World Mapping Center, 2014).

Hosner, Dominic, Mayke Wagner, Pavel Tarasov, Xiaocheng Chen, and Christian Leipe, "Spatiotemporal Distribution Patterns of Archaeological Sites in China during the Neolithic and Bronze Age: An Overview," *Holocene* 26.10 (2016), 1576–93.

Hou Ren-zhi, "Ancient City Ruins of the Inner Mongolia Region of China," *Journal of Historical Geography* 11.3 (1985), 241–52.

Hsu Cho-yun, "The Spring and Autumn Period," in Michael Loewe and Edward Shaughnessey, eds., *The Cambridge History of Ancient China: From the Origins of Civilization to 221 BC* (Cambridge: Cambridge University Press, 1990), 545–86.

Huang, Chun Chang, and Hongxia Su, "Climate Change and Zhou Relocations in Early Chinese History," *Journal of Historical Geography* 35.2 (2009), 297–310.

Huang, Chun Chang, Jiangli Pang, Hongxia Su, Shengli Li, and Benwei Ge, "Holocene Environmental Change Inferred from the Loess-Palaeosol Sequences Adjacent to the Floodplain of the Yellow River, China," *Quaternary Science Reviews* 28.25–26 (2009), 2633–46.

Huang, Chun Chang, Jiangli Pang, Xiaochun Zha, Yali Zhou, Hongxia Su, and Yuqing Li, "Extraordinary Floods of 4100–4000 a BP Recorded at the Late Neolithic Ruins in the Jinghe River Gorges, Middle Reach of the Yellow River, China," *Palaeogeography, Palaeoclimatology, Palaeoecology* 289.1–4 (2010), 1–9.

Huang, Chun Chang et al., "Charcoal Records of Fire History in the Holocene Loess-Soil Sequences over the Southern Loess Plateau of China," *Palaeography, Paleoclimatology, Palaeoecology* 239 (2006), 28–44.

Huanghe fanghongzhi bianzuan weiyuanhui 黃河防洪誌編纂委員會, *Huanghe fanghong zhi* 黃河防洪志 (Zhengzhou: Henan renmin chubanshe, 1991).

Huanghe shuili weiyuanhui Huanghezhi zongbian jishi 黃河水利委員會黃河志總編輯室 ed., *Huanghe dashi ji* 黃河大事記 (Zhengzhou: Henan renmin chubanshe, 2017).

Huanghe shuili weiyuanhui Huanghezhi zongbian jishi 黃河水利委員會黃河志總編輯室, *Lidai zhi Huang wenxuan* 歷代治黃文選 (2 vols.) (Zhengzhou: Henan renmin chubanshe, 1989), www.yrcc.gov.cn/hhwh/lszl/rw/gdzhrw/.

Huanghe shuili weiyuanhui Huanghezhi zongbian jishi 黃河水利委員會黃河志總編輯室, *Huanghe liuyu zongshu* 黃河流域綜述 (Zhengzhou: Henan renmin chubanshe, 2017).

Huanghe shuilishi shuyao bianxiezu 黃河水利史述要編寫組, *Huanghe shuilishi shuyao* 黃河水利史述要 (Zhengzhou: Huanghe shuili chubanshe, 2003).

Hucker, Charles, *A Dictionary of Official Titles in Imperial China* (Stanford, CA: Stanford University Press, 1985).

Hummel, Arthur W., Sr. (contributor), *Shandongsheng Huanghe tu.* Map, c. after 1881. https://www.loc.gov/item/gm71005026/.

Hymes, Robert P., and Conrad Schirokauer, ed., *Ordering the World: Approaches to State and Society in Sung Dynasty China* (Berkeley: University of California Press, 1993).

Ingram, Lynn B., and Frances Malamud Roam, *The West without Water: What Past Droughts, Floods and Other Climatic Clues Tell Us about Tomorrow* (Berkeley: University of California Press, 2013).

Jullien, François, *A Treatise on Efficacy: Between Western and Chinese Thinking* (Honolulu: University of Hawai'i Press, 2003).

Kidder, Tristam R., and Haiwang Liu, "Bridging Theoretical Gaps in Geoarchaeology, Archaeology, and History in the Yellow River Valley, China," *Archaeological and Anthropological Sciences* 9.8 (2017), 1585–1602.

Kidder, Tristam R., Haiwang Liu, Michael J. Storozum, and Qin Zhen, "New Perspectives on the Collapse and Regeneration of the Han Dynasty," in Ronald K. Faulseit, ed., *Beyond Collapse: Archaeological Perspectives on Resilience, Revitalization, and Transformation in Complex Societies* (Carbondale: Southern Illinois University Press, 2015), 70–99.

Kidder, Tristam R., and Yijie Zhuang, "Anthropocene Archaeology of the Yellow River, China, 5000–2000 BP," *Holocene* 25.10 (2015), 1627–39.

Klein, JoAnna, "Ritual Cleansing: A New Formula to Help Tame China's Yellow River," *New York Times*, June 6, 2017.

Klein, Naomi, *The Shock Doctrine: The Rise of Disaster Capitalism* (New York: Henry Holt and Co., 2008).

Kolbert, Elizabeth, "Under Water: Can Engineers Save Louisiana's Disappearing Coast?" *New Yorker*, April 1, 2019.

Kong Ping, Jun Jia, Yong and Zheng, "Time Constraints for the Yellow River Traversing the Sanmen Gorge," *Geochemistry, Geophysics, Geosystems* 15 (2014), 395–407.

Laflen, John M., *Soil Erosion and Dryland Farming* (Boca Raton, FL: CRC Press, 2000).

Lamouroux, Christian, "From the Yellow River to the Huai: New Representations of a River Network and the Hydraulic Crisis of 1128," in Mark Elvin and Liu Ts'ui-jung, eds., *Sediments of Time: Environment and Society in Chinese History* (Cambridge: Cambridge University Press, 1998), 545–84.

Lander, Brian, *The Nature of Political Power in Early China: From the First Farmers to the First Empire* (New Haven, CT: Yale University Press, forthcoming).

Lander, Brian, and Katherine Brunson, "The Sumatran Rhinoceros Was Extirpated from Mainland East Asia by Hunting and Habitat Loss," *Current Biology Magazine* 28 (March 19, 2018), R245–R253.

Lary, Diana, "Drowned Earth: The Strategic Breaching of the Yellow River Dike, 1938," *War in History* 8.2 (2001), 191–207.

Legge, James, trans., *The Chinese Classics* (1875; Taipei: SMC, 1994; digital edition at the Chinese Text Project, ctext.org).

Leonard, Jane, *Controlling from Afar: The Daoguang Emperor's Management of the Grand Canal Crisis, 1824–1826* (Ann Arbor: University of Michigan Press, 1996).

Lewis, Mark Edward, *The Early Empires: Qin and Han* (Cambridge, MA: Harvard University Press, 2010).

Lewis, Mark Edward, *The Flood Myths of Early China* (Albany: State University of New York Press, 2006).

Li Daoyuan 酈道元, *Shuijing zhu* 水經注 (Sibu congkan edition online at the Chinese Text Project, ctext.org).

Li Denan 李德楠, *Ming Qing Huang Yun diqu de hegong jianshe yu shengtai huanjing bianqian yanjiu* 明清黃運地區的河工建設與生態環境變遷研究 (Beijing: Zhongguo shehui kexue chubanshe, 2018).

Li Feng, *Landscape and Power in Early China: The Crisis and Fall of the Western Zhou, 1045–771 BC* (Cambridge: Cambridge University Press, 2006).

Li Huarui 李華瑞, *Song-Xia guanxi shi* 宋夏關係史 (Shijiazhuang: Hebei renmin chubanshe, 1998).

Li Liu, "The Products of Minds as Well as Hands: Production of Prestige Goods in the Neolithic and Early State Periods of China," *Asian Perspectives* 42.1 (2003), 1–40.

Li Ping, Sai Vanapalli, and Tonglu Li, "Review of Collapse Triggering Mechanism of Collapsible Soils Due to Wetting," *Journal of Rock Mechanics and Geotechnical Engineering* 8.2 (2016), 256–74.

Li Ronghua 李榮華, "Qingdai huangtu gaoyuan shuitu liushi ji shehui yingdui jizhi" 清代黃土高原水土流失及社會應對機制, *Lanzhou xuekan* 2015.05, 83–88.

Li Shixu 黎世序, *Xu xing shui jinjian* 續行水金鑑 (Beijing: Shangwu yinshuguan, 1937).

Li Tao 李燾, *Xu zizhi tongjian changbian* 續資治通鑑長編 (1137; Shanghai: Xinhua shudian, 1986).

Li Wang, Ming'an Shao, Quanjiu Wang, and William Gale, "Historical Changes in the Environment of the Chinese Loess Plateau," *Environmental Science and Policy* 9 (2006), 675–84.

Lin, Aiming, Zhenyu Yang, Zhiming Sun, and Tianshui Yang, "How and When Did the Yellow River Develop Its Square Bend?" *Geology* 29.10 (2001), 951–54.

Linqing 麟慶, *Hegong qiju tushuo* 河工器具圖說 (Taiwan: Taiwan shangwu yinshuguan, 1968).

Linton, Jamie, and Jessica Budds, "The Hydrosocial Cycle: Defining and Mobilizing a Relational-Dialectical Approach to Water," *Geoforum* 57 (2014), 170–80.

Liu An 劉安 and Xu Shen 許慎, *Huainanzi* 淮南子 (Shanghai: Sibu congkan) (also available at the Chinese Text Project, ctext.org).

Liu Heping, "Picturing Yu Controlling the Flood: Technology, Ecology, and Emperorship in Northern Song China," in Dagmar Schaefer, ed., *Cultures of Knowledge: Technology in Chinese History* (Sinica Leidensia 103) (Leiden: Brill, 2011).

Liu Jingchun 劉景純, *Mingdai jiubian shidi yanjiu* 明代九邊史地研究 (Beijing: Zhonghua shuju, 2014).

Liu Tianhe 劉天和, *Wenshui ji* 問水集 (ca. 1500; Nanjing: Zhongguo shuili gongcheng xuehui, 1936).

Liu Tungsheng, *Loess in China* (Beijing: China Ocean Press, 1988).

Liu Xu 劉昫 et al., *Jiu Tang shu* 舊唐書 (Beijing: Zhonghua shuju; online at Scripa Sinica).

Lorge, Peter, *War, Politics and Society in Early Modern China, 900–1795* (New York: Routledge, 2005).

Lowdermilk, Walter C, "Forest Destruction and Slope Denudation in the Province of Shanhsi," *China Journal of Science and Arts* 4.3 (Shanghai: China Society of Science and Arts, 1936), 127–36.

Lowdermilk, Walter C., *Forestry in Denuded China* (Philadelphia: American Academy of Political and Social Science, 1930).

Lü Buwei 呂不韋, *Lüshi Chunqiu* 呂氏春秋 (Sibu congkan edition online at the Chinese Text Project, ctext.org).

Ma Li 馬力, "Song Zhezong qinzhengshi dui Xi Xia de kaibian he Yuanfu xinjiangjie de queli" 宋哲宗親政時對西夏的開遍和元符新疆界的確立, in Deng Guangming 鄧廣銘 et al., eds., *Song shi yanjiu lunwen ji* 宋史研究論文集 (Shijiazhuang: Hebei jiaoyu chubanshe, 1989), 126–54.

Mallory, Walter Hampton, *China Land of Famine* (New York: American Geographical Society, 1926).

McGrath, Michael. "Frustrated Empires: The Song-Tangut War of 1038–44," in Don J. Wyatt, ed., *Battlefronts Real and Imagined: War, Border and Identity in the Chinese Middle Period* (London: Palgrave Macmillan, 2008), 151–90.

McMahon, Daniel, *Rethinking the Decline of China's Qing Dynasty: Imperial Activism and Borderland Management at the Turn of the Nineteenth Century* (New York: Routledge, 2014).

McNeill, J. R., and Peter Engelke, *The Great Acceleration: A History of the Anthropocene since 1945* (Philadelphia: University of Pennsylvania Press, 2016).

Meeks, Elijah, and Ruth Mostern, "The Politics of Territory in Song Dynasty China (960–1276 CE)," in Alistair Geddes and Ian Gregory, eds., *Rethinking Space and Place: New Directions with Historical GIS* (Indianapolis: Indiana University Press, 2012).

Mencius 孟子, *The Works of Mencius* translated by James Legge (Oxford: Clarendon Press, 1895), bilingual edition online at ctext.org.

Mengzi 孟子, *Mengzi* 孟子 (Wuyingdian shisanjing zhushu edition online at the Chinese Text Project, ctext.org).

Menzies, Nicholas, *Forest and Land Management in Imperial China* (New York: St. Martin's Press, 1994).

Menzies, Nicholas, "Forestry (Section 42b)," in Joseph Needham, ed., *Science and Civilization in China*, vol. 6, *Biology and Biological Technology, Part III Agro-Industries and Forestry* (Cambridge: Cambridge University Press, 1996).

Miller, Bryan K., "The Southern Xiongnu in Northern China: Navigating and Negotiating the Middle Ground," in J. Bemmann and M. Schumauder, eds., *Complexity of Interaction Along the Eurasian Steppe Zone in the First Millennium AD* (Bonn Contributions to Asian Archaeology 6) (Bonn: Bonn University Press), 127–98.

Miller, Ian, *Fir and Empire: The Transformation of Forests in Early Modern China* (Seattle: University of Washington Press, 2020).

Montgomery, David, *Dirt: The Erosion of Civilizations* (Berkeley: University of California Press, 2007).

Mostern, Ruth, "Cartography on the Song Frontier: Making and Using Maps in the Song-Xia Conflict, Evidence from *Changbian* and *Song huiyao*," *Proceedings of the Third International Symposium on Ancient Chinese Books and Records of Science and Technology*, edited by Hans Ulrich Vogel, Christine Moll-Murata, and Gao Xuan (Beijing: Daxiang chubanshe, 2004), 147–52.

Mostern, Ruth, *Dividing the Realm in Order to Govern: The Spatial Organization of the Song State (960–1276)* (Cambridge, MA: Harvard University Asia Center, 2011).

Mostern, Ruth, "From Battlefields to Counties: War, Border, and State Power in Southern Song Huainan," in Don J. Wyatt, ed., *Battlefronts Real and Imagined: War, Border, and Identity in the Chinese Middle Period* (New York: Palgrave Macmillan, 2008), 227–52.

Mostern, Ruth, "Mapping the Tracks of Yu: Yellow River Statecraft as Science and Technology, 1200–1600," in Patrick Manning and Abigail Owen, eds., *Knowledge in Translation: Global Patterns of Scientific Exchange, 100–1800 CE* (University of Pittsburgh Press, 2018), 134–46.

Mostern, Ruth, and Ian Johnson, "From Named Place to Naming Event: Creating Gazetteers for History," *International Journal of Geographic Information Science* 22.10 (2008), 1091–1108.

Mote, Frederic W., *Imperial China 900–1800* (Cambridge, MA: Harvard University Press, 1999).

Mukerji, Chandra, *Impossible Engineering: Technology and Territoriality on the Canal du Midi* (Princeton, NJ: Princeton University Press, 2009).

Muscolino, Micah, *Ecology of War in China: Henan Province, the Yellow River, and Beyond, 1938–1950* (Cambridge: Cambridge University Press, 2016).

Muscolino, Micah, "Soil and Society on Shaanxi's Loess Plateau: A History from the Bottom Up," unpublished paper, 2016.

Nanhe Huang Yun hu he xuxie jiyi tushuo 南河黃運湖河蓄洩機宜圖說 (Illustrated Explanations of Drainage and Storage Activities on the Lakes and Rivers of the Yellow River and the Grand Canal), ca. 1796–1820, in the collection of the National Central Library, Taiwan. https://www.wdl.org/en/item/11388/#q=huang+yun+hu+he+zong+tu.

NASA, "Shuttle Radar Topography Mission (SRTM) Version 3.0" (2020). https://www2.jpl.nasa.gov/srtm/cbanddataproducts.html.

Needham, Joseph, "Hydraulic Engineering (II): Control, Construction and Maintenance of Waterways," in *Science and Civilization in China*, vol. 4, *Physics and Physical Technology, Part 3, Civil Engineering and Nautics* (Cambridge: Cambridge University Press, 1971), 211–365.

Nixon, Rob, *Slow Violence and the Environmentalism of the Poor* (Cambridge, MA: Harvard University Press, 2013).

Obrusanszky, Borbala, "Tongwancheng, the City of Southern Huns." *Transoxiana* 14 (August 2009). Also published in the *Journal of Eurasian Studies* 1.1 (January–March 2009).

Osborne, Anne, "Highlands and Lowlands: Economic and Ecological Interactions in the Lower Yangzi Region under the Qing," in Mark Elvin and Liu Ts'ui-jung, eds., *Sediments of Time: Environment and Society in Chinese History* (Cambridge: Cambridge University Press, 1998), 203–34.

Ouyang Xiu, "Lun Xiuhe" 論修河 in Mao Kun 茅坤 ed., *Tang Song badajia wenchao* 唐宋八大家文鈔 38.1b–17a (Taibei: Taiwan shangwu, 1986), online at guoxuedashi.com.

Ouyang Xiu 歐陽修, *Xin Tangshu* 新唐書 (Beijing: Zhonghua shuju edition, online at Scripta Sinica, http://hanchi.ihp.sinica.edu.tw).

Pan Jixun 潘季馴, *Hefang yilan* 河防一覽 (Taibei: Wenhai chubanshe, 1970).

Pan Jixun 潘季馴, *Hefang yilan* 河防一覽, first edition held at Princeton University Library, call number TB197/685. Online at https://catalog.princeton.edu/catalog/4030712.

Parara-Caryannis, George, "Historical Earthquakes in China," unpublished manuscript (2013), http://www.drgeorgepc.com/EarthquakesChina.html.

Perdue, Peter, *China Marches West: The Qing Conquest of Central Eurasia* (Cambridge, MA: Harvard University Press, 2009).

Perry, Elizabeth, *Rebels and Revolutionaries in North China, 1845–1945* (Stanford, CA: Stanford University Press, 1980).

Pietz, David A., *Engineering the State: The Huai River and Reconstruction in Nationalist China, 1927–1937* (New York: Routledge, 2002).

Pietz, David A., *The Yellow River: The Problem of Water in Modern China* (Cambridge, MA: Harvard University Press, 2015).

Ping Kong, Jun Jia, and Yong Zheng, "Time Constraints for the Yellow River Traversing the Sanmen Gorge," *Geochemistry, Geophysics, Geosystems* 15 (2014), 395–407.

Pomeranz, Kenneth, "Introduction," in Edmund Burke III and Kenneth Pomeranz, eds., *The Environment and World History* (Berkeley: University of California Press, 2009).

Pomeranz, Kenneth, *The Making of a Hinterland: State, Society, and Economy in Inland North China, 1853–1937* (Berkeley: University of California Press, 1993).

Pomeranz, Kenneth, "The Transformation of China's Environment, 1500–2000," in Edmund Burke III and Kenneth Pomeranz, eds., *The Environment and World History* (Berkeley: University of California Press, 2009), 118–64.

Pottinger, Lori, "Why We Need Working Floodplains," Public Policy Institute of California (February 22, 2018). https://www.ppic.org/blog/need-working-floodplains/.

Qiao Shuqing, Xuefa Shi, Yoshiki Saito, Xiaoyan Li, Yonggui Yu, Yazhi Bai, Yanguang Liu, Kunshan Wang, and Gang Yang, "Sedimentary Records of Natural and Artificial Huanghe (Yellow River) Channel Shifts during the Holocene in the Southern Bohai Sea," *Continental Shelf Research* 31 (2011), 1336–42.

Ren Mei'e and Zhu Xianmo, "Anthropogenic Influences on Changes in the Sediment Load of the Yellow River, China, during the Holocene," *Holocene* 4.3 (1994), 314–20.

Richards, John, *The Unending Frontier: An Environmental History of the Early Modern World* (Berkeley: University of California Press, 2006).

Richter, Daniel, Allen Bacon, Zachary Brecheisen, and Megan Mobley, "Soil in the Anthropocene," *IOP Conference Series: Earth and Environmental Science* 25 (2015).

Rosen, Arlene, "The Impact of Environmental Change and Human Land Use on Alluvial Valleys in the Loess Plateau of China during the Middle Holocene," *Geomorphology* 101.1–2 (2008), 298–307.

Rosen, Arlene, Jinok Lee, Min Li, Joshua Wright, Henry Wright, and Hui Fang, "The Anthropocene and the Landscape of Confucius: A Historical Ecology of Landscape Changes in Northern and Eastern China during the Middle to Late Holocene," *Holocene* 10 (2015), 1640–50.

Rossabi, Morris, "The Ming and Inner Asia," in Denis Twitchett and Frederic Mote, eds., *The Cambridge History of China*, vol. 8, *The Ming Dynasty, Part 2: 1368–1644* (Cambridge: Cambridge University Press, 1998), 221–71.

Saito Yoshiki, Zuosheng Yang, and Kazuaki Hori, "The Huanghe (Yellow River) and Changjiang (Yangtze River) Deltas: A Review on Their Characteristics, Evolution, and Sediment Discharge During the Holocene," *Geomorphology* 41.2–3 (2001), 219–31.

Scott, James, *Against the Grain: A Deep History of the Earliest States* (New Haven, CT: Yale University Press, 2018).

Scott, James, *Seeing Like a State: How Certain Schemes to Improve the Human Condition Have Failed* (New Haven, CT: Yale University Press, 1999).

Sewell, William, Jr., *Logics of History: Social Theory and Social Transformation* (Chicago: University of Chicago Press, 2005).

Shangshu 尚書 James Legge, trans., *The Chinese Classics* (1875; Taipei: SMC, 1994; digital edition at the Chinese Text Project, ctext.org).

Shen Yi 沈怡, *Huanghe nianbiao* 黃河年表 (Shanghai: Junshi weiyuanhui ziyuan weiyuanhui, 1935).

Shen Zhifeng 申志鋒, *Jingguan, shengji, liebian: Yuzhong yan Huang quyu huanjing bianqian yu chengxiang jingji jiegou* 景觀、生計、裂變：豫中沿黃區域環境變遷與城鄉經濟結構, PhD dissertation, Department of History, Zhejiang University, 2019.

Shi Changxing, Zhang Dian, and Lianyuan You, "Changes in Sediment Yield of the Yellow River Basin of China during the Holocene," *Geomorphology* 46.3–4 (2002), 267–83.

Shi Naian 施耐庵, *Zhongyi shuihu quanshu* 忠義水滸全書. Rare book online at Harvard University Library (between 1621 and 1800), https://iiif.lib.harvard.edu/manifests/view/drs:53918403$107i.

Shi Nianhai 史念海, Cao Erqin 曹爾琴, and Zhu Shiguang 朱士光, *Huangtu gaoyuan senlin yu caoyuan de bianqian* 黃土高原森林的變遷 (Xi'an: Shaanxi renmin chubanshe, 1985).

Shiba Yoshinobu, *Commerce and Society in Sung China* (Ann Arbor: University of Michigan Press, 1970).

Shuilibu Huaihe shuili weiyuanhui 水利部淮河水利委員會, *Huaihe liuyu ditu ji* 淮河流域地圖集 (Beijing: Kexue chubanshe, 1999).

Sima Guang 司馬光, *Zizhi tongjian* 資治通鑑 (Shanghai: Shanghai guji chubanshe, 2017).

Sima Qian 司馬遷, *Shiji* 史記 (Beijing: Zhonghua shuju) (online version at Scripta Sinica, http://hanchi.ihp.sinica.edu.tw).

Skaff, Jonathan, *Sui-Tang China and Its Turko-Mongol Neighbors: Culture, Power, and Connections, 580–800* (Oxford: Oxford University Press, 2012).

Song Lian 宋濂, *Yuanshi* 元史 (Beijing: Zhonghua shuju) (online version at Scripta Sinica, http://hanchi.ihp.sinica.edu.tw).

Song Zhenhai 宋震海, *Zhongguo gudai ziran zaihai he yichan nianbiao zongji* 中國古代重大自然災害和異常年表總集 (Guangzhou: Guangdong jiaoyu chubanshe, 1992).

Sterckx, Roel, *The Animal and the Daemon in Early China* (Albany, NY: SUNY Press, 2018).

Sun Jinghao, "City, State, and the Grand Canal: Jining's Identity and Transformation, 1289–1937," PhD dissertation, University of Toronto, 2005.

Sun Wei 孫偉, "Bei Song shiqi huangtu gaoyuan diqu cheng, zhai, bao tixi yanbian yanjiu," 北宋時期黃土高原地區城寨堡體系演變研究, MA thesis in historical geography, Shaanxi Shifan Daxue, 2005.

Tackett, Nicholas, "The Great Wall and Conceptualization of the Border under the Northern Song," *Journal of Song-Yuan Studies* 38 (2008), 99–138.

Tan Qixiang 譚其驤, "Heyi Huanghe zai Dong Han yihou hui chuxian yige changqi anliu de jumian: Cong lishi shang lunzheng Huanghe zhongyou de tudi heli liyong shi xiaomixiayou shuihai de juedingxing yinsu" 何以黃河在東漢以後會出現一個長期安流的局面——從歷史上論證黃河中游的土地合理利用是消弭下游水害的決定性因素, *Xueshu yuekan* 1962.2 (1962), 23–35.

Tan Qixiang 譚其驤, *Huanghe shi luncong* 黃河史論叢 (Shanghai: Fudan daxue chubanshe, 1986).

Tan Qixiang 譚其驤, ed., *Zhongguo lishi ditu ji* 中國歷史地圖集 (Beijing: Zhongguo ditu chubanshe, 1982).

Tata, Samantha, "Dust from Chinese Storm Reaches California," NBC Los Angeles, April 1, 2013, https://www.nbclosangeles.com/news/local/gobi-desert-dust-china-southern-california-sierra-nevada-owens-valley/1945530/.

Taubes, Hannibal, "Some Early Chinese Descriptions of the Myer Ma Chen Range," May 18, 2016, at http://twosmall.ipower.com/blog/?p=4036.

Theobald, Ulrich, *Hefang yilan* (December 21, 2013), http://chinaknowledge.de/Literature/Science/hefangyilan.html.

Toqto'a 脫脫, *Jin shi* 金史 (Beijing: Zhonghua shuju edition online at Scripta Sinica, http://hanchi.ihp.sinica.edu.tw).

Toqto'a 脫脫, *Song shi* 宋史 (Beijing: Zhonghua shuju edition online at Scripta Sinica, http://hanchi.ihp.sinica.edu.tw).

Twitchett, Denis, ed., *The Cambridge History of China*, vol. 3, *Sui and T'ang China, 589–906 AD, Part One* (Cambridge: Cambridge University Press, 1979).

Vermeer, Edward, "P'an Chi-hsun's Solutions for the Yellow River Problem of the Late Sixteenth Century," *T'oung Pao* 73 (1987), 33–67.

Wagner, Donald B., *The State and the Iron Industry in Han China* (Copenhagen: NIAS, 2001).

Waldron, Arthur, *The Great Wall of China from History to Myth* (Cambridge: Cambridge University Press, 2003).

Wang Cong, Shuai Wang, Bojie Fu, Zongshan Li, Xing Wu, and Qiang Tang, "Precipitation Gradient Determines the Tradeoff Between Soil Moisture and Soil Organic Carbon, Total Nitrogen, and Species Richness in the Loess Plateau, China," *Science of the Total Environment* 575 (2017), 1538–45.

Wang Pu 王溥, Tang Huiyao 唐會要 (Shanghai: Shangwu yinshuguan, 1935).

Wang Qinruo 王欽若 et al., *Cifu yuangui* 冊府元龜 (Taibei: Dahua, 1984).

Wang Xinzeng, Feng Jiao, Xingping Li, and Shaoshan An, "The Loess Plateau," in L. Zhang and K. Schwarzel, eds., *Multifunctional Land-Use Systems for Managing the Nexus of Environmental Resources* (Spring 2017), 11–27.

Wang Yingjie and Su Yanjun, "The Geo-pattern of Course Shifts of the Lower Yellow River," *Journal of Geographic Science* 21.6 (2011), 1019–36.

Wang Zhengxiang, "Boundary Data of Loess Plateau Region," *Journal of Global Change Data & Discovery*, 1.1 (2017), 113. https://doi.org/10.3974/geodp.2017.01.17.

Watson, Burton, trans., *Records of the Grand Historian: Chapters from the Shih Chi of Ssu-ma Ch'ien* (New York: Columbia University Press, 1958).

Watts, Jonathan, "Provincial Tug-of-War Waters Down China's Yellow River Success Story," *The Guardian*, June 28, 2011. https://www.theguardian.com/environment/2011/jun/28/water-yellow-river-china.

Wei Huan 魏煥, *Huang Ming jiubian kao* 皇明九邊考 (1544), https://www.loc.gov/item/2001708572/.

Wheatley, Paul, *The Pivot of the Four Quarters: A Preliminary Inquiry into the Origins and Character of the Ancient Chinese City* (Edinburgh: Edinburgh University Press, 1971).

White, Gilbert Fowler, "Human Adjustment to Floods: A Geographical Approach to the Flood Problem in the United States" (University of Chicago Department of Geography Research Paper 29, 1945).

White, Richard, *The Organic Machine: The Remaking of the Columbia River* (New York: Hill and Wang, 1996).

Will, Pierre-Étienne, "Clear Waters versus Muddy Waters: The Zheng-Bai Irrigation System of Shaanxi Province in the Late-Imperial Period," in Mark Elvin and Liu Ts'ui-jung, eds., *Sediments of Time: Environment and Society in Chinese History* (Cambridge: Cambridge University Press, 1998), 283–343.

Will, Pierre-Étienne, "On State Management of Water Conservancy in Late Imperial China," *Papers on Far Eastern History* (1987) (Australian National University Department of Far Eastern History), 71–91.

Wilson, Ernest Henry, *A Naturalist in Western China* (1913; London: Cadogan, 1986).

Wittfogel, Karl, *Oriental Despotism: A Comparative Study of Total Power* (1957; New Haven, CT: Yale University Press, 2013).

Xiang Rong, Yang Zuosheng, Saito Yoshiki, Guo Zhigang, Fan Dejiang, Li Yunhai, Xiao Shangbin, Shi Xuefa, and Chen Muhong, "East Asian Winter Monsoon (EAWM) Changes Inferred from Environmentally Sensitive Grain Size Component Records during the Last 2300 Years in Mud Areas Southwest of Jeju Island," *Science in China: Series D Earth Sciences* 49.6 (2006), 604–14.

Xin Deyong 辛德勇, "You Yuanguang hejue yu suowei Wang Jing zhihe zhonglun Dong Han yihou Huanghe changqi anliu de yuanyin" 由元光河決與所謂王景治河重論東漢以後黃河長期安流的原因, *Wenshi* 2012.1 (2012), 1–39.

Xu Jiongxin, "Naturally and Anthropogenically Accelerated Sedimentation in the Lower Yellow River, China, over the Past 13,000 Years," *Geografiska Annaler Series A, Physical Geography* 80.1 (1998), 67–78.

Xu Jiongxin, "A Study of Long-Term Environmental Effects of River Regulation on the Yellow River of China in Historical Perspective," *Geografiska Annaler Series A, Physical Geography* 75.3 (1993), 61–72.

Xu Jiongxin, "A Study of the Accumulation Rate of the Yellow River in the Past 10000 Years," *Variability in Stream Erosion and Sediment Transport* (Proceedings of the Canberra Symposium, 1994) (International Association of Hydrological Sciences Publication 224, 1994), 421–30.

Xu Jiongxin, "The Wind-Water Two-Phase Erosion and Sediment-Producing Processes in the Middle Yellow River Basin, China," *Science in China Series D* 43.2 (2000), 176–86.

Xu Song 徐松, *Song huiyao jigao* 宋會要輯稿 (ca. 1809; Beijing: Zhonghua shuju, 1957).

Yang Lien-sheng, "Economic Aspects of Public Works in Imperial China," *Excursions in Sinology* (Cambridge: Cambridge University Press, 1969).

Yang, Soon-ja, "Shen Dao's Theory of *Fa* and His Influence on Han Fei," in Paul Goldin, ed., *Dao Companion to the Philosophy of Han Fei* (New York: Springer, 2013).

Yang Xiao and Han Zhaoqing, "The Change of the Gaobao Lakes and Its Driving Forces (1717–2011)," *Acta Geographica Sinica* 73.1 (2018), 127–37.

Yao Hanyuan 姚漢源, *Huanghe shuilishi yanjiu* 黃河水利史研究 (Zhengzhou: Huanghe shuili chubanshe, 2003).

Yi Sangheon, "Holocene Vegetation Responses to East Asian Monsoonal Changes in South Korea," in Juan Blanco and Houshang Kheradmand, eds., *Climate Change: Geophysical Foundations and Ecological Effects* (Rijeka, Croatia: InTech, 2011).

Yuan Zuliang 袁祖亮, ed., *Zhongguo zaihai tongshi* 中國災害通史 (Zhongzhou: Zhengzhou daxue chubanshe, 8 volumes, 2008–2009).

Yue Shi 樂史, *Taiping Huanyu ji* 太平寰宇記 (984; Taibei: Wenhai chubanshe, 1993).

Zhang Haiying 張海英 *Ming shi* 明史 (Beijing: Zhonghua shuju online at Scripta Sinica, http://hanchi.ihp.sinica.edu.tw).

Zhang Jin et al., "Qingkou Complex: An Engineering Masterpiece of the Grand Canal of China," *Engineering History and Heritage* 167.EH2 (2014), 74–87.

Zhang Ke, Zhao Yan, Zhou Aifeng, and Sun Huiling, "Late Holocene Vegetation Dynamics and Human Activities Reconstructed from Lake Records in Western Loess Plateau, China," *Quaternary International* 227 (2010) 38–45.

Zhang Ling, "Changing with the Yellow River: An Environmental History of Hebei, 1048–1128," *Harvard Journal of Asiatic Studies* 69.1 (2009), 1–36.

Zhang Ling, *The River, The Plain, and the State: An Environmental Drama in Northern Song China, 1048–1128* (Cambridge: Cambridge University Press, 2016).

Zhang Quanming 張全明, *Liang Song shengtai huanjing bianqian shi* 兩宋生態環境變遷史 (Beijing: Zhonghua shuju, 2015).

Zhang Zhengming 張正明, *Huanghe liuyu dituji* 黃河流域地圖集 (Beijing: Zhongguo ditu chubanshe, 1989).

Zheng Kunhe 張崑河, "Sui Yunhe kao" 隋運河考 *Yugong* 7.1–3 (1937), 201–11.

Zheng Rubi 鄭汝璧, *Yansui zhenzhi* 延绥镇志 (1608; Shanghai guji chubanshe, 2011).

Zhongguo kexueyuan huangtugaoyuan zonghe kexuekaocha duibian 中國科學院黃土高原綜合科學考察隊編, ed., *Huangtu gaoyuan diqu renkou wenti* 黃土高原地區人口問題 (Beijing: Zhongguo jingji chubanshe, 1990).

Zhou, W. J., J. Dodson, M. J. Head, B. S. Li, Y. J. Hou, X. F. Lu, D. J. Donahue, and A. J. T. Jull, "Environmental Variability within the Chinese Desert-Loess Transition Zone over the Last 20,000 Years," *Holocene* 12.1 (2002), 107–12.

Zhouli 周禮 (Sibu congkan edition online at Chinese Text Project, ctext.org).

Zhuang, Yijie, and Tristam R. Kidder, "Archaeology of the Anthropocene in the Yellow River Region, China, 8000–2000 cal. BP," *Holocene* 24.11 (2014), 1602–23.

Zou Yilin 鄒逸麟, *Huang Huai Hai pingyuan lishi dili* 黃淮海平原歷史地理 (Hefei: Anhui jiaoyu chubanshe, 1997).

INDEX

Page numbers in *italic* indicate material in figures or tables.

Abbasid Caliphate, 133
Acciaviati, Anthony, 3–4
alluvial plains, 5–6, 247, Plate 1; at start of Common Era, 66, 93–94, 99–100, 104, 110; in 5th, 6th century CE, 116–17; in 7th century CE, 137; in 10th century CE, 124; in 11th, 12th centuries CE, 139, 143; in 13th century CE, 175, *205*; in 18th, 19th century CE, 226; in 21st century CE, 245; chronicling of, *205*; effect of canals on, 137–39, 178; in Holocene, 62–63; human modifications to, 6–7, 62–63, 66; of Loess Plateau, 38, 41, 44–45, 62, 106, Plate 1; during Neolithic, Bronze ages, 62, 68–69; of Ordos region, 32, 97–98; recent scholarship on, 93; reduced resilience to floods, 11, 14; river as "sediment sorting machine," 3–4; sandification of, 178; "Three Theses" and, 106; Yellow River effect on, 45–46, 53
American visitors, advisors, 203, 240
An du ("Establishing the Capital"), 142

An Lushan and rebellion, 133, 137, 139, 176
anjing (stability), 164
Anthropocene, 239, 244
avulsions, 7, 51, *102*–103, *165*, 172, 175, 207, 234, 241

Bai Zhongshan, 202–3
Baima (White Horse) River, 117
bamboo, 102, 104–5, 115, 143, 186
barge traffic, 209, 211, 219, 243
Bayan Har Mountains, 29, Plate 4
Beijing, 137, 179–80 *(180)*, 237
Bian Canal, 114, 142, 158
Bohai Sea, 72, 179, Plate 19
Book of Jin, The (Jinshu), 115–16
breaches of other rivers, 230–31
breaches of Yellow River, 7; 132 BCE, 101–4 *(102)*; 3 and 11 CE, 108–9; 14, 109; 65 through 140, 114; 924 through 1029, 124, 140–41, 143, 163; 1034, 161, *165*; 1048, 159, 166–70; 1128, 170; 1168, 175; 1234 through 1368, 204–7 *(204, 206)*; 1344, 204, 206; 1855, 223, 225, 234–37, 242;

309

breaches of Yellow River (*continued*) Chanzhou embankments, 161; at Hongze Lake, 234–37, 240–43 *(241)*; of Huayuankou, 241–42 *(241)*; improving technology to handle, 162–64, 210–12; intentional breaches, 140, 170, 174–75; Jia Lu's solutions for, 207; Liu Tianhe explanation of, 213; mismanagement leading to, 108–9, 161, 241–42 *(242)*; Pan Jixun prevention system, 223–25, 227; as percentage of disasters, 8; of perched rivers, 54; post-Han, 115; repairing of, 105, 141, 143, 158, 161, 210, 220–21 *(221)*; on Shaanxi Province floodplain, 213; Tongwaxiang breach, 237; in TYDA database, 249–50, *262, 265*. See also Huai River

Bronze Age: charioteer combat, 75; developing water management theories, 83; environmental effects of war, 75; Erlitou erosion, 74; floods at start of, 71; forest on Loess Plateau, 59–60; intensive agriculture, 62; Mark Elvin on, 144; sediment deposition during, 68 *(68)*, 73; small floodplain population during, 79; written reports from, 78; Zhou era, 43

Budds, Jessica, 4
Bureau of Rivers and Canals, 169
Burtynsky, Edward, 1

canals. See drainage canals; feeder canals; Grand Canal; Huainan canals
cave dwellings, 198, Plate 32
Cen Zhongmian, 20
Cenozoic era, 57
Central Great Mountain (*zhongyue*), 49
Chahar Mongols, 188, 196
Chang'an, *35*, Plate 12; as capital city, 32, 34, 110, 118–19, 122, 145; capture, pillaging of, 133–34; decline of, 153, 178, 185; deforestation around, 109; intensive farming near, 94; silt accumulation near, Plate 12; timber requirements of, 40, 91, 98, 131; unreliable waterways to, 139–40, 142; as world's largest city, 128–31; as Xi'an, 32, 34, 38, 76, 185, 243; Zhou court moving from, 76

Changjiang delta. See Yangtze (Changjiang) delta

channelization: dangers of linearity in, 6; early examples of, 81; of floodplains, 22; perched channels in riverbeds, 54, 56; Qingkou channel, 219–20; of Yellow River, 78, 91, 113–14, 251

Chanzhou, 161, *165*, 166
Chao Cuo, 94, 96
charcoal: from biomass burning, 60, 98, 200; evidence from, 69, 71–74, 136; growing shortages of, 143; on Loess Plateau, 60; sources of, 82; at Taosi site, 72–73

check dams, 19, 202, 245, Plate 38
Chen Huang, 229–30
Chen Yunzhen, 12, *253*, 292n2
"China's Sorrow," Yellow River as, 10
Chinese Communist Party (CCP), 239, 242–44
Chinggis Khan, 188
Chongbo (Count of Chong), 3, 84, 105
Classic of Mountains and Oceans, The (Shanhaijing), 81, 113
Classic of Poetry, The (Shijing), 81
coal mining, 141, 143, 240, 245
Commentaries on River Defense (Chen), 230
Commentary of Zuo, The (Zuozhuan), 81
Commentary on the Water Classic, The (Shuijingzhu) (Li), 116–17
commodification of nature, 60, 141, 144, 183–84, 199
Compendium of Questions about Water (Wenshuiji), A, 213
corvée labor, 80, 89, 114, 140, 160, 169, 189, 211
course changes, 7, 51, *102–103, 165*, 172–75, 207, 234, 241
court actions, debates on hydrology, 18–19, 76, 194–95; Chang'an, 143;

Daoguang, 233–34; encouraging colonization, 184; Grand Canal system, 181–82; Han, 100–102, 108; Huazhou, 161; Jia Rang "Three Theses," 105–6; Jin, 116; Li Daoyuan, 116; Ming, 113, 188–89, 213–14; Qi, 86; Qin, 86; Qing, 184, 193, 201–2; Taizong, 158–59; Tang, 86, 140, 144–45; Yuan, 206–7; Zhou, 76, 86

currents, Plate 33; decreased, 3, 7, 161, 169; increased, 7, 106, 113–14, 159, 161–62, 214; keeping silt in, 162; losing control of, 165; in meanders, 54; Mencius on, 83; Pan Jixun on, 223–25; pelagic, 3, 7; in repair dikes, 220–21; of Sanmenxia Gorge, 36; and sediment, 6, 214, 219; as *shi*, 225; velocity of, 18; of Yellow River, 30

Da Cunha, Dilip, 3, 7
Dalu Marsh, 55
dams: during 132 BCE breach, 102; *Book of Jin* on, 115; check dams, 202, 245, Plate 38; defining "social reality," 14; Du Chong's breaching of, 170; endangering Yellow River, 244–45; Gaojiayan Dam (Hongze), *218*, 219, 222–23, 236; Gilbert F. White on, 5; Hu Ding proposal for check dams, 19; Jia Rang on, 107; Ming building of, 211; nineteenth century, 226, 233; Pan Jixun on, *224*; as part of "technology complex," 13–14; risks from, 3; Sanmenxia hydroelectric project, 36, 66, 242–43; and silt crisis, 232; with spillways, 162; TVA recommendations for, 240; in TYDA database, 251, 264–65; weirs, 114; Wu Guifang proposal, 214; Yu's understanding of, 3, 84

Daohe xingsheng shu (Li), 163
Daqing River, 237
database. *See* TYDA
Datong Prefecture, 190, 190f
David, Armand (Père David), 203
Dawen River, 52

Daye Marsh (Liangshan Marsh), 52, *53*, 55
declensionism, 17
deforestation, 39, 231, 240; around Chang'an, 109; around Yellow River, 143, 231; causing drought, 40; causing loess soil erosion, 19, 47, 121; contemporaneous assessments of, 194–95, 203; for embankment materials, 143, 168; by farmers clearing land, 63, 193–94, 198, 201; for firewood, 32; during "great acceleration," 73–74; in Great Bend (Hetao), 36, 200; in Guanzhong, 98, 109, 117, 122, 186; in Liupan Mountains, 129, 151, 155, 176, 200; *Mencius* on, 85–86; on Mount Tai massif, 79; Neolithic, 70; on Ordos, 93, 198; for paper, charcoal, construction, 40, 200; under Qing, 200–202; of Qinling Mountains, 129, 195, 200; regrowth, 196; in Shanxi Province, 143, 194, 199–200; Tan Qixiang on, 66. *See also* timber

delta. *See* Yangtze (Changjiang) delta
delta accretion and backwash, 225–26
desertification: climate causes of, 192; drought and, 122, 185, 193; human causes of, 28, 93, 136, 192 *(192)*; in Mu Us area, 136; in Ordos area, 93, 181–82, 185, 193, 198; sand dunes from, 30, 76, Plate 6
"developmentalist project," 14, 183
Diagrams of the Tribute of Yu, The (Yugongtu), 113
dike systems (see also levees), 219; breaches of, 54; Kaifeng dike overtopped, 234; need for regular maintenance, 161; to repair levee breaches, 220–21 *(221)*; splay fan evidence of, 56; in TYDA database, 262, 264; used for dredging, diversion, 224, 230
disaster–management ratio, 65, *124*, 183, *210*, 212
diversion techniques, 89, 138, 170, 209–10, 219, 221, 228–29
Dodgen, Randall, 12, 228–29

Dongping Lake, 52
Dongzhi Tableland, Gansu, 134
dou (measurement), 90, 107
drainage canals: built after Sizhou flood, 226; connecting Hongze, Gaobao, 221; under efficacy principle, 19; failures of, 55; Feng Qun's understanding of, 105–6; Grand Canal, 219–20; increasing flood hazard, 6; increasing usage of, 159; Jia Rang's understanding of, 106; *versus* levees, 120, 162, 207; under Yu the Great, 3, 84
dredging, 264, Plate 29; of Bian Canal, 158; corvée labor used for, 140, 160, 220; of Grand Canal, 211; Han Wudi comment on, 103; impossible after 1048 course change, 166–67; "iron dragon-claw silt dispersing wheel" for, 162; by Jia Lu, 207–8; mechanical, 220; Pan Jixun use of water for, 223–24; perpetual need for, 115; and "silt crisis," 232; weir system for, 114; by Yu the Great, 84
drought, 10–11, 28, 185, 188, 193. *See also* moisture on Loess Plateau historically
Du Chong, 170
Duke Huan (Qi Huan Gong), 79–80, 83
Dunnell, Ruth, 149
Dushuijian directorate, 161
dust storms, 41, 56, 76, 244

earthquakes, 24 *(24)*, 108, 145, 182, 185, 196, 205
earthworks, 13–14, Plates 24–25; breaches causing ruptures in, 204, 207; under CCP, 242; clogged with sediment, 158; debate over, 182; decreasing flooding frequency, 100; early examples of, 78–80; full buildout of, 10; on Huai River, 219; increasing salinity, 56; maintenance of, 7, 19, 210; management events involving, 108–9 *(108)*, 169; modern replacements for, 243; sediment undermining, 202, 219; under Song

regime, 147–49; stalk bundles for, *222*, 231, 243; Wang Jing's alternatives to, 113–14; Wu Guifang's 1577 proposal for, 214
eastern Asia elevation, hydrology, Plate 3
efficacy tradition *(shi)*, 2, 19, 83
"800-*li* Liangshan Marsh," 52
Elvin, Mark, 10, 144
embankments, Plate 25; around Hongze Lake, 106, 222; breach in Chanzhou, 161; collapse of from erosion, 19, 194; of concrete and steel, 243; deforestation for materials, 168; Feng Qun on, 105–6; and Gaojiayan dam, 106; Jia Liu on, 207; Jia Rang on, 106; laws requiring, 159; Ming improvements on, 210–11; Pan's Grand Canal project, 222–24, 237; with parallel levees, 159; repairs to, *265*; weirs, 163; *Zhouli* on, 83
emergency repairs, 155, 166, 221, 233–35, *265*, Plate 29
environmental protection as low priority, 19–20
epidemics, 188, 195–96
Erlitou cultural area, *72*, 74
erosion, 7, 11–13, Plate 30; along Wei River, 131; Chen Huang's suggested controls, 229–30; from deforestation, 19, 47, 121, 194, 198; from deforestation of Liupan Mountains, 129, 151, 155, 176, 200; early understanding of, 200–203; from farming, 18, 141, 155; and Grand Canal, 179; from herding, 73–74; Liu Tianhe understanding of, 213; of loess soil, 121; from loss of timber, 153, 200; from Loess Plateau, 181; Pan Jixun's suggested controls, 214, 223–26 *(224)*, 236; precipitation and, 18–19, 158; under Qin regime, 86; under Qing dynasty, 200–202; riverbank collapse from, 99–100, 158, 200, 213; timeline of human-caused, 6, *11*; upstream, 7, 11; Zhang Rong understanding of, 18
"Establishing the Capital" *(An du)*, 142

European comments on Yellow River, 203
event timelines, 11, *65*, *93*, *228*
events, defining, 16–18

famine, 110, 174, 195–96, 206, 226, 232
farming/farmers, *92*; after 1048 Yellow River breach, 166–68; after Wang Jing flood control, 114; army of defeating Wang Mang, 109; burning to clear fields, 60, 73, 101; causing erosion, 18, 141, 155; CCP initiatives for, 242–44; Chao Cuo plans for, 94–95; deforestation for, 193, 198, 200–201; digging canals, 79; displaced to marginal lands, 181; displacing pastoralists, 196; and drought, 122; early evidence of, 12, 69–70; effect of on river, 4, 12–13, 22, 70; eighth, ninth centuries, 131, 135–36; fertilizing with sediment, 99–100, 163; following moisture, 28; government contracts for, 198; Han regime and, 92, 112; in Heng Mountains, 152; during Holocene Climatic Optimum, 58, 63, 71; incentives for, 198–99, 201; Iron Age, 76, 79; irrigation lowering water table, 192; Jiajing regime and, 194; in Kaifeng, 142; levees causing difficulties for, 220; and loess soil, 41–42; during medieval period, 117; moisture levels affecting, 144; on Ordos, 97–99, 111, 130, 146–47, 153, 191; Paleolithic, Neolithic, Bronze, 59–60, 62, 68–71, 74–75; during Qin regime, 87; Qin regime and, 87, 90; Qing land reclamation for, 196–97; resettlement of, 13, 96; rice, 141, 178, 180; and sand infiltration, 51, 192, 223, 232, 236; and 1628 famine, 195–96; and soil erosion, 12, 18, 201–2; soil salinization, 13, 51, 181, 240; soldier-farmers *(tuntian)*, 146, 149; spillway flooding of, 219, Plate 34; switching to dryland crops, 181; Tan Qixiang on, 66; *versus* trees, 201; use of iron plows, 74–76, 91–92 *(92)*, 94; use of irrigation, 94; use of terracing, Plate 13; water managers flooding fields, 210; Yan Shengfang on damage from, 194; Yu on, 85; in Yulin, 191. *See also* desertification
fascines *(sao)*, 80, 101, 143, 162–63, 168, 174
feeder canals, 158, Plate 30
Fen River, Valley: agriculture in, 117; earthquake faults in, 24 *(24)*; flooding, breaching of, 230–31; forests, forestry in, 76, 98; Neolithic sites in, 69–70; Northern Han in, 140; as population center, 91, 210; Taosi city in, 72
fen sha qi shi, 175
Feng Qun, 105–6
fengshan rituals, 103
fire: charcoal from, 82, 98, 143; firewood collection, 143, 145, 151, 193, 195; human-induced, 85, 98, 136; in *The Mencius,* 85; used to clear farmland, 60, 73–75, 136
Five Dynasties era, 140
Five Great Mountains of China *(wuyue),* 49
flooding: as "acts of man," 5; becoming endemic, 240; caused by sedimentation, 13–14, 118–19; Chen Huang on, 230; choosing diversion over prevention, 231–32; dangers, costs of, 5–7, 11; diverting not preventing, 231–32, Plate 34; "flood and famine region," 232; Grand Canal and, 182–83, 209; as groundwater supply, 6; as part of ecosystem, 5; picking winners and losers, 13, 211–12; "technology complex" for, 13–14; and waterborne disease, 240; White on, 5; Wu Guifang scouring plan, 214
floodplains, Plate 23; awash (1048–1351), 166; defining success regarding, 182; lock-in, 19; reduction, channeling of, 22; residents of, 7, 13, 125; transportation canal system in, 13

forests. *See* deforestation
Fudan University Historical Geography Institute, 20–21

gabions, 105, 143
Gansu province, 30, 37, 110, 134, 147, 201, Plate 7
Gaobao lake system, 221–23, *241*, 242
Gaojiayan Dam, 218f, 219, 221, 223, 232, 236
geography of floodplain events 550–750 CE, 138
global cooling, 195
Gobi Desert, 24, 41, 133
Goryeo regime, 170
grain: cultivation, 85, 134–36; shipments, 36, 103, 137–40, 168, 206, 211–14, 219; as tribute, 212, 228–29, 237
Grand Canal (Dayunhe) network, 13–14, 179–81; commercial shipments through, 229; emperors' interest in, 226–27; failures of, 206, 208–9, 236; formal abandonment of, 237; Gaojiayan Dam and, 219; as "Great Transport Watercourse," 180; keeping sediment out of, 13, 181; locks to divert silt, *215–17*, Plates 33, 34; reconstruction of, 211; repair attempts, 206, 236
Great Bend (Hetao): around Ordos Block, 24, 29–33 *(31)*; deforestation around, 36, 200; fortification within, 110, 188, 190; Great Wall delimiting, 89; population of, 96–97; as territorial boundary, 92
Great River (Da he), 123. *See also* Yellow River
Great Wall, *88, 187*; construction of, 181; early stages of, 189; forest cover loss around, 193; under Qin regime, 89; settlements around, 196, *197*; at Yulin, 41, *190*, Plate 17
Greer, Charles, 24
groundwater, 6, 56, 82, 223, 244
Gu Yanwu, 223
Guangdong, 202
Guan Zhong, 80

Guanzhong, 128, Plates 12, 30; deforestation, reforestation in, 98, 109, 117, 122, 186; as longtime settlement, capital, 34–35 *(35)*, 77, 86–89, *91*, 146; population levels in, 110, 128, *129*, 186; rainfall in, *38*; Zhengguo Canal in, 89, 158, 195. *See also* Chang'an; Qin regime; Shaanxi province; Tang regime; Wei River
Gun, Count of Chong (Chongbo), 3, 84, 105
Guo Chang, 105
Guo River, 212
Guo Shoujing, 176
Guo Zhongyan, 117
Guomindang, 239–40, 242
Gyaring Lake, 30, Plate 6

Hai River, 53, 170, 179, 225, 237, 252, Plate 2
Han Dynasty, 119–20, 122, 158, Plate 12; deforestation by, 98; Eastern Han, 109; farming and iron industry, 92, *92*; flooding leading to migration, 110; founding of, 90; late sixth century, 120, 122; regulating river, 100; restorationist, 108; Western Han, 63–67, 94, *95*, *97*, 98, 115; Xiongnu conflict, 92–94
Han shu (The History of the Han Dynasty), 90
Han Wudi, 102–3
Haojing, 76
Hartwell, Robert, 137
He Xiubin, 47–48
headwaters of Yellow River, 23, 29–30, 94, Plate 5
Hebei, 143; after Jin invasion, 174; breaches at, near, 159, 161, 165–67, 170, 175; irrigation works, 80; Kidder excavation of, 99
Hekou village, 33, *34*
Helan Mountains, 32, Plates 8–10
Henan province, 48–52, Plates 20, 22; absence of trees in, 231; archeological sites in, 49; Bai Zhongshan on, 202; capital in, 77; irrigation canals in, 242; on Loess Plateau, 37, 231; Xu

Xiake on, 195; on Yellow River, 23, 33, 48, 52; in Yellow River floodplain, 49–51, 77, 168. *See also* Kaifeng and Zhengzhou

Heng Mountains, 152–53, 193, Plate 28

herders, 81; causing floods, erosion, 73–74; displaced, resettled, 12–13, 144–45, 199; effect of drought on, 122; in Ordos, 188, 199; sedentarization of, 183

Hetao. *See* Great Bend

High Tableland (Shaanxi), 134

historical events *versus* processes, 16–17

History of the Han Dynasty, The (Han shu), 90, 98

Holocene era: after Holocene Climatic Optimum, 73–74; desert boundary during, 28; during Holocene Climatic Optimum, 58–60, 63, 70–71; human modifications during, 12, 62–63, 67, 82; lakes and wetlands during, 55; Loess Plateau during, 39; sediment deposits during, 22; and TYDA, 248; Yellow River sediment deposition, *68*

Hong Mai, 143

Honggou Canal, 80

Hongze Lake, *247, 250*, Plates 33, 34; 1824 dike failures, 233; and canal system, 138, 221; disaster events, 174, 214, *215–18*, 233; and Gaobao lake, 221; Gaojiayan Dam, *218*, 219, 222–23, 236; and Huai River, 55, 138, 170, *215*, 219, 223, 233; Pan Jixun and, 214; sediment filling, 234–35; too low to scour Yellow River, 233; and Yellow River breaches, 233–37, 240–43 *(241)*

Hongze-Gaojiayan-Qingkou infrastructure, 223, 232

Hongze's dam failures, 222–23

Horne, Ryan M., 15, 248–56, 260

"horseheads" *(matou)*, 162

horses: in 1500–1600s, 193; and agriculture, 98; in Heng Mountains, 152; in Ordos frontier, 149; riding of, 76; ritual sacrifice of, 103; in Shaanxi, 191; in Song state, 144–45; Tang reserving pasture for, 130–31; as tribute, 86; warhorses, 13, 75, 135, 145

Hou Ren-zhi, 93

Hu Ding, 19, 202–3

Huai River, 23, Plates 2, 26; and 1128 breach, 170; and 1286 breach, 204; and 1855 Yellow River avulsion, 235–36; canals connecting, 137–38; engineering of, 55; in fourteenth century, 206–8; and Hongze Lake, *215*, 219, 223, 233; and Huayuankou breach, 241–42 *(241)*; population loss around, 210; sharing bed with Yellow River, 51–53, 101, 103, 109, 170–71, 207–8, 235–37; shifting streams across, 214; sources of, 49; splitting with Yellow River, 212; in TYDA database, 252; used for scouring Yellow River, 229; Yellow River seepage into, 225

Huainan canals, 160, 221

Huang Chao Rebellion, 134

Huang he as name for Yellow River, 123

Huanghe nianbiao (Yellow River Annals), 247, 249, 251, 255

Huashan (Mount Hua), 49, 51

Huayuankou breach, 241–42, 241f

Hukou Waterfall (Hukou pubu), 33, 197

hunting, 59, 69–70, 76, 78, 81, 115, 130

Hutuo River, 231

Huzi breach, 103

"hydraulic contradictions" within river, 12

hydraulic engineering, 163, 208, 229, 237

hydrosocial defined, 4, 6

imperial capital: Beijing as, 179, 208; Chang'an (Xi'an) as, 34, 89, 109, 122, 142, 185, Plate 12; Kaifeng as, 50 *(50)*, 142; Luoyang as, 109; Nanjing as, 208

imperial floodplain, 214, 223–31, 238

"In one *shi*" saying, 90, 230

Indian tectonic plate, 57

Inner Mongolia, 29, 33, 37, 94

Invention of Rivers, The (Da Cunha), 3
Iron Age, 60, 62, 67, 74–79, 83
"iron dragon-claw silt dispersing wheel" *(tie longzhua yangni che),* 162

Japanese army in China, 239, 241–42
Ji Chaoding (Chi Ch'ao-ting), 20
Jia Lu, 207–8
Jia Rang, 105–6, 113
Jiajing era, 194
Jiangnan, 180
Jianxin Cui, 192
jiedushi (hereditary warlords), 133
Jin dynasty, 115–16, *119, 154,* 166, 170–77, 186
Ji'nan, 54, Plate 25
Jing River, Plate 30; in 2 CE, 90; on arable land, 32; as clear, 81, 117; cultivation effects on, 89–91; feeder canal from, 158; flooding of, 70–71; reforestation around, 245; saying about, 230; watershed erosion, 201; and Wei River, 30, 89–90
Jin-Yuan conflict, 176
Jullien, François, 2, 83
Jurchen people, 153, 166, 170, 184, 196
Juye Marsh, 103

Kaifeng, Plates 22, 34–36; canal network and, 142, 169; captured by Jurchens, 170; complexity of breach repairs to, 143; flood disasters of 1841, 1842, 1855, 234–37; flood-control system destroyed, 226–27; floods in, around, 50, *(50, 51),* 142; Japanese occupation of, 241; Liu Tianhe on, 213; as Song capital, 145, 158
Ke Zhang, 136
Khalkas, 188
Khitan (Qidan) Liao, 159–61, 170
Khubilai Khan, 175–76
Kidder, T. R., 66–67, 73, 99
Klein, Naomi, 16
Kolbert, Elizabeth, 239, 245
Kunlun Mountains, 29

lake district, Yellow/Yangtze Rivers, Plate 26
Lankao, 51–52
Lanzhou, 30, 175
Leonard, Jane Kate, 234
levees (see also dike systems): alternatives to, 120, 209; blocking tributaries, 18; causing subsidence, 7; costs of repair, 232–33; *versus* drainage canals, 120, 162, 207; flooding from failures of, 3, *8,* 13, 212–13, 234–36; lifting riverbeds, 100, 105; moisture accumulation alongside, 106; Pan Jixun interventions, 223–26; parallel, 159; raising level of Yellow River, 225; repairing breaches in, 220–21 *(221);* and river course changes, 7; sand and, 54, 162, 220; technology of, 162–64, 220–21 *(221)*
Li Chui, 163
Li Daoyuan, 116
Li Mountain Village, 198, Plate 32
Li Yizhi, 20
Li Zicheng, 195, 226
Liangshan Marsh, 52, *53,* 53f, 55
Liao, Khitan (Qidan), 159–60, 170
life spans of Yellow River, 9–10
Lingzhou, 32
Linton, Jamie, 4
Little Ice Age, 192, 288n38
Liu Daxia, 212
Liu Tianhe, 213–14
Liupan Mountains: 1279 expedition through, 175; deforestation, erosion of, 129, 151, 155, 176, 200; forests in, 98; fortifications, battles in, 123, 144–47 *(147),* 149; and Great Bend, 30–31; settlement of, 96; Tianchi Lake in, 43
liusha ("flow of sand"), 136
locks, 142, 163, 181, 211, 219–20, 226, 232, Plate 33
Loess Plateau, Plates 1, 2, 16
loess soil, Plates 13, 18; deforestation causing erosion of, 121; Li Yizhi on hydrology of, 20; and Loess Plateau, 38; as sediment, 41, 79, 119;

Shen Menglan on, 230; used in wall manufacture, 189

lower course of Yellow River, 22; attempts to channelize, 113–14; civil engineering along, 74–75, 91; complete reconstruction of, 214, 226; deforestation issues around, 143; Han activity around, 98; mismanagement leading to breaches, 108–9; as separate from middle course, 14, 19, 29, 52; silt buildup in, 165, 213; as unruly, 9; YRCC control over, 240

Lüliang Mountains, Plate 13; agriculture in, 37; forests in, 81, 98, 117, 129, 142–43, 193; and Great Bend (Hetao), 31–32 *(31)*; Qikou at foot of, 198

Luo River/Yiluo River, 32, 49–50, 69–70, *72*, 74, 76–77, 81, 109

Luo Valley, 91

Luochuan Tableland, 134

Luoyang (Luoyi), 49, 51, 76, 109–10, 114, 133, 137–38, 142–43

Luoyi valley, 96

maize, 60, 182, 199

management and disaster ratio, *65, 124, 182, 183, 210, 212*

Manchu Qing dynasty, 182, 184, 196, 226

Mang Mountains, 109, 242

Mao Zedong, 242

Map of Yu's Traces (Yujitu), 3, *4*

meanders *(zuowan)*, 7, 54–55, 172, 219–20, Plate 24

Mencius (aka Mengzi, Meng Ke), 1–2, 6, 83, 86–87

Mencius, The, 85–86

Meng Tian, General, 89

metaphor in Chinese tradition, 1–2, 82, 86, 280n68

middle course of Yellow River, 22, 33–34, 34f; and 1048 course change, 166–67; considered separate from lower course, 14, 29, 52; construction projects on, 100; eastern edge of, 36; Henan province and, 48–49; impact of irrigation works on, 244–45; mid-Holocene changes to, 12; new towns along (1750–1860), 197; population of, 60; resettlement of residents, 13; sediment transportation in, 44; as separate from lower course, 14, 19, 29, 52; Tan Qixiang on, 66; TYDA database on, 248–49; YRCC control over, 240

military farm system, 193

Miller, Ian, 144

millet, 70, 74, 99, 231

Ming regime, 184–88, 210; deforestation during, 193; destruction of flood-control system, 226; fall of, 185; forests during, 186; and Great Wall, *187*, 196, Plate 17; Jia Lu system under, 208; and Loess Plateau, 196–98; moisture mean during, 192 *(192)*; on Ordos, 184–85, 188, 190, 192; Pan Jixun, 214, 223–26, 224f, 236; political geography under, *187*, 188–90; population under, 185 *(185)*, 210, *211*; retreat of, 192; subsidizing farmers, 198; water management by, 208–13, 219, 223; Yulin garrison, 190–93, Plate 31; Zhu Yuanzhang, 188, 208

"modern" Yellow River, 47, 52, 58–59, 210, 242, Plate 4

moisture, *26, 32, 56,* 105–6, 144

moisture on Loess Plateau historically, 27 *(27)*, 134; 750–1300 CE, 122–25 *(122, 125)*; 1300 to 1911 CE, 181–82 *(182)*; 1500 to 1650 CE, 189, *190*, 192 *(192)*; 1650 to 2012 CE, 198 *(198)*; by ecological zone, 37–38 *(38)*; forests and, 40–41; land clearance and, 75–76; micro-landforms and, 56; moisture gradient on, 24, *38*, 122; by season, 44

Mongolia, Inner, 29, 33, 37, 94

Mongols, 111, 153; Chinese colonization and, 181; under Manchu dynasty, 196, 199; on Ordos, 188–89; Toqto'a as consultant, 207; Yuan regime, 175–76, *177*, 184–85, 204

Monograph of the River's Origin, A (Heyuan zhi) (Pan Angxiao), 175
monsoons: 200–0 BCE, 90; Asian summer, 277n88; and dam openings, 49; East Asian system, 12–13; El Niño influencing, 27; during Holocene Climatic Optimum, 58–59, 63, 70; Indian summer, 277n88; and Mu Us Desert, 192; Ordos and, 22; sand transported by, 135; summer, 25, *26*, 28–29, 49, 58–59; winter, 25, *26*
Mount Hua (Huashan), 49, 51
Mount Song (Songshan), 49–51, 69, 195
Mount Tai: deforestation of, 79; directing course changes, 52; erosion on, 99; and Huai River, 55, 109, 212, 237; Neolithic sites in, 69 *(69)*; rituals on, 103; tributaries of, 55
Mu Us Desert (Maowusu shamo), Plate 14; climate variability in, 28; desertification causes in, 93, 135–36; early settlement of, 96–97 *(97)*, 118, 131; Great Wall construction near, *187*, *190*; moisture gradient in, *38*; shifting of, 59; Wuding River along, 134, *135*; Xi Xia settlement near, 146; Xiazhou City near, 130; Yulin city near, *190*, 191–92
Muddy River, 117

Needham, Joseph, 20
Neolithic era, 57, 59–60, 62, 68–74 *(69)*, 79
Ngoring Lake, 30, Plate 5
Nihe river, 117
Nine Defense Areas (Jiubian), *189*, 190–91
Nine Provinces (Jiuzhou), 3, 84
"nine rivers," separating, 83, 209
Ningxia prefecture, 190, *190*
nitrogen-deficient soil, 240
Nixon, Rob, 17
nomadic herders, 14, 28, 59, 81, 86, 183
North China Plain, 63; dust storms on, 41; early settlement of, 70, 80, 136–37; geography of, 29, 33, 36, 49; shifting control of, 115, 125; soil-core study of, 45–46; as symbol of misery, 10, 237; water system on, 137, 179, *180*
Northern Han, 140
Northern Song, 50, 249; Bian Canal, 114, 142, 158; cultivated land of, 174; deforestation by, 143; flooding-related disasters under, 50, 125, 142; Jurchen Jin defeat of, 184, 186; Tangut resistance to, 144–45

Ordos Plateau area, 22, 130; alluvial plains, 32, 97–98; commercial towns on, 197; deforestation, desertification in, 93, 181–82, 185, 193, 198–99; erosion, flooding in, 121; ethnic cleansing on, 191; farming/farmers, 97–99, 110–11, 130, 146–47, 153, 191; Great Bend (Hetao), 24, 29–33 *(31)*; horses, herders in, 119, 149, 188, 199; increasing drought in, 71; Loop, 29, 58, 94; military retreat from, 181–82; Ming regime on, 184–85, 188, 190, 192; Mongols on, 188–89; monsoons and, 22; peasant rebellions on, 195; population of, 128, 185; Qin wall across, 87; Song losing control of, 123; supplying Chang'an, 128–29; Tongwancheng as capital, 118; tripartite balance of power in, 133–34; Tümeds, Mongols, Han on, 188
Organic Machine, The (White), 8–9
Ouyang Xiu, 19, 151, 153, 169
Ouyang Xuan, 207–8
overflow basins, 19

paleodeltas, 45, Plate 19
Pan Angxiao, 175
Pan Jixun, 214, 223–26 *(224)*, 236
pastoralism on Loess Plateau, 12–13; around Tongwancheng, 135; in competition with agriculturalists, 59–60, 71–72, 76, 87, 119, 136, 144; consigned to north, 144, 199; displacing Neolithic agriculture, 59–60; and Guanzhong, 77; and horses, 76; under Jurchen, Mongols, 153; under Ming, 188, 196; under Qin, 86–89;

sustaining trees, grasses, 117, 119; temporary elimination of, 183–86; in Xiazhou, 130; Xiongnu, 92, 94, 110–11
perched riverbeds, 54, 56, 100–101, 104–6, 155, 161
Père David (Armand David), 203
phosphate-deficient soil, 240
pine, 43, 73, 130, 143–44
Pingliang, 37, Plates 16, 37, 38
pirates, 206
plague, 205–6, 226
Pleistocene era, 55, 58
plows, plowing, 75, 91–92 *(92)*, 141
political geography on Loess Plateau, 3, *186–87*, 189
Pomeranz, Kenneth, 183–84
population: in 1550, 211f; Bronze Age, 79; erosion increasing with, 11–12; Great Bend (Hetao), 96–97; in Guanzhong, 110, 128, *129*, 186; in Loess Plateau, 40; middle course of Yellow River, 60; Ming regime, 185 *(185)*; Qin regime, *119*; Qing regime, *185*; Shaanxi Province, 118–19 *(119)*, 128, 153, 155, 185 *(185)*, 195; Shanxi Province, 210; southern shift in, 137; Western Zhou, 119t; Yangtze (Changjiang) delta, *211*; Yuan regime, *185*
pounded earth structures, 13, 41, 72, 189, Plates 17, 31
precipitation: contents of runoff from, 44, 99; drought periods, 10–11, 28, 185, 188, 193; and erosion, 18–19, 158; estimating floods from, 210; in TYDA database, 262

Qi, state of, 79–80, 83, 86–88, 194
Qikou, 197–98, Plate 32
Qin regime, 63, *88*, *95*; China unified under, 88; erosion under, 86; forced migrations, colonization under, 88–89, 96; Han defeat of, 90; population under, *119*; saying on river sediment, 230; Shang Yang advice to, 86–87; as state monopoly, 87; Tan Qixiang on, 66; walls, 89, 92, 94–96, 145 (See *also* Great Wall); Zhengguo Canal, 89–90, 158, 195, 201, Plate 30
Qin river, 212, 230
Qing dynasty, 10, *77*, 226; attempts to manage sediment, 203, 219, 225–30; cave houses, 198, Plate 32; Chen Huang's suggested erosion controls, 229–30; deforestation, erosion under, 200–202; disagreements over Yellow River management, 202–3; fall of, 239; floodplain management by, *183*, 226–27; Manchu Qing dynasty, 182, 184, 196, 226; offering agricultural incentives, 182, 184; Pan Jixun's suggested erosion controls, 214, 223–26, 224f, 236; population under, *185*; promoting migration to Loess Plateau, 196; rise, fall of, 226, 239; ruling house, 10, 199–200, 227, 239; salt monopolies under, 198; shifting, balancing of resources, 199; UNESCO waterworks site, 243, Plate 35; use of spillways, Plate 34; White Lotus Rebellion against, 199
Qinghai Province, 29–30, Plate 6
Qingkou: in 18th century, 214, Plate 33; and 1841 flood, 236; hydrological architecture around, *215–18*; and Pan Jixun system, 223; "silt crisis" at, 232; spillways to protect, Plate 34; water diverted to, 228; water flow through, 219
Qinling Mountains: deforestation of, 129, 195, 200; immigrants crossing, 112; location of, 24, 30; Mount Hua (Huashan) in, 49, 51; only alternative to canal travel, 140; White Lotus Rebellion in, 199; and Yellow River, 35–36
Qinzhou prefecture, 145, *146*
Quaternary period, 41, 46, 58

rainstorms. *See* precipitation
Records of Counties and Prefectures from the Yuanhe Era (Yuanhe junxian zhi), 116–17
Records of River Defense from the Zhizheng Era (Zhizheng hefang ji), 207–8

Records of the Historian, The, 85
Records of the Historian (Shiji), The (Sima Qian), 80, 85
reforesting plans, 201, 243, Plate 37
Ren Mei-e, 42, 48
reservoirs, 13, 118–20, 181; CCP Sanmenxia project, 242–43; check dams to create, 202; on Grand Canal, 211, 214; Guo Shoujing and, 176; Hongze Lake, 214; Jia Rang and, 113; loess landscape as, 41; Pan Jixun on, *224*; preindustrial, 162; silt release techniques for, 49–50, Plate 21; Wang Jing and, 114–15
reversing water flow, 99, 225
rice, 70, 74, 128, 141, 178, 180, 223
Richards, John, 274n28
Rituals of Zhou (Zhouli), 83
River, the Plain, and the State, The (Zhang), 159
River and Canal Monograph, The (Hequshu), 113
river basins, 4–7
river course changes/avulsions, 7, 51, *102*–103, *165*, 172–75, 207, 234, 241
River Defense Abstracts (Hefang zhaiyao) (Chen), 230
River Elegy (Heshang) documentary, 244
riverbanks: erosion, collapse of, 99–100, 200, 213; farming near, 72; housing near, 184; meander currents and, 54; rising, 7, 48
riverbeds, 223; controlled releases to lower, 49–50; course changes, 44, 52, 212, 214; desiccation of, 244; erosion of, 158; high-sediment era, 54; Jia Lu repair work, 207; Jia Rang "Three Theses" on, 106–7; Kaifeng repair work, 235–36; levees lifting, 100, 105; Liu Tianhe treatise on, 213; Pan Jixun repair work, 223; perched channels in, 54, 56; Qingkou channel, 219–20; sediment causing rise in, 6–7, 18, 155, 204, 225, 233; "silt crisis," 232; silt washing, dredging of, 114–15; Wang Anshi silt dispersal wheel, 162; Wei Yuanyu on, 233
rivers: biographic approach to, 8–9; central authority management of, 12; as human inventions, 3; not seen as ecological systems, 12; as sediment sorting machines, 3–4
river course changes, *209*, 214, Plate 1, Plate 22; from alluvial deposits, 6; CCP report on, 243; course reversals, 225, 232; impact of levees on, 7; during imperial times, 13–14; minor changes, 219, 225; near Mount Song, 50; primary location of, 52–53; in TYDA database, 263; visible indications of, 56; during Wang Mang regime, 108–9; of year 14 and 17 CE, 161; of year 1034, 161; of year 1048, 143, 159, 166–70; of year 1128, 143; of year 1279, 55; of year 1391, 208; of year 1492, 212; of year 1855, 236–37, 242; *Yellow River Annals* listing of, 247, 255–56. *See also* lower course of Yellow River; middle course of Yellow River; upper course of Yellow River
Rosen, Arlene, 66, 70, 79

salt marshes, ponds, flats, 56, 58, 145, 152, 188, 191, 236, 244
salt mining, trade, 38; metropolises built around, 211; sediment interfering with, 204; smuggling, 134; state, clan control of, 13, 80, 82, 130, 144–45, 184, 198
sand: after Huayuankou Breach, 241 (*241*); breaches depositing, 161, 166–68, 171; and declining soil fertility, 17, 236; deposition over time, 67, *68*; dunes from desertification, 30, 76, Plate 6; encircling Yulin Garrison, 191–93; filling lakes, wells, 221, 223, 232; in first, second centuries BCE, 92, 94, 99; in Gaobao lakes, 221; Hu Ding's report on, 202; hydrology becoming circular, 227–28; in Lankao, 51–52, Plate 22; leading to 1755 breach, 236–37; le-

vees and, 54, 162, 220; *liusha* ("flow of sand") 841 CE, 136; loss of ground cover exposing, 151; on Loess Plateau, 37–38, 41, 56; Mao's initiative in 1950s, 242; mobile dunes, 76; Pan Jixun's solution to, 223–26, *224*, 236; sandification of alluvial plains, 178, 181; sandstorms, 56, 168; as sediment component, 44, 181; splay fans of, 56; transported by monsoons, floods, 7, 25, *26*, 135, 200; use of drainage basins for, 210; water table and, 191, 232; and yearly silt release, 49, Plate 21; and Yellow River, 20, 202–3; Yellow River tributaries as source of, 33; in Zhengguo Canal, 201. *See also* sediment; silt

Sanmenxia Gorge hydroelectric dam, 36 *(36)*, 66, 242–43

Sanyangzhuang, 99

"sawtooths" *(juya)*, 162

saying about silt, 90, 230

scholarship: on alluvial plains, 93; on river history, 8–10, 20–21; on Yellow River, 6

Scott, James, 273n9

scouring of sediment, 49, 83, 107, 162, 214, 223–25, 229, 233

sediment: attempts to slow, 202; burying villages, 237; crisis caused by, 225; critical to river systems, 5–6; deposition from Loess Plateau, 6, 120, 181; deposition over time, 8f, 11–12, *68*, 212, 223; early understanding of erosion issues, 200–203; erodes upstream, accretes downstream, 17; factors affecting flow of, 25; flow inhibited by weak currents, 214; and human lifespan, 18; lack of consensus on handling, 212; long-distance effects on, 4, 6–7; as opportunity, 90; pushed out to sea, 13; in Qingkou channel, 219; in river mouth, 214, 218, 225, 236; sand, gravel as, 44, 181, 201–2; source of nutrients, 6, 22; wind-borne, 28, 37, 41, 44, 76

seeing as empowered illumination, 5, 273n9

settlement geography of Loess Plateau, 196–97 *(197)*

Sewell, William, 16

Shaanxi Province, Plate 32; breaches on floodplain, 213; break up of tableland, 134; colonist farmers in, 188, 191; earthquakes in, 145; famine in, 195; fortification in, 147, 193; on Loess Plateau, 37; Mongols in, 188; population over time, 118–19 *(119)*, 128, 153, 155, 185 *(185)*, 195; poverty leading to rebellion, 199; salt-mining in, 130; sediment loss from, 202; Shen Menglan on, 230; timber overcutting, 200; in TYDA database, 249

Shandong Peninsula, 52, 100, 166, 170–71, 204, 234

Shandong Province, 79, 130, 140, 143, 155, 231, Plate 25

Shang Yang, 86–87

Shanhaijing (The Classic of Mountains and Oceans), 81, 113

Shanxi Province: effects of deforestation in, 143, 194, 199–200; Li Daoyuan in, 116; in loess region, 37; on Loess Plateau, 37; under Northern Han, 140; population of, 210; sediment loss from, 202; Shen Menglan on, 230–31; under Southern Xiongnu, 110–11; waterworks system in, 139; Yan Shengfang treatise on, 194

Shaobai Xiong, 252, 254

Shen Dao, 83

Shen Kua, 143–44, 151–53, 163

Shen Menglan, 230

Shenzong, 169

shi (power), 2–3, 19, 83, 169, 225

Shi Nianhai, 20

shifting baseline theory, 18

Shijing (The Classic of Poetry), 81

"shocks," 16

Shun, 85

shusan ("loose and crumbling") soils, 202

Si River, 103, 207

Sichuan, 112, 136, 195

silt, 168; agriculture abandoned due to, 236; blocking Yellow River outlet, 236; current speed and, 106, 114; dispersing wheel for, 162; drainage basins for, 210; in Gaobao lakes, 222; increasing amounts of, 124, 195; levee height and, 120; Liu Tianhe on origins of, 213–14; on Loess Plateau, 41, 44; nineteenth century, 10, 51; Ouyang Xuan on, 208; Pan Jixun focus on, 223–25; replaced by sand, gravel, 202; "silt crisis," 232; twentieth century, 51; Xiaolangdi Dam releases of, 49, Plate 21; in Yellow River, 30, 33, 37, 44, 52, 56–57. *See also* sand; sediment

Sima Qian, 80, 85, 102–4

Sino-Tibetan treaty (821), 133

Sizhou inundation, 226

slackwater lakes, Plate 35

slow violence, 17, 232

sluices: Gaojiayan collapse, 223; Jia Rang's guidance, remediation of, 106, 113, 207; limits of, 232, 243; Ming improvements on, 232, 243; Pan Jixun on, *224*; proper, improper use of, 7, 13, 210–12, 219, 232, 264; Su Shi and Shen Kua on, 163; Wang Jing construction of, 113–14

soil: declining fertility, 17; erosion requiring innovation, 12–13; farm erosion, 12, 18; North China Plain core study, 45–46; salinization of, 13, 232; *shusan* ("loose and crumbling") soils, 202; topsoil, 7, 74, 92, 193. *See also* loess soil

Song History, The (Song shi), 143–44

Song Jiang outlaw band, 55. *See also* Northern Song

Song regime, 123, 141, 234; banning tree felling, 153; conflict with Jin-Yuan, 174–76; conflict with Jin, 176; conflict with Xi Xia, 145, 176, 200; emperor site inspections, 158; favoring colonization, 144–45; fortification campaign, 147–49 (148f); inventions under, 128; population distribution, 153, 174, *185*; river sacrificial ceremonies, 159; and steppe-facing regimes, 153; Tang transition to, 128, 177; war in 1040s, 144–45; water engineering under, 160–61, 163–64, 167–71 *(167)*, 174–75, *205*. *See also* Northern Song

sorghum stalks, 101, 115, 162, *222*, 231, 243

splay fans, 56, 244

Spring and Autumn Period (Chunqiu shidai), 77–81

Su River, 50

Su Shi, 139, 163

subsidence of land, 7, 25, 223

Sui regime, 118, 122, 128, 143

sunspot cycle and floods, 28

tablelands, highland plains *(yuan)*, 43, 130, 134, *154*, Plate 35

Taihang Mountains, 32, 81, 143–44, 194

Taizong emperor, 158

Taizu emperor, 158

Tan Qixiang, 20, 66

Tang regime, 123; disasters during, 64–66 *(65)*, 123, *124*, 139–40; dry climate during, 122; militarization under, 118, 129–30, 133–34; population under, 32, 128, *129*, 153; Tang-Song transition, 128, 177–78. *See also* Chang'an; Guanzhong

Tangut people, 123, 133–36, 140, 144–53, 169

Tao Mo, 201

Taosi, 72–73 *(72)*

taxation, 150, 183–84; Han difficulties with, 110; and Huang Chao Rebellion, 134; under Qing, 196, 198; resistance to, 161, 181, 195, 199; tied to agriculture, 75, 89, 182–83, 189, 198, 212; tied to earthworks, 210; tied to Grand Canal, 181, 237; tied to river management, 2, 212, 229–30, 233, 236

tectonic plates, 23, 31, 52, 57–58

Teh, Ian, 30

terraces *(titian)*, 141; under Mao Zedong, 242; near Li Mountain Village, 198, Plate 32; near Lüliang, Plate 13; near Pingliang, Plate 37; used for agriculture, 198, 200, 202; used for forestry, 245
"Three Theses on River Regulation" *(Zhihe sance)* (Jia), 105–6, 113
Tianchi Lake, 43, 76, 136, 200
Tibetan Plateau, 5, 23, Plates 3; declining glacial volume on, 244; as Earth's third pole, 25; floodplain connected to, 137; seismic activity of, 24, 24f, 58; Yellow River headwaters on, 29–30
Tibetans, 110–11, 133–34, 136, 145
timber, 32, 168, 186, 200; in 1600s, 196; banning of firewood, 143; for canal support structures, 158; from central, western Loess Plateau, 131; Chang'an requirements of, 40, 142–43; commercial market in, 123, 143–44, 198, 200; commodification of, 60, 184, 197, 200; earmarked for military, 145, 150, 153; erosion from loss of, 153, 200; following 1048 flood, 168; on Guanzhong Plain, 186; Hong Mai on, 143; from logging, 123, 143, 198; loss of in middle Ming period, 193; for mining, smelting, 75; rebounding 200–600 CE, 117; replacements for, 162, 231; streamlining removal of, 81, 92, 98; transported by river, 76. *See also* deforestation
timelines, vii; breaches as percentage of disasters, *8*; compared to human lifespan, 18; disaster and moisture correlation, 700 to 1400 CE, 125; disaster to management ratio, 750 BCE to 750 CE, *65*, *124*; disaster to management ratio, 1300 to 1911 CE, *183*; disasters to events, 1815 to 1915, *235*; erosion and settlement history, 11; eventful era on floodplain, 920–1165, 156–57; events, 100 to 900 CE, 117; events, 150 BCE to 150 CE, 65; events, 300 BCE to 14, 93; events, 750 BCE to 750 CE, 65; events and moisture levels, 1300 to 1911, 182; extension of Yellow River Delta, 57; floodplain events (1034–1127), 167; floodplain events (1128–1234), 172; floodplain events (1234–1368), 177; floodplain events (1820–1911), 235; historical floods around Kaifeng, 51; Loess Plateau political geography, 64, 112, 186; management and disaster events (2–220), 108; management to disaster ratio (1368–1590), 210; moisture index (700–1400), 122; moisture mean (1500–1650), 192; moisture mean (1650–2012), 198; moisture on Loess Plateau, 27; Qing Dynasty events, 228; repairs to new construction (1300–1850), 227; sedimentation on Yellow River, 47; Yuan floodplain events (1234–1368), 205
Tongwancheng, 118, 134–36, *135*, Plate 27
Tongwaxiang, 236–37, 242
Toqto'a, 207
tree planting attempts, 201–2
tree-ring records, 27 *(27)*, 277n88
tribute goods, 130, 212, 228–29, 237
Tribute of Yu (Yugong), 20–21, 83–84, 113
tubers from Americas, 60, 182, 199
Tümeds, 188
turbidity, 202, 213, 230, 236
Turks/Turkics, 111, 129–30, 133
turning points, 9, 63, 66, 119, 123, 176, 249
TVA (Tennessee Valley Authority), 240
Twelve Canals of Yinzhang/Zhang River, 80
TYDA (Tracks of Yu Digital Atlas), 15–17, 247; compilation of data, 248–52; designing and validating, 252–55; contents of, 255–56; Jupyter Notebook for, 256–61; event types in, 262–65

Uighurs, 133–34, 145
UNESCO World Heritage Site, 243
Unified View of River Management, A (Hefang yilan) (Jixun), 224
upper course of Yellow River, 30, 100, 123, Plates 5–7

Viers, Josh, 5, 7

walls, fortified, 190–93. *See also* Great Wall
Wang Anshi, 162
Wang Jing, 113–14
Wang Mang, 108–9
Wang Yanshi, 105
war analogy for flood control, 5
Warring States Period (Zhanguo shidai), 2, 77–78, 80–82, 85–87
"wasteland" *(huangdi),* 191, 193
water as metaphor, 1–2
Water Margin (Shuihu zhuan), 52, 55
water tables: falling, 7, 82, 98, 184, 191–92, 232; high, 243–44
waterworks, 21, 34, *224,* Plates 12; Bronze Age, 71; and channelization, 78; destroyed in war, 100–101, 115, 139; eighteenth, nineteenth centuries, 200–203, 219–20, 226; eleventh century, 139–40, 143, 155, *156,* 160–61; fourteenth century, 177; management, regulation of, *8,* 80, *138;* poem on, 90; program for whole floodplain, 113; in Shanxi, 139; and *shi,* 83; spending on, 12; twelfth, thirteenth centuries, 172–74 *(172), 177;* twentieth century, 240, 243; in TYDA database, 249; Wang Jing, 114–15; Yiluo River valley, 109; Yu example, 84–85. *See also* Grand Canal
weather effects on conflict, 11
Wei River, 30, *35,* Plate 12; called clear-flowing, 130; canal debris in, 201–2; and Chang'an foodshed, 90, 129; Chinese settlement along, 146; confluence with Yiluo, Su, 50; cultivation effects on, 89–90; debris from timber in, 200; earthquake (1556), 185; effects of flooding of, 230–31; Emperor Han Wudi on, 104; erosion along, 131; following fault lines, 24 *(24);* on Guanzhong Plain, 32, 146, 195, Plate 12; and Jing River, 30, 89–90; navigability of, 36; Neolithic sites near, 69–70; outposts on tributaries, 131; pasturage and hunting around, 69–70, 80–81; plains around, 32, 43; sediment into and from, 34, 107, 200–201, 230, 243; and Song, Xi Xia conflict, 146, 146f; timber transport down, 76, 98; Zhang Rong on, 107; Zhengguo Canal near, 89, 195; Zhou kingdom and, 76
Wei Yuanyu, 233
weirs, 114, 163, 219
Western Han, 63–67, 94, *95, 97,* 98, 115, 119. *See also* Han Dynasty
White, Gilbert F., 5–7, 19
White, Richard, 8–9
White Horse river, 117
White Lotus Rebellion, 199
Will, Pierre-Étienne, 12
Wittfogel, Karl, 20
"wooden dragons" *(mulong),* 162
Writings of Master Guan (Guanzi) (Zhong), 80
Writings of Master Xun (Xunzi), 81
Writings of the Huainan Masters, The (Huainanzi), 98
Wu Guifang, 214
Wu Liang shrine, *84,* 85
Wuding River, 32, 90–91, 130, 134–36, 151–52, 190–91, 198
wuyue cosmology, 49, 55

Xi Xia regime, 123, 131, 158, 176, 193; conflict with Song regime, 144–53 *(146, 147),* 176, 200
Xi'an (Chang'an), 32, 34, 38, 76, 185, 243
Xiaolangdi Dam, Reservoir, 49, 114, Plate 21
Xiazhou, 130
Xihe jurisdiction, 112
Xin Deyong, 66

xing (nature), 2, 164n4
Xing County, 199
Xiongnu, 87, 89, 92–96, 101, 105, 110–12, 118
Xu Jiongxin, 45, *57*, *253*
Xu Xiake, 195

Yan Shengfang, 194
Yangtze (Changjiang) delta, Plate 26; and Bian Canal, 114; canals linking to, 137–40; cost of grain transport, 212; floodplain connecting to, 53; as foodshed for Beijing, 180; hard labor on, 160; as population center, *211*; Southern Song control of, 205; and Yellow River breach, 161, 241–42 *(242)*
Yangtze River, 24, 29, 55, 128, 205, 208, 223, 245
Yao, Emperor, 3, 85
yaodong (cave houses), 198, Plate 32
Yellow Earth (Huang tudi) movie, 244
Yellow River, 19, Plates 1–4; course change region of, Plate 22; elevation and hydrology, Plate 3; floodplain, 52–57, Plate 23; headwaters of, 23, 29–30, 94; modern course as conscious choice, 183–84, 210; multiple channels era of, 204–14 *(206, 209)*, 226; naming of, 123; paleodeltas, Plate 19; scholarship on, 6, 229; watershed, Plate 2. *See also* lower course of Yellow River; middle course of Yellow River; upper course of Yellow River
Yellow River Annals, The (Huanghe nianbiao), 247, 249, 251, 255
Yellow River Dredging Commission, 162
Yellow Sea, 23, 29
Yi, 85
Yi River, 49
Yiluo River/Luo River, 32, 49–50, 69–70, 72f, 74, 76–77, 81, 109
Yinchuan Plain, 15, 32, 37, 96, 123, 131, 134, 186, *190*, Plates 8–11
Yongding River, 231

YRCC (Yellow River Conservancy Commission), 46, 240, 242
Yu the Great (Da Yu): *Book of Jin* on, 115; building drainage canals, 84; emperors commemorating, 158; floodplain management by, 3, 5, 84; founding Nine Provinces (Jiuzhou), 3, 84; legend of, 2–3, 71, 78, 85, 280n66; *Map of Yu's Traces*, *4*; on need for study before action, 83; Ouyang Xiu on, 169; separating nine rivers technique, 3, 5, 84, 209; Tracks of Yu Digital Atlas, 248; using river's power, 50, 225; Yugong *(Tribute of Yu)* journal, 20–21, 113; Zhou Enlai invoking, 242
yuan (tableland, highland plains), 43, 130, 134, *154*, Plate 35
Yuan regime: destruction from Jin-Yuan conflict, 176; distribution of settlements, 186; floodplain events under, *177*, 204, *205*–6; money system breakdown, 205; Mongol Yuan conquest, 175–76, 184; political geography under, *187*; population under, *185*; Yuan-Ming transition, 208
Yugong (Tribute of Yu), 20–21, 83, 113
Yugongtu (The Diagrams of the Tribute of Yu), 113
Yulin: desertification in region, 191–93; garrison at, 191–93, Plate 31; and Great Wall, *190*, 192, Plate 17; pounded-earth Great Wall fragment, 41, Plate 17; terraces, gullies, 37, 130, Plate 14; Xiazhou, 130

Zhang Ling, 159, 164, 168
Zhang River, 80
Zhang Rong, 18–20, 106
Zhang Xianzhong, 195
Zhengguo Canal, 89–90, 158, 195, 201, Plate 30
Zhengzhou, 33, 51–52, 220, 236, 240–41, 244
Zhenzong, 159
Zhili province, 212, 231, 237
Zhou Enlai, 242

Zhou kingdom, 76–77; Eastern Zhou period, 79; highland plains *(yuan)* in, 43; later period (950s), 141; Qin tributary state to, 86; Western Zhou population, *119*
Zhouli (Rituals of Zhou), 83
Zhu Xianmo, 42, 48
Zhu Yuanzhang, 188, 208
Zhuang, Yijie, 67
Zou Yilin, 20
Zuozhuan (The Commentary of Zuo), 81

The Agrarian Studies Series at Yale University Press seeks to publish outstanding and original interdisciplinary work on agriculture and rural society—for any period, in any location. Works of daring that question existing paradigms and fill abstract categories with the lived experience of rural people are especially encouraged.

—James C. Scott, *Series Editor*

James C. Scott, *Seeing Like a State: How Certain Schemes to Improve the Human Condition Have Failed*
Steve Striffler, *Chicken: The Dangerous Transformation of America's Favorite Food*
James C. Scott, *The Art of Not Being Governed: An Anarchist History of Upland Southeast Asia*
Timothy Pachirat, *Every Twelve Seconds: Industrialized Slaughter and the Politics of Sight*
Edward Dallam Melillo, *Strangers on Familiar Soil: Rediscovering the Chile-California Connection*
Kathryn M. de Luna, *Collecting Food, Cultivating People: Subsistence and Society in Central Africa through the Seventeenth Century*
James C. Scott, *Against the Grain: A Deep History of the First Civilizations*
Loka Ashwood, *For-Profit Democracy: Why the Government Is Losing the Trust of Rural America*
Jonah Steinberg, *A Garland of Bones: Child Runaways in India*
Hannah Holleman, *Dust Bowls of Empire: Imperialism, Environmental Politics, and the Injustice of "Green" Capitalism*
Johnhenry Gonzalez, *Maroon Nation: A History of Revolutionary Haiti*
Christian C. Lentz, *Contested Territory: Điện Biên Phủ and the Making of Northwest Vietnam*
Dan Allosso, *Peppermint Kings: A Rural American History*
Jamie Kreiner, *Legions of Pigs in the Early Medieval West*
Christian Lund, *Nine-Tenths of the Law: Enduring Dispossession in Indonesia*
Shaila Seshia Galvin, *Becoming Organic: Nature and Agriculture in the Indian Himalaya*
Michael Dove, *Bitter Shade: The Ecological Challenge of Human Consciousness*
Japhy Wilson, *The Reality of Dreams: Post-Neoliberal Utopias in the Ecuadorian Amazon*
Aniket Aga, *Genetically Modified Democracy: The Science and Politics of Transgenic Crops in Contemporary India*
Ruth Mostern, *Following the Tracks of Yu: The Ecological and Imperial World of the Yellow River*
Brian Lander, *The King's Harvest: A Political Ecology of China from the First Farmers to the First Empire*
For a complete list of titles in the Yale Agrarian Studies Series, visit yalebooks.com/agrarian.